Probability and Mathematical Statistics (Continued)

ROHATGI • An Introduction to Probability Theory and Mathematical Statistics
SCHEFFE • The Analysis of Variance
SEBER • Linear Regression Analysis
SERFLING • Approximation Theorems of Mathematical Statistics
WILKS • Mathematical Statistics
WILLIAMS • Diffusions, Markov Processes, and Martingales, Volume I: Foundations
ZACKS • The Theory of Statistical Inference

Applied Probability and Statistics

ANDERSON, AUQUIER, HAUCK, OAKES, VANDAELE, and WEISBERG • Statistical Methods in Comparative Studies
BAILEY • The Elements of Stochastic Processes with Applications to the Natural Sciences
BAILEY • Mathematics, Statistics and Systems for Health
BARNETT and LEWIS • Outliers in Statistical Data
BARTHOLEMEW • Stochastic Models for Social Processes, *Second Edition*
BARTHOLOMEW and FORBES • Statistical Techniques for Manpower Planning
BECK and ARNOLD • Parameter Estimation in Engineering and Science
BELSLEY, KUH, and WELSCH • Regression Diagnostics: Identifying Influential Data and Sources of Collinearity
BENNETT and FRANKLIN • Statistical Analysis in Chemistry and the Chemical Industry
BHAT • Elements of Applied Stochastic Processes
BLOOMFIELD • Fourier Analysis of Time Series: An Introduction
BOX • R. A. Fisher, The Life of a Scientist
BOX and DRAPER • Evolutionary Operation: A Statistical Method for Process Improvement
BOX, HUNTER, and HUNTER • Statistics for Experimenters: An Introduction to Design, Data Analysis, and Model Building
BROWN and HOLLANDER • Statistics: A Biomedical Introduction
BROWNLEE • Statistical Theory and Methodology in Science and Engineering, *Second Edition*
BURY • Statistical Models in Applied Science
CHAMBERS • Computational Methods for Data Analysis
CHATTERJEE and PRICE • Regression Analysis by Example
CHERNOFF and MOSES • Elementary Decision Theory
CHOW • Analysis and Control of Dynamic Economic Systems
CLELLAND, deCANI, and BROWN • Basic Statistics with Business Applications, *Second Edition*
COCHRAN • Sampling Techniques, *Third Edition*
COCHRAN and COX • Experimental Designs, *Second Edition*
CONOVER • Practical Nonparametric Statistics, *Second Edition*
COX • Planning of Experiments
DANIEL • Biostatistics: A Foundation for Analysis in the Health Sciences, *Second Edition*
DANIEL • Applications of Statistics to Industrial Experimentation
DANIEL and WOOD • Fitting Equations to Data: Computer Analysis of Multifactor Data, *Second Editon*
DAVID • Order Statistics
DEMING • Sample Design in Business Research

continued on back

The Statistical Analysis of Failure Time Data

E. Arseven.

The Statistical Analysis of Failure Time Data

J. D. KALBFLEISCH

University of Waterloo, Ontario

R. L. PRENTICE

The Fred Hutchinson Cancer Research Center
Seattle, Washington
and
Department of Biostatistics
University of Washington

JOHN WILEY AND SONS

New York Chichester Brisbane Toronto

Library of Congress Cataloging in Publication Data

Kalbfleisch, J D
The statistical analysis of failure time data.

(Wiley series in probability and mathematical statistics)
Bibliography: p.
Includes indexes.
1. Failure time data analysis. 2. Regression analysis. I. Prentice, Ross L., joint author. II. Title.
QA270.K34 519.3 79-21889
ISBN 0-471-05519-0

Printed in the United States of America

10 9 8 7 6 5 4 3 2 1

To

Sharon and Didi

Preface

The main purpose of this book is to collect and unify some statistical models and methods that have been proposed for analyzing failure time data. Special attention has been paid to problems arising in the biomedical sciences, though many of the models and procedures apply also to industrial and engineering applications.

Some knowledge of the theory of statistics is assumed, though many of the principal statistical results used are reviewed. The book is intended to serve not only as a reference for persons involved in the analysis of failure data but also as a text for a graduate course in statistics or biostatistics.

We consider failure time data that arise when items are placed at risk under varying experimental conditions. Special features of such data are stress variables which vary with time and the presence of complicated censoring patterns. Recent books by Mann et al. (1974) and Gross and Clark (1975) deal extensively with estimation procedures for failure time data from homogeneous populations. Accordingly, these methods have not been extensively reviewed. Our main focus is on regression problems with survival data, specifically the estimation of regression coefficients and distributional shape in the presence of censoring. Many recent developments in this area, including the important contribution of Cox (1972), have made this book timely.

The choice of subject matter was made more difficult by the rapid pace with which methodology in the analysis of failure time data has been developing. Of course, the material covered is to some extent a reflection of our own interests in this field and there are some fairly extensive but specialized developments that have not been covered in any depth. Our attempt, however, has been to present a coherent background to the direction in which we think the field is moving.

Chapter 1 is introductory and deals with the basic formulation of survival models along with elementary methods of analysis. Chapter 2 presents many common survival models for homogeneous populations and regression problems, and Chapter 3 deals with parameter estimation within such models. The proportional hazards model introduced by Cox (1972) is considered in Chapter 4. Chapters 5 to 8 deal with more specialized topics: Chapter 5 considers more advanced topics in the proportional hazards model. Rank procedures in regression problems are considered in Chapter 6. The analysis of multivariate failure time data and competing risks data is the subject of Chapter 7. Chapter 8 considers the adaptation of the previous methods to such special topics as paired failure time data and case control studies. Bayesian methods of estimation are also given. The final section gives a detailed analysis of a data set using several of the methods developed in preceding chapters.

References are kept to a minimum throughout, but readers are referred to the bibliographic notes following each chapter which give a brief historical review of the material covered and related references.

We would like to express our thanks to Dr. Art Peterson for his careful reading of the manuscript and for his extensive comments. The comments and criticisms of graduate students at the University of Washington and the University of Waterloo also have been most helpful.

J. D. KALBFLEISCH
R. L. PRENTICE

Waterloo, Ontario, Canada
Seattle, Washington
January 1980

Contents

The Statistical Analysis of Failure Time Data

CHAPTER 1

Introduction

1.1 FAILURE TIME DATA

This book deals with the modeling and analysis of data that have as a principal end point the time until an event occurs. Such events are generically referred to as failures though the event may, for instance, be the performance of a certain task in a learning experiment in psychology or a change of residence in a demographic study. Major areas of application, however, are medical studies on chronic diseases and industrial life testing.

We assume that observations are available on the failure time of n individuals usually taken to be independent. The principal problem examined is that of developing methods for assessing the dependence of failure time on explanatory variables. Typically such explanatory variables will describe pre-study heterogeneity in the experimental material or differential alterations of treatments resulting from the study design. A secondary problem involves the estimation of and the specification of models for the underlying failure time distribution.

It is convenient to illustrate some of the distinguishing features of failure time data through the following examples.

1.1.1 Carcinogenesis

Table 1.1, from Pike (1966), gives the times from insult with the carcinogen DMBA to mortality from vaginal cancer in rats. Two groups were distinguished by a pretreatment regime.

We might consider comparing pretreatment regimens using the t-test (presumably to transformed data) or one of several nonparametric tests. Such procedures cannot immediately be applied, however, because of a feature very prevalent in failure time studies. Specifically, four failure times in Table 1.1 are "censored." For these four rats, we can see that the failure times exceed 216, 244, 204, and 344 days, respectively, but we do not know the failure times exactly. In this example, the (right) censoring

Table 1.1 DAYS TO VAGINAL CANCER MORTALITY IN RATS

Group 1	143,	164,	188,	188,	190,	192,	206,	209,	213,	216,
	220,	227,	230,	234,	246,	265,	304,	216,*	244*	
Group 2	142,	156,	163,	198,	205,	232,	232,	233,	233,	233,
	233,	239,	240,	261,	280,	280,	296,	296,	323,	204,*
	344*									

*Censored.

may have arisen because these four rats died of causes unrelated to the application of the carcinogen and were free of tumor at death, or they may simply not have developed tumor at the time of data analysis.

The necessity of obtaining methods of analysis that accommodate censoring is probably the most important reason for developing specialized models and procedures for failure time data.

A larger set of animal carcinogenesis data, from Hoel (1972), is given in Appendix 1. Two groups of male mice were given 300 rads of radiation and followed for cancer incidence. One group was maintained in a germ-free environment. The new feature of these data is that more than one failure mode occurs. It is of interest, for example, to evaluate the effect of a germ-free environment on the incidence rate of reticulum cell sarcoma while adjusting for the competing risks of developing thymic lymphoma or other causes of failure.

1.1.2 A Randomized Clinical Trial

Appendix 1 gives the data for a part of a large clinical trial carried out by the Radiation Therapy Oncology Group in the United States. The full study included patients with squamous carcinoma of 15 sites in the mouth and throat, with 16 participating institutions, though only data on three sites in the oropharynx reported by the six largest institutions are considered here. Patients entering the study were randomly assigned to one of two treatment groups, radiation therapy alone or radiation therapy together with a chemotherapeutic agent. One objective of the study was to compare the two treatment policies with respect to patient survival.

Approximately 30% of the survival times are censored owing primarily to patients surviving to the time of analysis. Some patients were lost to follow-up because the patient moved or transferred to an institution not participating in the study, though these cases were relatively rare. From a statistical point of view, the main feature of these data that distinguishes this example from the preceding one is the considerable lack of homogeneity between individuals being studied. Of course, as part of the study design, certain criteria for patient eligibility had to be met which eliminated extremes in the extent of disease, but still many factors are not controlled. This study included measurements of many covariates which would be expected to relate to survival experience. Six such variables are given in the data in Appendix 1 (sex, T staging, N staging, age, general condition, and grade). The site of the primary tumor and possible differences between participating institutions require consideration as well.

The TN staging classification gives a measure of the extent of the tumor at the primary site and at regional lymph nodes. T_1 refers to a small primary tumor, 2 centimeters or less in largest diameter, whereas T_4 is a massive tumor with extension to adjoining tissue. T_2 and T_3 refer to intermediate cases. N_0 refers to there being no clinical evidence of a lymph node metastasis and N_1, N_2, N_3 indicate, in increasing magnitude, the extent of existing lymph node involvement. Patients with classifications T_1N_0, T_1N_1, T_2N_0, or T_2N_1 or with distant metastases were excluded from study.

The variable general condition gives a measure of the functional capacity of the patient at the time of diagnosis (1 refers to no disability whereas 4 denotes bed confinement; 2 and 3 measure intermediate levels). The variable grade is a measure of the degree of differentiation of the tumor (the degree to which the tumor cell resembles the host cell) from 1 (well differentiated) to 3 (poorly differentiated).

In addition to the primary question whether the combined treatment mode is preferable to the conventional radiation therapy, it is of considerable interest to determine the extent to which the several covariates relate to subsequent survival. It is also imperative in answering the primary question to adjust the survivals for possible imbalance that may be present in the study with regard to the other covariates. Such problems are similar to those encountered in the classical theory of linear regression and the analysis of covariance. Again, the need to accommodate censoring is an important distinguishing point. In many situations it is also important to develop nonparametric and robust procedures since there is frequently little empirical or theoretical work to support a particular family of failure time distributions.

1.1.3 Heart Transplant Data

Crowley and Hu (1977) give survival times of potential heart transplant recipients from their date of acceptance into the Stanford heart transplant program. These data are reproduced in Appendix 1. One problem of considerable interest is to evaluate the effect of heart transplantation on subsequent survival.

For each study subject the explanatory variables age and prior surgery were recorded. There were also donor–recipient variables that may be predictive of post-transplant survival time. The main new feature here is that patients change treatment status during the course of the study. Specifically, a patient is part of the control group until a suitable donor is located and transplantation takes place, at which time he joins the treatment group. Correspondingly, some explanatory variables such as waiting time for transplant are observed during the course of the study and depend on the time elapsed to transplant. This study is examined in some detail in Chapter 5 using the ideas of time dependent covariates.

The existence of covariates that change over time is yet another problem unique to the analysis of failure time data. The heart transplant study is one example where such covariables arise because of the nature of the treatment. Alternatively, we can imagine a system operating under stress where the stress factor is varied as time elapses. In such a situation it would be common to examine the relationship between the stress applied and the instantaneous risk of failure. Other examples may arise in clinical studies: for instance, a measure of immune function may be recorded at regular intervals for leukemia patients in remission. One may wish to study the relationship between changes in immune function and corresponding propensity to relapse.

1.1.4 Accelerated Life Test

Nelson and Hahn (1972) presented data on the number of hours to failure of motorettes operating under various temperatures. The name "accelerated life test" for this type of study derives from the use of a stress factor, in this case temperature, to increase the rate of failure over that which would be observed under usual operating conditions. The data are presented in Table 1.2 and exhibit severe censoring, with only 17 out of 40 motorettes failing. Note that the stress (temperature) is constant for any particular motorette over time. The principal interest in such a study involves determination of the relationship between failure time and temperature for the purpose of extrapolating to usual running temperatures. Of course the validity of such extrapolation depends on the constancy of certain relationships over a very wide range of temperatures. For the present study, the failure time distribution at the design temperature of 130°C was of interest.

As in earlier examples, the censoring here is Type I or time censoring; that is, censored survival times were observed only if failure had not

Table 1.2 HOURS TO FAILURE OF MOTORETTES

150°C	All 10 motorettes without failure at 8064 hours
170°C	1764, 2772, 3444, 3542, 3780, 4860, 5196 3 motorettes without failure at 5448 hours
190°C	408, 408, 1344, 1344, 1440 5 motorettes without failure at 1680 hours
220°C	408, 408, 504, 504, 504 5 motorettes without failure at 528 hours

occurred prior to termination of the study. Experiments of this type, where considerable control is available to the experimenter, offer the possibility of other censoring schemes. For instance, in the above study it might have been decided in advance to continue the study until specified numbers of motorettes had failed at each of the temperatures (e.g., until one, three, five, and seven motorettes had failed at 150°C, 170°C, 190°C, and 220°C, respectively). Such censoring is usually referred to as Type II or order statistic censoring in that the study terminates as soon as certain order statistics are observed. With certain models, some inferential procedures (e.g., significance tests) are simpler for Type II than Type I censoring. It should be noted that Type II censoring usually does not allow an upper bound to be placed on the total duration of the study.

Some of the above examples are considered further throughout the book. We turn now, however, to mathematical representations of failure times and consider the very simplest case of an independent sample from a homogeneous population (no explanatory variables) with a single failure mode.

1.2 FAILURE TIME DISTRIBUTIONS

Let T be a nonnegative random variable representing the failure time of an individual from a homogeneous population. The probability distribution of T can be specified in many ways, three of which are particularly useful in survival applications: the survivor function, the probability density function, and the hazard function. Interrelations between these three representations are given below for both discrete and continuous distributions.

The survivor function is defined for both discrete and continuous distributions as the probability that T is at least as great as a value t; that is,

$$F(t) = P(T \geq t), \qquad 0 < t < \infty.$$

Note that F in some settings refers to the cumulative distribution function and therefore gives the probabilities in the left tail rather than the right tail of the distribution. The notation used here, however, is more convenient for the incorporation of censoring. Clearly $F(t)$ is a monotone nonincreasing left continuous function with $F(0) = 1$ and $\lim_{t\to\infty} F(t) = 0$. The probability density and hazard functions are most easily specified separately for discrete and continuous T.

1.2.1 *T* (Absolutely) Continuous

The probability density function (p.d.f.) of T is

$$f(t) = \lim_{\Delta t \to 0^+} \frac{P(t \leq T < t + \Delta t)}{\Delta t}$$

$$= \frac{-dF(t)}{dt}.$$

Conversely, $F(t) = \int_t^\infty f(s)\, ds$ and $f(t) \geq 0$ with $\int_0^\infty f(t)\, dt = 1$. The range of T is $[0, \infty)$, and this should be understood as the domain of definition for functions of t.

The hazard function specifies the instantaneous rate of failure at $T = t$ conditional upon survival to time t and is defined as

$$\lambda(t) = \lim_{\Delta t \to 0^+} \frac{P(t \leq T < t + \Delta t | T \geq t)}{\Delta t}$$

$$= \frac{f(t)}{F(t)}. \tag{1.1}$$

It is easily seen that $\lambda(t)$ specifies the distribution of T since, from (1.1),

$$\lambda(t) = \frac{-d \log F(t)}{dt}$$

so that integrating and using $F(0) = 1$, we obtain

$$F(t) = \exp\left(-\int_0^t \lambda(u)\, du\right). \tag{1.2}$$

The p.d.f. of T can be written

$$f(t) = \lambda(t) \exp\left(-\int_0^t \lambda(u)\, du\right). \tag{1.3}$$

Examination of (1.2) indicates that $\lambda(t)$ is a nonnegative function with

$$\int_0^s \lambda(u)\,du < \infty$$

for some $s > 0$ and $\int_0^\infty \lambda(u)\,du = \infty$.

Other representations of the failure time distribution are occasionally useful. An example is the expected residual life at time t,

$$r(t) = E(T - t \mid T \geq t), \qquad 0 \leq t \leq \infty,$$

which uniquely determines a continuous survival distribution with finite mean, as can be seen by noting that

$$\begin{aligned} r(t) &= \int_t^\infty \frac{(u-t)f(u)\,du}{F(t)} \\ &= \int_t^\infty \frac{F(u)\,du}{F(t)}, \end{aligned} \tag{1.4}$$

upon integration by parts. Equivalently,

$$-\frac{1}{r(t)} = \frac{d}{dt} \log \int_t^\infty F(u)\,du.$$

Substituting $t = 0$ in (1.4) gives $r(0) = \int_0^\infty F(u)\,du$ and hence

$$\int_0^t \frac{du}{r(u)} = -\log \int_t^\infty F(u)\,du + \log r(0)$$

which leads finally to

$$F(t) = \frac{r(0)}{r(t)} \exp\left(-\int_0^t \frac{du}{r(u)}\right).$$

1.2.2 *T* Discrete

If T is a discrete random variable taking values $x_1 < x_2 < \cdots$ with associated probability function

$$f(x_i) = P(T = x_i), \qquad i = 1, 2, \ldots,$$

then the survivor function is

$$\begin{aligned} F(t) &= \sum_{j \mid x_j \geq t} f(x_j) \\ &= \sum f(x_j) H(x_j - t), \end{aligned}$$

where $H(x)$ is the Heaviside function

$$H(x) = \begin{cases} 0, & x < 0 \\ 1, & x \geq 0. \end{cases}$$

The hazard at x_j is defined as the conditional probability of failure at x_j,

$$\lambda_j = P(T = x_j | T \geq x_j)$$
$$= \frac{f(x_j)}{F(x_j)}, \qquad j = 1, 2, \ldots .$$

Corresponding to (1.2) and (1.3), the survivor function and the probability function are given by

$$F(t) = \prod_{j | x_j < t} (1 - \lambda_j) \tag{1.5}$$

and

$$f(x_j) = \lambda_j \prod_{1}^{j-1} (1 - \lambda_i). \tag{1.6}$$

1.2.3 Discussion

More generally, the distribution of T may have both discrete and continuous components, in which case the density function is a sum of discrete and continuous parts. A hazard function can be defined as above. The survivor function obtained from it involves products of terms like (1.2) and (1.5). Specifically, if λ_c denotes the hazard function for the continuous part and mass points occur at $x_1 < x_2 < \cdots$, then the overall survivor function can be written

$$F(t) = \exp\left[-\int_0^t \lambda_c(u)\, du \right] \prod_{j | x_j < t} (1 - \lambda_j).$$

The discrete, mixed, and continuous cases can be combined. We write the hazard function as

$$\lambda(t)\, dt = \lambda_c(t)\, dt + \sum \lambda_j \delta(t - x_j)\, dt$$

where $\delta(t - x_j)$ is the Dirac delta function defined so that

$$\delta(x)\, dx = \begin{cases} 1, & x = 0 \\ 0, & \text{otherwise.} \end{cases}$$

The cumulative hazard is now defined as

$$\Lambda(t) = \int_0^t \lambda(u)\, du = \int_0^t \lambda_c(u)\, du + \sum_{j | x_j < t} \lambda_j$$

where the Dirac delta components define the discrete contributions to the integral. The survivor function in the discrete, continuous, or mixed cases can then be written as

$$F(t) = \mathop{\mathscr{P}}_0^t [1 - d\Lambda(u)]$$

where the product integral $\mathscr{P}$ is defined by

$$\mathop{\mathscr{P}}_0^t [1 - d\Lambda(u)] = \lim \prod_1^r \{1 - [\Lambda(u_k) - \Lambda(u_{k-1})]\}.$$

Here $0 = u_0 < u_1 < \cdots < u_r = t$ and the limit is taken as $r \to \infty$ and $u_k - u_{k-1} \to 0$. Alternatively, we might write

$$F(t) = \mathop{\mathscr{P}}_0^t [1 - \lambda(u)\, du]$$

where again the Dirac delta functions handle the discrete contributions. It is easily seen that

$$\mathop{\mathscr{P}}_0^t [1 - \lambda_c(u)\, du] = \exp\left[-\int_0^t \lambda(u)\, du\right].$$

This unification shows that failure time data can be considered to arise in both the discrete and the continuous cases in essentially the same way. The product representation is similar to a coin tossing experiment in which the head probability varies over time. The coin is tossed repeatedly and failure corresponds to the first occurrence of a tail. Thus in general the survival probability at time t is obtained by taking the product of the conditional survival probabilities for infinitesimal intervals up to time t. This way of viewing a failure mechanism has led to many of the recent developments in the area. In effect, it is possible to examine survival experience by looking at the survival experience over each interval conditional upon the experience to that point. The construction of the likelihood for censored survival data in Section 5.2 depends on this as do the partial likelihood analysis of the proportional hazards model in Section 5.4, the multivariate failure time models of Chapter 7, and the Bayesian analyses of Section 8.4.

Note that $f(t)$ and $F(t)$ [or more usually the cumulative distribution function $P(T \leq t)$] are common representations of the distribution of a random variable. The hazard function $\lambda(t)$ is a more specialized characterization but is particularly useful in modeling survival time data. In many instances, information is available as to how the failure rate will change with the amount of time on test. This information can be used to model $\lambda(t)$ and easily translated into implications for $F(t)$ and $f(t)$ using the

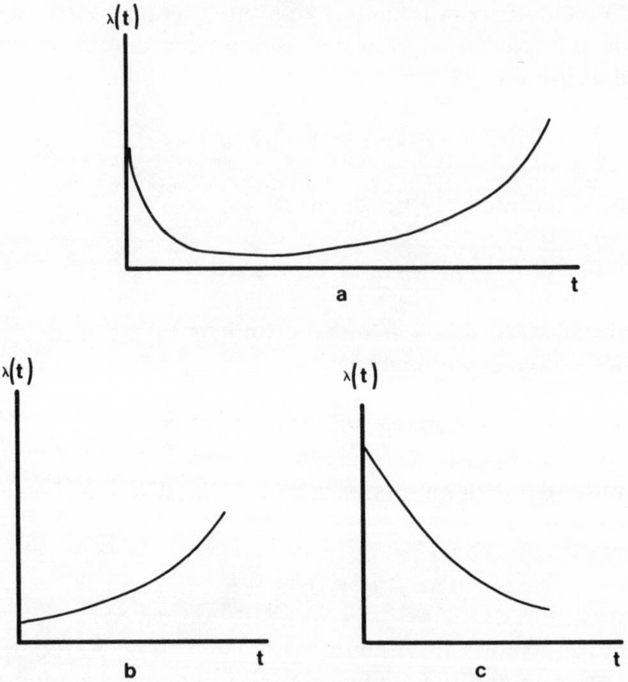

Figure 1.1 Some types of hazard functions: (a) hazard for human mortality; (b) positive aging; (c) negative aging.

above formulas. For example, in modeling age at death of human populations, it is clear that $\lambda(t)$ is initially large owing to infant mortality and childhood diseases. This is followed by a period of relatively low mortality after which the mortality rate increases very rapidly (c.f. Figure 1.1a). In other applications, monotone increasing hazards (positive aging) or decreasing hazards (negative aging) may be suggested (Figures 1.1b and c). Such qualitative information on $\lambda(t)$ is often useful in selecting a family of probability models for T. Chapter 2 discusses and examines some commonly used models for failure time and their associated hazard functions.

1.3 ESTIMATION OF THE SURVIVOR FUNCTION

The sample cumulative distribution function (c.d.f.)

$$\bar{F}_n(x) = \frac{\text{No. of sample values} \leq x}{n}$$

is a familiar and convenient way to summarize and display data. A plot of $\bar{F}_n(x)$ versus x visually represents the sample and provides full information on the percentile points, the dispersion, and the general features of the sample distribution. Besides these obvious descriptive uses, it is an indispensable aid in studying the distributional shape of the population from which the sample arose; in fact, the sample distribution function can serve as a basic tool in constructing formal tests of goodness of fit of the data to hypothesized probability models (c.f., for example, Cox and Hinkley, 1974, pp. 69ff.).

In the analysis of survival data, it is very often useful to summarize the survival experience of particular groups of patients in terms of the sample c.d.f., or more usually in terms of the sample survivor function. If an uncensored sample of n distinct failure times is observed from a homogeneous population, the sample survivor function is a step function decreasing by n^{-1} immediately following each observed failure time. A simple adjustment accommodates any ties present in the data. However, as noted earlier, survival data very often involve right censoring and a methodology for handling this with convenience is required.

Let $t_1 < t_2 < \cdots < t_k$ represent the observed failure times in a sample of size n_0 from a homogeneous population with survivor function $F(t)$. Suppose that d_j items fail at t_j $(j = 1, \ldots, k)$ and m_j items are censored in the interval $[t_j, t_{j+1})$ at times $t_{j1}, \ldots, t_{jm_j}$ $(j = 0, \ldots, k)$, where $t_0 = 0$ and $t_{k+1} = \infty$. Let $n_j = (m_j + d_j) + \cdots + (m_k + d_k)$ be the number of items at risk at a time just prior to t_j. The probability of failure at t_j is

$$F(t_j) - F(t_j + 0),$$

where $F(t_j + 0) = \lim_{x \to 0^+} F(t_j + x)$ $(j = 1, \ldots, k)$. We assume that the contribution to the likelihood of a survival time censored at t_{jl} is

$$P(T > t_{jl}) = F(t_{jl} + 0).$$

In effect, we are assuming that the observed censoring time t_{jl} tells us only that the unobserved failure time is greater than t_{jl}. Conditions under which such an assumption would be appropriate are discussed in Sections 3.2 and 5.2, but note that this would be appropriate, for example, if censoring times were fixed in advance for each item. Thus we obtain

$$\mathscr{L} = \prod_{j=0}^{k} \left\{ [F(t_j) - F(t_j + 0)]^{d_j} \prod_{l=1}^{m_j} F(t_{jl} + 0) \right\}$$

which, given the data, can be viewed as a likelihood function on the space of survivor functions $F(t)$. The maximum likelihood estimate is the survivor function $\hat{F}(t)$ that maximizes $\mathscr{L}$. This definition of the maximum likelihood estimate is a generalization of the usual concept used in parametric models. The results of such problems where many parameters are involved in the maximization should be treated with care. There are dangers associated with maximizing likelihoods of many parameters since such techniques may lead to inefficient or inconsistent estimates. The results of such maximizations require some investigation to assure they are reasonable.

Clearly $\hat{F}(t)$ is discontinuous at the observed failure times since otherwise $\mathscr{L} = 0$. Further, subject to $t_{jl} \geq t_j$, $F(t_{jl} + 0)$ is maximized by taking $F(t_{jl} + 0) = F(t_j + 0)$ $(j = 1, \ldots, k,\ l = 1, \ldots, m_j)$ and $F(t_{0l}) = 1$ $(l = 1, \ldots, m_0)$. The function $\hat{F}(t)$ is then a discrete survivor function with hazard components $\hat{\lambda}_1, \ldots, \hat{\lambda}_k$ at $t_1, \ldots, t_k$, respectively. Thus

$$\hat{F}(t_j) = \prod_{l=1}^{j-1} (1 - \hat{\lambda}_l) \tag{1.7}$$

$$\hat{F}(t_j + 0) = \prod_{l=1}^{j} (1 - \hat{\lambda}_l) \tag{1.8}$$

where the $\hat{\lambda}_l$'s are chosen to maximize the function

$$\prod_{j=1}^{k} \left\{ \left[\prod_{l=1}^{j-1} (1 - \lambda_l)^{d_j} \right] \lambda_j^{d_j} \left[\prod_{l=1}^{j} (1 - \lambda_l)^{m_j} \right] \right\} = \prod_{j=1}^{k} \lambda_j^{d_j} (1 - \lambda_j)^{n_j - d_j} \tag{1.9}$$

obtained by substitution of (1.7) and (1.8) in $\mathscr{L}$. Clearly $\hat{\lambda}_j = d_j / n_j$ $(j = 1, \ldots, k)$ and the *product limit estimate* of the survivor function is

$$\hat{F}(t) = \prod_{j \mid t_j < t} \left(\frac{n_j - d_j}{n_j} \right). \tag{1.10}$$

In obtaining the product limit estimate, we are in effect making the conditional probability of failure at each t_j agree exactly with the observed conditional relative frequency of failure at t_j given by d_j / n_j. A slightly problematic point is that $\hat{F}(t)$ never reduces to zero if $m_k > 0$. For this reason $\hat{F}(t)$ is usually taken to be undefined for $t > t_{km_k}$.

The "nonparametric maximum likelihood estimate" (1.10) needs a word of explanation since measure theoretic difficulties arise in interpreting $\mathscr{L}$. If one considers the data to have arisen from a continuous distribution with a small measurement error associated with each observed failure time (as is the actual situation), the first factor of $\mathscr{L}$ becomes $\Pi[F(t_j - \delta_j) - F(t_j + \delta_i)]^{d_j}$ where the interval $[t_j - \delta_j, t_j + \delta_j) = A_j$ is defined by the precision of measurement. In order to avoid difficulties with censoring times tied with failure times, we adopt the convention of moving such censoring times a small amount Δ to the right. Consider now $\delta_j \to 0$ and suppose that δ_j is small enough that all censoring points are excluded from A_j. With δ_j so chosen, standard maximum likelihood arguments show that any continuous survivor function $\tilde{F}_\delta$ with $\tilde{F}_\delta(t_j - \delta_j)$ and $\tilde{F}_\delta(t_j + \delta_j)$ given by (1.7) and (1.8), respectively, is a maximum likelihood estimate. As the $\delta_j \to 0$, any maximum likelihood estimate $\tilde{F}_\delta$ converges to the product limit estimate (1.10).

The estimate $\hat{F}(t)$ is the direct generalization of the sample survivor function for censored data. It was derived by Kaplan and Meier (1958) and as a consequence, is often referred to as the *Kaplan–Meier estimate.* Table 1.3 and Figure 1.2 exemplify the Kaplan–Meier estimate (1.10) for the carcinogenesis data of Section 1.1.1.

A heuristic derivation of an asymptotic variance formula for the product limit estimator can be obtained by regarding (1.9) as an ordinary parametric likelihood in the parameters $\lambda_1, \ldots, \lambda_k$. Standard likelihood methods, as reviewed in Section 3.4, would yield an estimator $n_j^{-3}[d_j(n_j - d_j)]$ for the asymptotic variance of λ_j and hence, for

$$\log \hat{F}(t) = \sum_{j|t_j<t} \log(1 - \hat{\lambda}_j),$$

an asymptotic variance estimator of

$$\begin{aligned}\widehat{\mathrm{Var}}[\log \hat{F}(t)] &= \sum_{j|t_j<t} (1 - \hat{\lambda}_j)^{-2}\, \widehat{\mathrm{Var}}(1 - \hat{\lambda}_j) \\ &= \sum_{j|t_j<t} \frac{d_j}{n_j(n_j - d_j)}.\end{aligned}$$

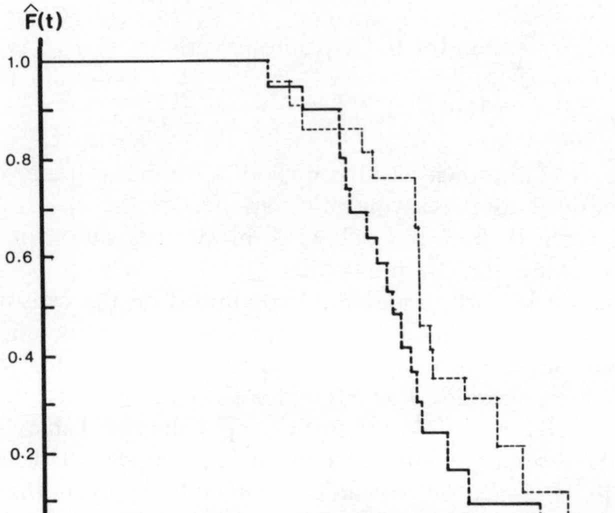

Figure 1.2 Kaplan–Meier survivor function estimates for the carcinogenesis data: solid line, group 1; broken line, group 2.

Table 1.3 KAPLAN–MEIER SURVIVOR FUNCTION ESTIMATE FOR CARCINOGENESIS DATA

Group 1					Group 2				
t_j	d_j	n_j	$\hat{F}(t_j+0)$	$\widehat{\text{Var}}(\hat{F})$	t_j	d_j	n_j	$\hat{F}(t_j+0)$	$\widehat{\text{Var}}(\hat{F})$
143	1	19	.947	.00262	142	1	21	.952	.00216
164	1	18	.895	.00496	156	1	20	.905	.00410
188	2	17	.789	.00875	163	1	19	.857	.00583
190	1	15	.737	.01021	198	1	18	.810	.00734
192	1	14	.684	.01137	205	1	16	.759	.00885
206	1	13	.632	.01225	232	2	15	.658	.01109
209	1	12	.579	.01283	233	4	13	.455	.01240
213	1	11	.526	.01312	239	1	9	.405	.01208
216	1	10	.474	.01312	240	1	8	.345	.01148
220	1	8	.414	.01311	261	1	7	.304	.01067
227	1	7	.355	.01264	280	2	6	.202	.00814
230	1	6	.296	.01170	296	2	4	.101	.00459
234	1	5	.237	.01029	323	1	2	.051	.00243
246	1	3	.158	.00873					
265	1	2	.079	.00530					
304	1	1	.000						

The induced expression for the asymptotic variance of $\hat{F}(t)$ is then

$$\widehat{\text{Var}}[\hat{F}(t)] = \hat{F}^2(t) \sum_{j|t_j<t} \frac{d_j}{n_j(n_j - d_j)}. \tag{1.11}$$

Expression (1.11), known as Greenwood's formula (Greenwood, 1926), was first derived as the asymptotic variance of the classical life table estimator, which is discussed below. A proper derivation of (1.11) with continuous failure time requires that (1.10) be viewed as a stochastic process in t. Such work, under mild conditions on the censoring, shows (1.10) to be asymptotically a Gaussian process so that one may calculate, for example, an approximate 95% confidence interval for the survivor function at some specified t as $\hat{F}(t) \pm 1.96[\widehat{\text{Var}}\,\hat{F}(t)]^{1/2}$. At extreme values of t (e.g., $t \le 188$ or $t > 246$ for the Group 1 data of Table 1.3) such an approximate confidence interval may include impossible values outside the range $[0, 1]$. This problem can be avoided by applying the asymptotic normal distribution to a transformation of $F(t)$ for which the range is unrestricted. For example, the asymptotic variance of

$$\hat{v}(t) = \log[-\log \hat{F}(t)]$$

is, from Greenwood's formula and asymptotic maximum likelihood theory (Section 3.4), estimated by

$$\hat{s}^2(t) = \frac{\sum_{j|t_j<t} \frac{d_j}{n_j(n_j - d_j)}}{\left[\sum_{j|t_j<t} \log\left(\frac{n_j - d_j}{n_j}\right)\right]^2}.$$

An asymptotic 95% confidence interval of $\hat{v}(t) \pm 1.96\hat{s}(t)$ for $v(t) = \log[-\log F(t)]$ gives a corresponding asymptotic 95% confidence interval for $F(t)$ of

$$\hat{F}(t) \exp[\pm 1.96\hat{s}(t)] \tag{1.12}$$

which takes values in [0, 1]. Application of this method to the Group 1 data of Table 1.1 gives an approximate 95% interval for $F(t)$ at $t = 150$ of (.679, .992). A normal approximation to the distribution of $\hat{F}(150)$, in contrast, gives (.847, 1.047).

In addition to the avoidance of impossible values, parameter transformations can improve the adequacy of normal approximations. Chapter 3 gives some further discussion of this topic in some simpler settings. A detailed evaluation of parameter transformations for survivor function estimation does not seem to have been published.

Many other estimators of the survivor function have also been considered. The oldest is that formed from the life table (see, for example, Chiang, 1968). The life table is a summary of the survival data grouped into convenient intervals. In some applications (e.g., actuarial), the data are often collected in such a grouped form. In other cases, the data might be grouped to get a simpler and more easily understood presentation. If, for example, the data are grouped into intervals $I_1, \ldots, I_k$ such that $I_j = [b_0 + b_1 + \cdots + b_{j-1}, b_0 + \cdots + b_j)$ is of width b_j with $b_0 = 0$ and $b_k = \infty$, the life table would present the number of failures and censored survival times falling in each interval.

Suppose that m_j censored times and d_j failure times fall in the interval I_j and let $n_j = \Sigma_{l \geq j}(d_l + m_l)$ be the number of individuals at risk at the start of the jth interval. The standard life table estimator of the conditional probability of failure in I_j given survival to I_j is $\hat{q}_j = 1$ if $n_j = 0$ and

$$\hat{q}_j = \frac{d_j}{n_j - m_j/2}$$

otherwise. The $m_j/2$ term in the denominator is used in an attempt to adjust for the fact that not all the n_j individuals are at risk for the whole of I_j. The corresponding life table estimator of the survivor function at the end of I_j is

$$\tilde{F}(b_1 + \cdots + b_j) = \prod_{i=1}^{j} (1 - \hat{q}_i). \tag{1.13}$$

Greenwood's formula (1.11), with n_j replaced by $n_j - m_j/2$, provides an estimator of the variance of $\tilde{F}$.

The life table method is primarily designed for situations in which actual failure and censoring times are unavailable and only the d_j's and m_j's are given for the jth interval. A simple modification of the life table method utilizes the additional information when the survival times are known. Suppose, for example, that $t_{j1}, \ldots, t_{jr_j}$ are the observed times in I_j of which m_j are censored and d_j are failures, $r_j = d_j + m_j$ $(j = 1, \ldots, k)$. Suppose the hazard function $\lambda(t)$ is taken to be a step function having the constant value λ_j in the interval I_j. It is in this case an easy exercise to show that the maximum likelihood estimate of λ_j is

$$\hat{\lambda}_j = \frac{d_j}{S_j}$$

where

$$S_j = \sum_{l=1}^{r_j} \left(t_{jl} - \sum_{0}^{j-1} b_i \right) + n_{j+1} b_j$$

is the total observed survival time in the interval I_j. The corresponding estimator of the survivor function is for $t \in I_j$,

$$\hat{\hat{F}}(t) = \exp\left[-\hat{\lambda}_j \left(t - \sum_{0}^{j-1} b_l \right) - \sum_{1}^{j-1} \hat{\lambda}_i b_i \right]. \tag{1.14}$$

This, unlike the preceding estimators, is a continuous function of t. There is, however, an arbitrariness in the choice of intervals.

The problem of estimating the survivor function in the presence of covariates is discussed in Sections 4.3 and 4.8.

1.4 COMPARISON OF SURVIVAL CURVES

Often it is of interest to determine whether two or more samples could have arisen from identical survivor functions. One approach would involve the use of the asymptotic results for $\hat{F}(t)$ mentioned above to devise a test for equality of the survivor functions at some prespecified time t. Such a procedure, however, would not usually make efficient use of the available data, and attention in recent years has turned instead to test statistics that attempt to summarize differences between survivor function estimators over the whole of the study period. The most commonly used statistics of this type can be viewed as censored data generalizations of such familiar rank tests as the Wilcoxon test and the Savage (exponential scores) test. In this chapter, only a heuristic derivation of the generalized Savage (1956) or *log-rank* test is given. This test is particularly good when the ratio of hazard functions in the populations being compared is approximately constant. It can also be advocated on the basis of ease of presentation to nonstatistical personnel since the test statistic is the difference between the observed number of failures in each group and a quantity that, for most purposes, can be thought of as the corresponding expected number of failures under the null hypothesis.

Suppose one wishes to test equality of the survivor functions $F_1(t), \ldots, F_r(t)$ on the basis of samples from each of r populations. Let $t_1 < t_2 < \cdots < t_k$ denote the failure times for the sample formed by pooling the r individual samples. Suppose d_j failures occur at t_j and that n_j study subjects are at risk just prior to t_j $(j = 1, \ldots, k)$ and let d_{ij} and n_{ij} be the corresponding numbers in sample i $(i = 1, \ldots, r)$. The data at t_j are in the form of a $2 \times r$ contingency table with d_{ij} failures and $n_{ij} - d_{ij}$ survivors in the ith row $(i = 1, \ldots, r)$. Conditional on the failure and censoring experience up to time t_j the distribution of $d_{1j}, \ldots, d_{rj}$ is simply the product of binomial distributions

$$\prod_{i=1}^{r} \binom{n_{ij}}{d_{ij}} \lambda_j^{d_{ij}} (1 - \lambda_j)^{n_{ij} - d_{ij}}$$

where λ_j is the conditional failure probability at t_j, which is common for each of the r samples under the null hypothesis. The conditional distribution for $d_{1j}, \ldots, d_{rj}$ given d_j is then the hypergeometric distribution

$$\frac{\prod_1^r \binom{n_{ij}}{d_{ij}}}{\binom{n_j}{d_j}}. \tag{1.15}$$

The mean and variance of d_{ij} from (1.15) are, respectively,

$$w_{ij} = n_{ij} d_j n_j^{-1}$$

and

$$(V_j)_{ii} = n_{ij}(n_j - n_{ij}) d_j (n_j - d_j) n_j^{-2} (n_j - 1)^{-1}$$

The covariance of d_{ij} and d_{lj} is

$$(V_j)_{il} = -n_{ij} n_{lj} d_j (n_j - d_j) n_j^{-2} (n_j - 1)^{-1}.$$

Thus the statistic $\mathbf{v}_j' = (d_{ij} - w_{ij}, \ldots, d_{rj} - w_{rj})$ has (conditional) mean zero

and Grunkemeier (1975) investigate the nominal significance levels using simulations and also consider a generalized likelihood ratio test. Peto et al. (1977) suggest that the variance estimator (1.11) is inaccurate in the tail of the distribution and give an alternative estimator. The use of transformations to improve the normal approximation does not seem to have been considered.

The discussion of the log-rank test in Section 1.4 is similar to that by Mantel (1966). The log-rank test is discussed further in Chapter 4. Chapter 6 deals more completely with the derivations and applications of censored data rank tests, and additional references are given there.

Censoring mechanisms, although mentioned briefly above, are considered in more detail in Chapters 3 and 5, where references are given.

CHAPTER 2

Failure Time Models

2.1 INTRODUCTION

Even though our main interest in this book concerns the relationship between failure time and explanatory variables, it is necessary to consider briefly failure time distributions for homogeneous populations. Throughout the literature on failure time data, certain parametric models have been used repeatedly; exponential and Weibull models, for example, are often used. These distributions admit closed form expressions for tail area probabilities and thereby simple formulas for survivor and hazard functions. Log-normal and gamma distributions are generally less convenient computationally but are still frequently applied.

Section 2.2 discusses some of the standard failure time models for homogeneous populations. The properties and the theoretical bases of these distributions are considered here only briefly. These have been discussed in some detail by Johnson and Kotz (1970a, 1970b) and Mann et al. (1974) for many of the models introduced. Section 2.3 generalizes the failure time models to include regressor variables. Section 2.4 considers discrete regression models.

2.2 SOME (CONTINUOUS) PARAMETRIC FAILURE TIME MODELS

As before, $T > 0$ is a random variable representing failure time and t represents a typical point in its range. We use $Y \equiv \log T$ to represent the log failure time and summarize the survival distributions in terms of both T and Y. Shape comparisons among the parametric models are often simpler in terms of Y than T.

2.2.1 The Exponential Distribution

The one parameter exponential distribution is obtained by taking the hazard function to be a constant, $\lambda(t) = \lambda > 0$, over the range of T. The instantaneous failure rate is independent of t so that the conditional chance of failure in a time interval of specified length is the same regardless of how long the individual has been on trial; this is referred to as the memoryless property of the exponential distribution. The survivor function and density functions of T are, respectively,

$$F(t) = e^{-\lambda t} \qquad \text{and} \qquad f(t) = \lambda e^{-\lambda t}$$

from (1.2) and (1.3). Figure 2.1 gives sketches of the functions $\lambda(t)$, $f(t)$, and $F(t)$ for the exponential distribution. An empirical check of the appropriateness of the exponential model for a set of survival data is provided by plotting the log of a survivor function estimate versus t. Such a plot should approximate a straight line through the origin. As can be deduced from Table 1.1, the carcinogenesis data of Section 1.1.1 is not well described by a single parameter exponential model, primarily owing to the absence of tumor incidence within the first 140 days.

The p.d.f. of $Y \equiv \log T$ is

$$\exp(y - \alpha - e^{y-\alpha}), \qquad -\infty < y < \infty,$$

where $\alpha = -\log \lambda$. Letting $Y = \alpha + W$, the p.d.f. of W is

$$\exp(w - e^{w}), \qquad -\infty < w < \infty, \tag{2.1}$$

which is an extreme value (minimum) distribution. This distribution derives its name from its appearance as the limiting distribution of a standardized form of the minimum of a sample selected from a continuous distribution with support on $(-\infty, a)$ for some $a \leq \infty$. Details of

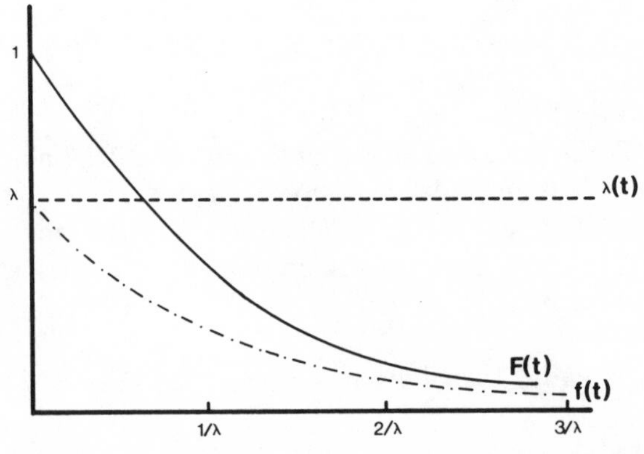

Figure 2.1 Hazard function, density function, and survivor function for the single parameter exponential model. Note λ may exceed 1.

the derivation are given, for example, by Johnson and Kotz (1970a, Chapter 2.1). The exponential distribution arises, also, as the limiting form of the distribution of a minimum of samples from some densities with support on $(0, \infty)$. (See Problem 6 in Appendix 2 for details.) This can sometimes be taken as theoretical justification for its use in survival studies in which a complex mechanism fails when any one of its many components fails. The extreme value distribution (2.1) is a unimodal distribution with skewness -1.14 and kurtosis 2.4. The mean of (2.1) is $-\gamma$ ($\gamma = .5722\ldots$ is Euler's constant) and the variance is $\pi^2/6 = 1.6449\ldots$. The moment generating function is

$$M_W(\theta) = E(e^{\theta W}) = \Gamma(\theta + 1), \qquad \theta > -1,$$

where

$$\Gamma(k) = \int_0^\infty x^{k-1} e^{-x}\, dx$$

is the gamma function.

2.2.2 The Weibull Distribution

An important generalization of the exponential distribution allows for a power dependence of the hazard on time. This yields the two parameter Weibull distribution with hazard function

$$\lambda(t) = \lambda p (\lambda t)^{p-1}$$

for λ, $p > 0$. This hazard (see Figure 2.2) is monotone decreasing for

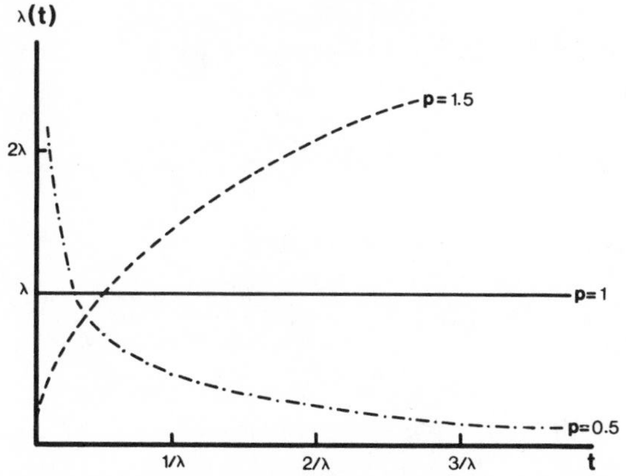

Figure 2.2 Hazard functions for the two parameter Weibull model.

$p<1$, increasing for $p>1$, and reduces to the constant exponential hazard if $p=1$. The probability density function is

$$f(t)=\lambda p(\lambda t)^{p-1}\exp[-(\lambda t)^p]$$

and the survivor function is

$$F(t)=\exp[-(\lambda t)^p].$$

Clearly

$$\log[-\log F(t)]=p(\log t+\log\lambda)$$

so that an empirical check for the Weibull distribution is provided by a plot of $\log[-\log \hat{F}(t)]$ versus $\log t$ where $\hat{F}(t)$ is a sample (Kaplan–Meier) estimate of the survivor function. The plot should give approximately a straight line, the slope of which provides a rough estimate of p and the $\log t$ intercept an estimate of $-\log\lambda$.

The p.d.f. of the log failure time Y is

$$\sigma^{-1}\exp\left(\frac{y-\alpha}{\sigma}-e^{(y-\alpha)/\sigma}\right),\qquad -\infty<y<\infty,$$

where $\sigma=p^{-1}$ and $\alpha=-\log\lambda$. More simply, we can write $Y=\alpha+\sigma W$, where W has the extreme value p.d.f. (2.1) as before. The shape of the density for Y is fixed because λ and p affect only the location and the scaling of the distribution. The Weibull distribution can also be developed as the limiting distribution of the minimum of a sample from a continuous distribution with support on $[0, u)$ for some u $(0<u<\infty)$.

2.2.3 The Log-Normal Distribution

Again, the model for $Y=\log T$ is of the form $Y=\alpha+\sigma W$, where W is a standard normal variate with density

$$\phi(w)=\frac{e^{-w^2/2}}{\sqrt{2\pi}}. \tag{2.2}$$

The density function for T can be written

$$f(t)=(2\pi)^{-1/2}pt^{-1}\exp\left(\frac{-p^2(\log\lambda t)^2}{2}\right)$$

where as before $\alpha=-\log\lambda$ and $\sigma=p^{-1}$. The survivor and hazard functions involve the incomplete normal integral

$$\Phi(w)=\int_{-\infty}^{w}\phi(u)\,du.$$

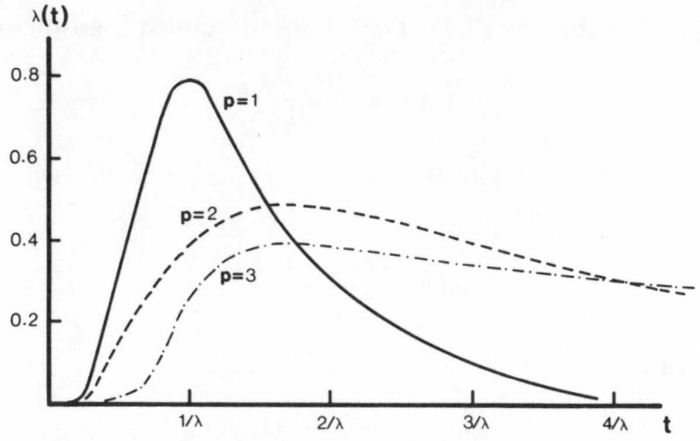

Figure 2.3 Hazard functions of log-normal distributions.

The survivor function is

$$F(t) = 1 - \Phi(p \log \lambda t)$$

and the hazard function is $f(t)/F(t)$. The hazard function has value 0 at $t = 0$, increases to a maximum and then decreases, approaching zero as t becomes large. The log-normal hazard is illustrated in Figure 2.3. The log-normal model is particularly simple to apply if there is no censoring, but with censoring the computations quickly become formidable. The log-logistic distribution of Section 2.2.6 provides a good approximation to the log-normal distribution and may frequently be a preferable survival time model.

2.2.4 The Gamma Distribution

As noted above, the Weibull distribution is a two parameter generalization of the exponential model; another such generalization is the gamma distribution with density function

$$f(t) = \frac{\lambda(\lambda t)^{k-1} e^{-\lambda t}}{\Gamma(k)}$$

where k, $\lambda > 0$. The model for $Y = \log T$ can be written $Y = \alpha + W$, where W has the density

$$\frac{\exp(kw - e^w)}{\Gamma(k)}. \tag{2.3}$$

The error quantity W has a negatively skewed distribution with skewness decreasing with increasing k. At the exponential model $k = 1$, W has the extreme value distribution (2.1). The moment generating function of W is

$$M(\theta) = \frac{\Gamma(\theta + k)}{\Gamma(k)}$$

from which it is easily shown that the mean and variance of (2.3) are the digamma function

$$\psi(k) = \frac{d \log \Gamma(k)}{dk}$$

and the trigamma function

$$\psi^{(1)}(k) = \frac{d^2 \log \Gamma(k)}{dk^2},$$

respectively. These are tabulated, for example, in Abramowitz and Stegun (1965).

It is of interest to note that W, suitably standardized, has a limiting normal distribution as $k \to \infty$: from Stirling's formula,

$$\log \Gamma(k) = -k + (k - \tfrac{1}{2}) \log k + \log \sqrt{2\pi} + O\left(\frac{1}{k}\right),$$

it is easily seen that $\psi(k) = \log k + O(k^{-1})$ and $\psi^{(1)}(k) = k^{-1} + O(k^{-2})$. We consider, then, the standardized variate

$$W^* = \sqrt{k}(W - \log k)$$

and obtain

$$\lim_{k \to \infty} M_{W^*}(\theta) = \exp\left(\frac{\theta^2}{2}\right)$$

where $M_{W^*}(\theta)$ is the moment generating function of W^*. Thus as $k \to \infty$, the distribution of W^* converges to that of a standard normal variate.

The survivor and hazard functions of the gamma distribution involve the incomplete gamma integral

$$I_k(s) = \frac{\int_0^s x^{k-1} e^{-x}\, dx}{\Gamma(k)}$$

and are, respectively,

$$F(t) = 1 - I_k(\lambda t)$$

and

$$\lambda(t) = \frac{\lambda(\lambda t)^{k-1} \exp(-\lambda t) \Gamma(k)^{-1}}{1 - I_k(\lambda t)}.$$

The hazard function is monotone increasing from 0 if $k > 1$, monotone decreasing from ∞ if $k < 1$, and in either case approaches λ as t becomes large. If $k = 1$, the gamma distribution reduces to the exponential distribution. With integer k, the gamma distribution is sometimes called a special Erlangian distribution.

The gamma distribution with integer k (and the exponential distribution, $k = 1$) can be derived as the distribution of the waiting time to the kth emission from a Poisson source with intensity parameter λ. As a side result, it is apparent that the sum of k independent exponential variates with failure rate λ has the gamma distribution with parameters λ and k.

2.2.5 Generalized Gamma Distribution

The gamma family of Section 2.2.4 can be generalized by incorporating a scale parameter σ in the model for $Y = \log T$ to give $Y = \alpha + \sigma W$ where W has the distribution with density (2.3). This three parameter model was introduced by Stacy (1962) and includes as special cases the exponential ($\sigma = k = 1$), the gamma ($\sigma = 1$), and the Weibull ($k = 1$). The log-normal is also a limiting special case as $k \to \infty$.

The p.d.f. for T can be written

$$f(t) = \frac{\lambda p(\lambda t)^{pk-1} \exp[-(\lambda t)^p]}{\Gamma(k)}, \qquad t > 0,$$

where $\lambda = e^{-\alpha}$ and $p = \sigma^{-1}$. The hazard function incorporates a variety of shapes as indicated by the special cases. The distribution is most easily visualized in terms of Y, the log survival time.

2.2.6 Log-Logistic Distribution

Other failure time models can be constructed by selecting different distributions for the error variable W in $Y = \alpha + \sigma W$. One such is the log-logistic distribution for T obtained if W has the logistic density

$$\frac{e^w}{(1 + e^w)^2}. \tag{2.4}$$

This is a symmetric density with mean 0 and variance $\pi^2/3$ with slightly heavier tails than the normal density function, the excess in kurtosis being 1.2. The probability density function of t is then

$$f(t) = \lambda p(\lambda t)^{p-1}[1 + (\lambda t)^p]^{-2}$$

where again $\lambda = e^{-\alpha}$ and $p = \sigma^{-1}$.

Although this model has been used only occasionally in life testing applications, it has the advantage (like the Weibull and exponential models) of having simple algebraic expressions for the survivor and hazard functions. It is therefore more convenient in handling censored data than the log-normal distribution while providing a good approximation to it except in the extreme tails. The survivor and hazard functions are, respectively,

$$F(t) = \frac{1}{1+(\lambda t)^p}$$

and

$$\lambda(t) = \frac{\lambda p(\lambda t)^{p-1}}{1+(\lambda t)^p}.$$

This hazard function is identical to the Weibull hazard aside from the denominator factor $1+(\lambda t)^p$; it is monotone decreasing from ∞ if $p<1$ and is monotone decreasing from λ if $p=1$. If $p>1$, the hazard resembles the log-normal hazard in that it increases from zero to a maximum at $t=(p-1)^{1/p}/\lambda$ and decreases toward zero thereafter.

2.2.7 Generalized *F* distribution

The final parametric model to be discussed incorporates all the above distributions as special cases; the primary value of this model may be its use for discriminating between such distributions as the Weibull and log-logistic distributions as models for a given set of data (c.f. Section 3.9).

Once again we consider a location and scale model for the log failure time Y in which the error distribution is now assumed to be that of the logarithm of an F variate on $2m_1$ and $2m_2$ degrees of freedom. That is, $Y=\mu+\sigma W$ where the p.d.f. of W is

$$\frac{(m_1/m_2)^{m_1}\exp(wm_1)[1+m_1e^w/m_2]^{-(m_1+m_2)}}{B(m_1, m_2)}, \tag{2.5}$$

where $B(m_1, m_2)=\Gamma(m_1)\Gamma(m_2)/\Gamma(m_1+m_2)$ is the beta function. The resulting model for T is the generalized F distribution. It is easily seen that the distributions discussed in Sections 2.2.1–2.2.6 are special cases of the generalized F. If $(m_1, m_2)=(1,1)$, then (2.5) reduces to the logistic model and T has the log-logistic distribution. The Weibull model is obtained as $(m_1, m_2)=(1,\infty)$ for which (2.5) is the extreme value error density (2.1); if in addition $\sigma=1$, the exponential model is obtained. The generalized gamma distribution corresponds to $m_2=\infty$ and as before reduces to the gamma distribution for $\sigma=1$. The log-normal distribution arises by allowing $(m_1, m_2) \to (\infty, \infty)$, as is discussed below. The densities (2.5) are positively skewed for $m_1 > m_2$, negatively skewed for $m_2 > m_1$, and symmetric along $m_1 = m_2$. The log F distribution is tabulated in the Fisher–Yates tables (1938).

The moment generating function can be used to develop the limiting special cases noted above:

$$M_W(\theta) = E(e^{\theta W}) = \frac{1}{B(m_1, m_2)} \left(\frac{m_2}{m_1}\right)^{\theta} \int_0^{\infty} \frac{x^{\theta + m_1 - 1}}{(1+x)^{m_1 + m_2}} \, dx$$

$$= \frac{\Gamma(m_1 + \theta)\Gamma(m_2 - \theta)}{\Gamma(m_1)\Gamma(m_2)} \left(\frac{m_2}{m_1}\right)^{\theta}$$

where a change of variables to $x = (m_1/m_2)e^W$ and

$$B(a, b) = \int_0^{\infty} \frac{x^{a-1}}{(1+x)^{a+b}} \, dx, \qquad a, b > 0,$$

have been used. The mean and variance of W are easily obtained as

$$E(W) = \frac{d}{d\theta} \log M_W(\theta)|_{\theta=0} = \psi(m_1) - \psi(m_2) + \log\left(\frac{m_2}{m_1}\right)$$

$$\text{Var}(W) = \frac{d^2}{d\theta^2} \log M_W(\theta)|_{\theta=0} = \psi^{(1)}(m_1) + \psi^{(1)}(m_2)$$

as before, ψ and $\psi^{(1)}$ are the digamma and trigamma functions.

All the special cases of the generalized F distribution listed above are apparent upon inserting the Stirling approximation into $M_W(\theta)$ except the convergence to the normal as $(m_1, m_2) \to (\infty, \infty)$. It is easily seen that $E(W) \to 0$ as $(m_1, m_2) \to (\infty, \infty)$ and further that

$$\text{Var}(W) \doteq m_1^{-1} + m_2^{-1} + O(m_1^{-2}) + O(m_2^{-2})$$

by Stirling's approximation. If the variable $W^* = \sqrt{m_1 m_2/(m_1 + m_2)}W$ is considered, then $\text{Var}(W^*) \to 1$ as $(m_1, m_2) \to (\infty, \infty)$ and it can further be shown (with rather tedious algebra) that as $(m_1, m_2) \to (\infty, \infty)$, $M_{W^*}(\theta) \to e^{\theta^2/2}$, the moment generating function of a standard normal variate.

The generalized F distribution is discussed in more detail in Section 9 where parameter transformations are considered to obtain the properties of regular estimation on the boundaries $m_1 = \infty$ or $m_2 = \infty$. Its use in discrimination is also considered there.

2.2.8 Other Distributions and Generalizations

There are of course many other distributions that have been or could be used as models for survival data. We have attempted only to outline some of the more commonly used models along with some of their extensions and generalizations.

In modeling adult human mortality a more rapidly increasing hazard function than that represented by, say, the Weibull distribution is necessary. In fact, a relationship in which the hazard function is an exponential function of failure time (age at death) has been found to be descriptive in many investigations, at least for ages greater than 35. Such a relation leads to the Gompertz (1825) hazard $\lambda(t) = \lambda \exp(\gamma t)$. Sometimes the exponential term is generalized to a polynomial function of t. The Makeham (1860) generalization adds a constant to the hazard to give $\lambda(t) = \alpha + \lambda \exp(\gamma t)$.

Failure time is sometimes modeled to include an initial threshold parameter (or guarantee parameter) Δ before which it is assumed that failure cannot occur. The models given above could all be modified in this way simply by replacing T with $T' = T - \Delta$. When such a Δ is known to exist, it should be incorporated in the modeling. But it would be rare that Δ would be known to exist without its value being known. For this reason, and also because of analytical difficulties in estimating Δ, such a threshold is usually not included as a free parameter to be estimated in the methods of Chapter 3. The problems in Appendix 2 outline some of the standard results for models with thresholds.

All the models discussed above are appropriate for continuous failure time variables. As noted earlier, however, failure time can be discrete and, correspondingly, discrete models are required. Some examples of discrete failure time models are given in the problems of Appendix 2, and some discrete models in a regression framework are discussed in Section 2.4.

2.3 REGRESSION MODELS

In Section 2.2, several survival distributions were introduced for modeling the survival experience of a homogeneous population. Usually, however, there are explanatory variables upon which failure time may depend. It therefore becomes of interest to consider generalizations of these models to take account of concomitant information on the individuals sampled.

Consider failure time $T > 0$ and suppose a vector $\mathbf{z} = (z_1, \ldots, z_s)$ of explanatory variables (or covariates) has been observed. Note that $\mathbf{z}$ may include both quantitative variables and qualitative variables such as treatment group; the latter can be incorporated through the use of indicator variables. The principal problem dealt with in this book is that of modeling and determining the relationship between t and $\mathbf{z}$. Certain of the covariates are usually of primary interest such as those specifying the treatment groups. One then wishes to evaluate treatment effects while accounting for heterogeneity in the individuals sampled.

2.3.1 Exponential and Weibull Regression Models

The exponential distribution can be generalized to obtain a regression model by allowing the failure rate to be a function of the covariates $\mathbf{z}$. The hazard at time t for an individual with covariates $\mathbf{z}$ can be written

$$\lambda(t; \mathbf{z}) = \lambda(\mathbf{z}).$$

Thus the hazard for a given $\mathbf{z}$ is a constant characterizing an exponential failure time distribution, but the failure rate depends on $\mathbf{z}$. The $\lambda(\mathbf{z})$ function may be parametrized in many ways. If the effect of the components of $\mathbf{z}$ is only through a linear function, $\mathbf{z}\boldsymbol{\beta}$, one has

$$\lambda(t; \mathbf{z}) = \lambda c(\mathbf{z}\boldsymbol{\beta})$$

where $\boldsymbol{\beta}' = (\beta_1, \ldots, \beta_s)$ is a vector of regression parameters, λ is a constant, and c is a specified functional form. The choice of c may depend on the particular data being considered. Three specific forms have been used (e.g., Feigl and Zelen, 1965): (1) $c(x) = 1 + x$, (2) $c(x) = (1 + x)^{-1}$, and (3) $c(x) = \exp(x)$. The first two of these correspond to (1) the failure rate and (2) the mean survival time being linear functions of $\mathbf{z}$. They both suffer from the disadvantage that the set of $\boldsymbol{\beta}$ values considered must be restricted to guarantee $c(\mathbf{z}\boldsymbol{\beta}) > 0$ for all possible $\mathbf{z}$. In many ways, (3) is the most natural of the above forms since it takes only positive values. We use the form $c(x) = e^x$ here and elsewhere, though it should be kept in mind that other forms may be more appropriate in specific examples and could be used without adding unduly to numerical or analytical computations.

Consider then the model with hazard function

$$\lambda(t; \mathbf{z}) = \lambda e^{\mathbf{z}\boldsymbol{\beta}}. \tag{2.6}$$

The conditional density function of T given $\mathbf{z}$ is then

$$f(t; \mathbf{z}) = \lambda e^{\mathbf{z}\boldsymbol{\beta}} \exp(-\lambda t e^{\mathbf{z}\boldsymbol{\beta}}).$$

In words, the model (2.6) specifies that the log failure rate is a linear function of the covariates $\mathbf{z}$. In terms of the log survival time, $Y = \log T$, the model (2.6) can be written

$$Y = \alpha - \mathbf{z}\boldsymbol{\beta} + W \tag{2.7}$$

where $\alpha = -\log \lambda$ and W has the extreme value distribution (2.1). The model (2.6) is a log-linear model; it is a linear model for Y with the error variable W having a specified distribution.

The Weibull distribution can be generalized to the regression situation in essentially the same way. If the conditional hazard is

$$\lambda(t; \mathbf{z}) = p(\lambda t)^{p-1} e^{\mathbf{z}\boldsymbol{\beta}},$$

then the conditional density of T is

$$f(t; z) = \lambda p(\lambda t)^{p-1} e^{\mathbf{z}\boldsymbol{\beta}} \exp[-(\lambda t)^p e^{\mathbf{z}\boldsymbol{\beta}}]. \tag{2.8}$$

The effect of the covariates is again to act multiplicatively on the Weibull hazard. Alternatively, in terms of $Y = \log T$, the model (2.8) is the linear model

$$Y = \alpha + \mathbf{z}\boldsymbol{\beta}^* + \sigma W \tag{2.9}$$

where $\alpha = -\log \lambda$, $\sigma = p^{-1}$, and $\boldsymbol{\beta}^* = -\sigma\boldsymbol{\beta}$.

The forms of the exponential and Weibull regression models suggest two distinct generalizations. First, the effect of the covariates in either (2.6) or (2.8) is to act multiplicatively on the hazard function. This relationship suggests a general model called the *proportional hazards model.* Second, both of these models are log-linear models; that is, the covariates act additively on Y (or multiplicatively on T). From this we obtain a general class of log-linear models called the *accelerated failure time model.*

2.3.2 The Proportional Hazards Model

Again, let $\lambda(t; \mathbf{z})$ represent the hazard function at time t for an individual with covariates $\mathbf{z}$. The proportional hazards model (Cox, 1972) specifies that

$$\lambda(t; \mathbf{z}) = \lambda_0(t) e^{\mathbf{z}\boldsymbol{\beta}} \tag{2.10}$$

where $\lambda_0(t)$ is an arbitrary unspecified base-line hazard function for continuous T. In this model, the covariates act multiplicatively on the hazard function. If $\lambda_0(t) = \lambda$, (2.10) reduces to the exponential regression model (2.6); the Weibull model (2.8) is the special case $\lambda_0(t) = \lambda p(\lambda t)^{p-1}$.

The conditional density function of T given $\mathbf{z}$ corresponding to (2.10) is

$$f(t; \mathbf{z}) = \lambda_0(t) e^{\mathbf{z}\boldsymbol{\beta}} \exp\left[-e^{\mathbf{z}\boldsymbol{\beta}} \int_0^t \lambda_0(u)\, du\right]. \tag{2.11}$$

The conditional survivor function for T given $\mathbf{z}$ is

$$F(t; \mathbf{z}) = [F_0(t)]^{\exp(\mathbf{z}\boldsymbol{\beta})} \tag{2.12}$$

where

$$F_0(t) = \exp\left[-\int_0^t \lambda_0(u)\,du\right].$$

Thus the survivor function of t for a covariate value, $\mathbf{z}$, is obtained by raising the base-line survivor function $F_0(t)$ to a power. The class of models produced by this process is sometimes referred to as the class of Lehmann alternatives (Lehmann, 1953).

If $\lambda_0(\cdot)$ is arbitrary, this model is sufficiently flexible for many applications. There are, however, two important generalizations that do not substantially complicate the estimation of $\boldsymbol{\beta}$. First, the nuisance function $\lambda_0(t)$ can be allowed to vary in specific subsets of the data. Suppose the population is divided into r strata and that the hazard $\lambda_j(t; \mathbf{z})$ in the jth stratum depends on an arbitrary shape function $\lambda_{0j}(t)$ and can be written

$$\lambda_j(t, \mathbf{z}) = \lambda_{0j}(t) \exp(\mathbf{z}\boldsymbol{\beta}), \tag{2.13}$$

for $j = 1, \ldots, r$. Such a generalization is useful, for instance, if some explanatory variable or variables do not appear to have a multiplicative effect on the hazard function. The range of such variables can then be divided into strata with only the remaining regression variables contributing to the exponential factor in (2.13).

The second important generalization allows the regression variable $\mathbf{z}$ to depend on time itself. Such regression variables arise in the heart transplant example of Section 1.1.3 where treatment group itself is "time dependent," as are certain donor recipient matching variables. The use and the analysis of time dependent covariables are examined in Chapter 5.

2.3.3 The Accelerated Failure Time Model

The multiplicative effect of the regression variables on the hazard as specified in (2.10) has a clear and intuitive meaning. Without restriction on $\lambda_0(\cdot)$, however, this model postulates no direct relationship between $\mathbf{z}$ and t itself. In Section 2.2.1, it was noted that the exponential and Weibull regression models are linear in $Y = \log T$ [see (2.7) and (2.9)]. In this section we obtain a second class of survival models, the class of log-linear models for T.

Suppose that $Y = \log T$ is related to the covariate $\mathbf{z}$ via a linear model $Y = \mathbf{z}\boldsymbol{\beta} + W$, where W is an error variable with density f. Exponentiation gives $T = \exp(\mathbf{z}\boldsymbol{\beta})T'$ where $T' = e^W > 0$ has hazard function $\lambda_0(t')$, say, that is independent of $\boldsymbol{\beta}$. It follows that the hazard function for T can be written in terms of this base-line hazard $\lambda_0(\cdot)$ according to

$$\lambda(t; \mathbf{z}) = \lambda_0(te^{-\mathbf{z}\boldsymbol{\beta}})e^{-\mathbf{z}\boldsymbol{\beta}}. \tag{2.14}$$

The survivor function is

$$F(t; \mathbf{z}) = \exp\left[-\int_0^{t\exp(-\mathbf{z}\boldsymbol{\beta})} \lambda_0(u)\, du\right], \tag{2.15}$$

and the density function is the product of (2.14) and (2.15).

Although the interpretation of the model (2.14) in terms of log T is straightforward, it is also easily seen that this model specifies that the effect of the covariable is multiplicative on t rather than on the hazard function. That is, we assume a base-line hazard function to exist and that the effect of the regression variables is to alter the rate at which an individual proceeds along the time axis. Equivalently, it is supposed that the role of $\mathbf{z}$ is to accelerate (or decelerate) the time to failure. The accelerated failure time model (2.14) is discussed further in Chapters 3 and 6.

All the parametric models discussed in Section 2.2 lead to linear models for Y so that, in many ways, the log-linear regression model is their most natural generalization. The exponential and Weibull regression models can be considered as special cases of either the accelerated failure time model (2.14) or the proportional hazards model (2.10). Note, however, that log-linear models derived from the other parametric models are not special cases of (2.10). For example, log-normal hazard functions (Section 2.2.3) with different location parameters α_1 and α_2, are generally not proportional to one another. In the next section, it is shown that the only log-linear models that are also proportional hazards models are the exponential and Weibull regression models of Section 2.3.1.

2.3.4 Comparison of the Regression Models

Consider now the intersection of the proportional hazards and accelerated failure time models; or equivalently, consider the subset of log-linear models in which the regression variable acts multiplicatively on the hazard function. Using subscripts 1 and 2 for the respective models, we require that

$$\lambda_{01}(t)\exp(\mathbf{z}\boldsymbol{\beta}_1) = \lambda_{02}[t\exp(\mathbf{z}\boldsymbol{\beta}_2)]\exp(\mathbf{z}\boldsymbol{\beta}_2)$$

for all $(t, \mathbf{z})$. The value $\mathbf{z} = \mathbf{0}$ gives $\lambda_{01}(\cdot) = \lambda_{02}(\cdot) = \lambda_0(\cdot)$, say, while $z = (-\log t/\beta_{21}, 0, \ldots, 0)$ gives, at that t,

$$\lambda_0(t)t^{\beta_{11}\beta_{21}^{-1}} = \lambda_0(1)t^{-1}$$

where β_{11} and β_{21} are the first components of $\boldsymbol{\beta}_1$ and $\boldsymbol{\beta}_2$, respectively. It follows that for all t

$$\lambda_0(t) = \lambda p(\lambda t)^{p-1}$$

where $p = \beta_{11}\beta_{21}^{-1}$ and $\lambda = \{\lambda_0(1)/p\}^{1/p}$. Note also that $\boldsymbol{\beta}_1 = -p\boldsymbol{\beta}_2$. The Weibull (and exponential) log-linear regression models are then the only log-linear models in (2.10). Also, the above discussion leads to a characterization of the two parameter Weibull model as the unique family that is closed under both multiplication of failure time and multiplication of the hazard function by an arbitrary nonzero constant.

Both the general classes of models described would provide sufficient flexibility for most purposes if methods for estimating $\boldsymbol{\beta}$ were available that did not require undue restrictions on the nuisance function $\lambda_0(\cdot)$. The remarkable feature of the proportional hazards model is that suitable methods of inference are available without any restriction on $\lambda_0(\cdot)$. This is discussed in Chapter 4. Parametric procedures for estimation in the accelerated failure time family are given in Chapter 3. Chapter 6 gives less model dependent methods of inference based on (2.14) using rank regression techniques.

2.4 DISCRETE FAILURE TIME MODELS

2.4.1 General

All the models discussed to this point are appropriate for failure time data arising from continuous distributions. As remarked earlier, however, failure time data is sometimes discrete either through the grouping of continuous data due to imprecise measurement, or because time itself is discrete.

Any of the continuous failure time models discussed in Section 2.2 can be used to generate a discrete model by introducing a grouping on the time axis. For example, if the underlying continuous failure time X has a Weibull distribution with survivor function

$$\exp[-(\lambda x)^p]$$

and times are grouped into unit intervals so that the discrete observed variable is $T = [X]$, where $[X]$ represents "integer part of X," the probability function of T can be written

$$\begin{aligned} f(t) &= P(T = t) = P(t \le X < t+1) \\ &= \theta^{t^p} - \theta^{(t+1)^p}, \qquad t = 0, 1, 2, \ldots, \end{aligned} \tag{2.16}$$

where $0<\theta=\exp(-\lambda^p)<1$. The special case $p=1$ is the geometric distribution with probability function $\theta^t(1-\theta)$. The hazard function corresponding to (2.16) is

$$\begin{aligned}\lambda(t) &= P(T=t|T\geq t)\\ &= 1-\theta^{(t+1)^p-t^p}\end{aligned}$$

which is monotone increasing, monotone decreasing, or constant for $p>1$, $p<1$, or $p=1$, respectively. This can be generalized to a regression model by applying the same grouping to the Weibull regression model with density (2.8).

In many applications, there is no theoretical justification for adopting particular parametric models for discrete failure time data. The next two sections discuss general discrete regression models which, like the continuous proportional hazards model, allow an arbitrary base-line hazard function which can be estimated from the data.

2.4.2 Discrete Proportional Hazards Model

A discrete analogue of the proportional hazards model can be obtained by applying the survivor function relationship (2.12) directly to a discrete model. Let the failure time T given covariates $\mathbf{z}$ have a discrete distribution with mass points at $0\leq x_1<x_2<\cdots$ and let $F_0(t)$ represent the base-line survivor function for $\mathbf{z}=\mathbf{0}$. The corresponding survivor function for covariates $\mathbf{z}$ is

$$F(t;\mathbf{z})=F_0(t)^{\exp(\mathbf{z}\boldsymbol{\beta})} \tag{2.17}$$

as in (2.12). If the hazard function corresponding to F_0 has contribution λ_i at x_i, then

$$F_0(t)=\prod_{i|x_i<t}(1-\lambda_i)$$

and, from (2.17),

$$F(t;\mathbf{z})=\prod_{i|x_i<t}(1-\lambda_i)^{\exp(\mathbf{z}\boldsymbol{\beta})}. \tag{2.18}$$

The hazard at x_i for covariate $\mathbf{z}$ is then

$$1-(1-\lambda_i)^{\exp(\mathbf{z}\boldsymbol{\beta})}. \tag{2.19}$$

It is of some interest to note that the discrete model (2.19) can also be obtained by grouping the continuous model (2.10). Thus if continuous failure times arising from the proportional hazards model (2.10) are grouped into disjoint intervals $[0 = a_0, a_1), [a_1, a_2), \ldots, [a_{k-1}, a_k = \infty)$, the hazard of failure in the ith interval for an individual with covariate $\mathbf{z}$ is

$$P\{T \in [a_{i-1}, a_i) | T \geq a_{i-1}\} = 1 - (1 - \lambda_i)^{\exp(\mathbf{z}\boldsymbol{\beta})}$$

where $\lambda_i = \exp[-\int_{a_{i-1}}^{a_i} \lambda_0(u)\, du]$. This discrete model is then the uniquely appropriate one for grouped data from the continuous proportional hazards model.

If the discrete base-line hazard function is written as

$$\lambda_d(t)\, dt = \sum \lambda_i \delta(t - x_i)\, dt \tag{2.20}$$

as in Section 2.3.1, we see that the hazard function for covariates $\mathbf{z}$ is

$$\lambda(t; \mathbf{z})\, dt = 1 - [1 - \lambda_d(t)\, dt]^{\exp(\mathbf{z}\boldsymbol{\beta})}. \tag{2.21}$$

Note that if $\lambda_d(t)$ is replaced with a continuous hazard $\lambda(t)$ in (2.21), the relationship is precisely $\lambda(t; \mathbf{z}) = \lambda_0(t) \exp(\mathbf{z}\boldsymbol{\beta})$ as in (2.10). Thus if $\lambda_0(t)$ is the base-line hazard function ($\mathbf{z} = \mathbf{0}$) for a discrete, continuous, or mixed random variable, the relationship between the survivor and hazard functions is

$$\begin{aligned} F(t; \mathbf{z}) &= \mathop{\mathscr{P}}_0^t [1 - \lambda(u; \mathbf{z})\, du] \\ &= \mathop{\mathscr{P}}_0^t [1 - \lambda_0(u)\, du]^{\exp(\mathbf{z}\boldsymbol{\beta})} \end{aligned} \tag{2.22}$$

where $\mathscr{P}$ is the product integral of Section 1.2.3. The Dirac delta components of λ_0 account for any discrete contributions to the survivor function. The expression (2.22) reduces to (2.18) in the discrete case and to (2.12) in the continuous case.

2.4.3 An Alternative Discrete Model

A second discrete failure time regression model was proposed by Cox (1972) and specifies a linear log odds model for the hazard probability at each potential failure time. Thus if $\lambda_d(t)\, dt$ is an arbitrary discrete hazard as defined in (2.19), the hazard for an arbitrary $\mathbf{z}$ is $\lambda(t; \mathbf{z})dt$, where

$$\frac{\lambda(t; \mathbf{z})\, dt}{1 - \lambda(t; \mathbf{z})\, dt} = \frac{\lambda_d(t)\, dt}{1 - \lambda_d(t)\, dt} \exp(\mathbf{z}\boldsymbol{\beta}). \tag{2.23}$$

This is a linear logistic model with an arbitrary logistic location parameter corresponding to each discrete point (or to each time interval). Since the denominator terms approach 1, this model again reduces to the continuous proportional hazards model (2.10) as the intervals become infinitesimal. The two discrete models (2.20) and (2.23) are therefore very similar if all of the discrete hazard contributions λ_i are small. In general, however, the hazard relationship (2.23) applies only approximately to grouped data from the proportional hazards model.

BIBLIOGRAPHIC NOTES

Properties of the exponential, Weibull, log-normal, gamma, generalized gamma, and log-logistic distributions are discussed by Johnson and Kotz (1970a, b), who also give extensive bibliographies on these distributions. Some of these distributions are also discussed by Cox (1972), Mann et al. (1974), and Gross and Clark (1975). The generalized gamma distribution was introduced by Stacy (1962) and has been discussed by Stacy and Mihram (1965), Parr and Webster (1965), Harter (1967), Hagar and Bain (1970), and Prentice (1974). The generalized F distribution is discussed by Prentice (1975).

Exponential, Weibull and log-normal regression models have received considerable use in the literature. Since most of this work has been concerned with estimation, the list of references for these and other parametric models is deferred to the bibliographic notes at the end of Chapter 3.

The proportional hazards model was introduced by Cox (1972); the two sample special case with censored data was considered by Peto and Peto (1972). This model has been extensively discussed with regard to inferential problems, and references are given in Chapter 4. The discrete analogue to the proportional hazards model was given by Kalbfleisch and Prentice (1973), and Cox (1972) considered the linear log odds model in the discrete case. The accelerated failure time model was introduced by Cox (1972) and considered by Prentice (1978). Methods of inference for discrete failure time distributions are considered in Section 4.6, and additional references are given there.

CHAPTER 3

Inference in Parametric Models and Related Topics

3.1 INTRODUCTION

As mentioned previously, one important reason for specialized statistical models and methods for failure time data is the need to accommodate right censoring in the data. It is usually the case that censoring greatly complicates the distribution theory for the estimators even when the censoring mechanism is simple and well understood. In other cases, complex censoring mechanisms may make such computations impossible. This fact leads, in most instances, to a reliance on asymptotic methods for estimation and testing.

The principal purpose of this chapter is to consider estimation and testing with parametric regression models. Discussions of censoring mechanisms and asymptotic likelihood methods are, however, given as necessary background for this and the following chapters. Inference procedures are illustrated throughout this chapter by showing their application to the carcinogenesis data of Section 1.1.1. The use of the generalized F model for both estimation and discrimination is also considered.

3.2 CENSORING MECHANISMS

We consider survival studies in which n items are put on test and data of the form $(t_i, \delta_i, \mathbf{z}_i)$, $i = 1, \ldots, n$, are collected. Here δ_i is an indicator variable ($\delta_i = 0$ if the ith item is censored; $\delta_i = 1$ if the ith item failed) and t_i is the failure time ($\delta_i = 1$) or the censoring time ($\delta_i = 0$). As before, $\mathbf{z}_i$ is a row vector of s covariates associated with the ith individual.

Suppose that the survival time model is specified up to a parameter vector $\boldsymbol{\theta}$ and that the survivor function for the ith individual is $F(t_i^0; \boldsymbol{\theta}, \mathbf{z}_i)$ with the corresponding density $f(t_i^0; \boldsymbol{\theta}, \mathbf{z}_i)$. In order to avoid confusion, we use the variable t_i^0 to indicate the failure time of the ith individual which, because of censoring, may or may not be observed.

In order to obtain the likelihood function for $\boldsymbol{\theta}$, it is necessary to consider the nature of the censoring mechanism acting on the data. For the moment, we make the rather restrictive assumption of *random censorship*. Specifically we assume that the censoring time C_i for the ith individual is a random variable with survivor and density functions $G_i(c)$ and $g_i(c)$, respectively $(i = 1, \ldots, n)$, and further that $C_1, \ldots, C_n$ are stochastically independent of each other and of the failure times $T_1^0, \ldots, T_n^0$. Note that the random censorship model includes the special case of *type I censoring*, where the censoring time of each individual is fixed in advance, as well as the case where items enter the study at random over time and the analysis is carried out at some prespecified time. This latter situation occurs in some medical studies. It should be noted, however, that the formulation is not sufficiently general to include some censoring schemes that are commonly used in certain areas of application as is discussed below.

For random censorship,

$$\begin{aligned} P[T_i \in (t, t+dt), \delta_i = 1; \mathbf{z}_i, \boldsymbol{\theta}] &= P[T_i^0 \in (t, t+dt), C_i > t; \mathbf{z}_i, \boldsymbol{\theta}] \\ &= G_i(t+0) f(t; \boldsymbol{\theta}, \mathbf{z}_i)\, dt \end{aligned}$$

and

$$P[T_i \in (t, t+dt), \delta_i = 0; \mathbf{z}_i, \boldsymbol{\theta}] = g_i(t) F(t+0; \boldsymbol{\theta}, \mathbf{z}_i)\, dt, \tag{3.1}$$

where, as before, $F(t+0; \boldsymbol{\theta}, \mathbf{z}_i) = \lim_{u \to t^+} F(u; \boldsymbol{\theta}, \mathbf{z}_i)$. If F is continuous, $F(t+0; \boldsymbol{\theta}, \mathbf{z}_i) = F(t; \boldsymbol{\theta}, \mathbf{z}_i)$. Thus if the censoring is noninformative [i.e., $G_i(t)$ does not involve $\boldsymbol{\theta}$], the likelihood on the data $(t_i, \delta_i, \mathbf{z}_i)$, $i = 1, \ldots, n$, is

$$L(\boldsymbol{\theta}) \propto \prod_{i=1}^{n} f(t_i; \boldsymbol{\theta}, \mathbf{z}_i)^{\delta_i} F(t_i + 0; \boldsymbol{\theta}, \mathbf{z}_i)^{1-\delta_i} \tag{3.2}$$

and further the pairs (t_i, δ_i), $i = 1, \ldots, n$, are (given $\mathbf{z}_1, \ldots, \mathbf{z}_n$) independent. In effect the likelihood is formed as $L(\boldsymbol{\theta}) = \prod L_i(\boldsymbol{\theta})$ where $L_i(\boldsymbol{\theta})$ is $f(t_i; \boldsymbol{\theta}, \mathbf{z}_i)$ for a failure and $F(t_i + 0; \boldsymbol{\theta}, \mathbf{z}_i)$ for a censored data point. A consequence of this model for the censoring is that the contribution to the likelihood of an individual censored at t_i is just $P\{T_i^0 > t_i; \mathbf{z}_i, \boldsymbol{\theta}\}$.

In fact the likelihood (3.2) is much more generally correct than the above discussion would suggest. As is shown in Section 5.2, there is a class of censoring mechanisms called *independent censoring* for which this likelihood is appropriate. Briefly, what we require is that, at time t and for any given $\mathbf{z}$, individuals cannot be censored because they appear to be at unusually high (or low) risk of failure. Some restriction of this type is necessary since it would clearly be impossible to obtain meaningful survival data if, for example, individuals were withdrawn from study whenever they appeared to be in imminent danger of failure. Many of the usual censoring schemes are special cases of independent censoring. For example, the study may continue until the dth smallest failure time occurs, at which time all surviving items are censored (*type II censoring*), or a specific fraction of individuals at risk may be censored at each of several ordered failure times (*progressive type II censoring*). More generally, the censoring procedure may depend arbitrarily during the course of the study on previous failure times, previous censoring times, or values of covariates included in the model.

3.3 SAMPLING FROM A CENSORED EXPONENTIAL DISTRIBUTION

A simple example illustrates the relationship between the censoring mechanism and the complexity of exact (frequentist) inferences. Suppose failure times arise from an exponential distribution with failure rate λ. In the above notation $\boldsymbol{\theta} = \lambda$, $F(t; \mathbf{z}, \boldsymbol{\theta}) = F(t; \lambda) = e^{-\lambda t}$ and $f(t; \mathbf{z}, \boldsymbol{\theta}) = \lambda e^{-\lambda t}$. With no censoring the likelihood function is given by

$$L(\lambda) = \lambda^n \exp(-\lambda V),$$

where $V = \Sigma\, t_i$. The likelihood function is determined by V, or equivalently by the maximum likelihood estimate $\hat{\lambda} = n/V$, so that V is sufficient for λ. That is, inference on λ can be based on the value of V and its distribution. The distribution of V is simply obtained from its moment generating function (m.g.f.). The m.g.f. of λt_i with parameter ξ is $(1-\xi)^{-1}$, so that of λV is $(1-\xi)^{-n}$, which is the m.g.f. of a gamma variate (2.2.4) with $k = n$, $\lambda = 1$. Equivalently, $2\lambda V$ has a χ^2 distribution with $2n$ degrees of freedom. This result leads to simple significance tests and confidence intervals for λ.

Inference on λ is equally simple with type II censoring. Suppose that n individuals are placed simultaneously on test and the study terminates when the dth failure occurs. Denote the ordered failure times by $t_{(1)} < t_{(2)} < \cdots < t_{(d)}$. As in the general expression (3.2), the likelihood contribution from each of the $(n-d)$ items censored at $t_{(d)}$ is $\exp(-\lambda t_{(d)})$, since the corresponding failure times are known only to exceed $t_{(d)}$, and the contribution of the failure at $t_{(i)}$ is $\lambda \exp(-\lambda t_{(i)})$, $i = 1, \ldots, d$. The likelihood is then

$$L(\lambda) = \lambda^d \exp(-\lambda V), \tag{3.3}$$

where the total survival time $V = \Sigma_1^d t_{(i)} + (n-d)t_{(d)}$, or the maximum likelihood estimate $\hat{\lambda} = d/V$, is again sufficient for λ. The joint density of $t_{(1)}, \ldots, t_{(d)}$ is required in order to derive the density for V. A rather informal derivation involves dividing the time axis into intervals $[0, t_{(1)})$, $[t_{(1)}, t_{(1)} + dt_{(1)})$, $[t_{(1)} + dt_{(1)}, t_{(2)})$, $\ldots$, $[t_{(d)}, t_{(d)} + dt_{(d)})$, $[t_{(d)} + dt_{(d)}, \infty)$ and appealing to the multinomial probability of obtaining frequencies 0, 1, 0, $\ldots$, 1, $(n-d)$ in the respective intervals. This gives a probability element

$$\frac{n!}{(n-d)!} \exp[-(n-d)\lambda t_{(d)}] \prod_1^d [\lambda \exp(-\lambda t_{(i)})\, dt_{(i)}]$$

and p.d.f. for $t_{(1)}, \ldots, t_{(d)}$

$$\frac{n!\lambda^d \exp(-\lambda V)}{(n-d)!}$$

which, of course, gives rise to (and amounts to a derivation of) the likelihood function (3.3) above. The change of variables to

$$u_i = (n-i+1)(t_{(i)} - t_{(i-1)}), \qquad i = 1, \ldots, d$$

where $t_{(0)} = 0$, has Jacobian $(n-d)!/n!$. It follows that the joint density of the u_i's is

$$\prod_1^d (\lambda e^{-\lambda u_i}), \qquad u_1, \ldots, u_d > 0,$$

and the so-called normalized spacings, u_i, have independent exponential distributions with failure rate λ. As above, it then follows that $2\lambda V = 2\lambda \Sigma_1^d u_i$ has a χ^2 distribution with $2d$ degrees of freedom and inference proceeds as in the uncensored case. Note that, with exponential data, the same estimating efficiency is achieved by following d items until all have failed or a larger number, n, until the dth failure.

The transformation to the u_i's gives the expected exponential order statistics. We can write

$$t_{(i)} = \frac{u_1}{n} + \frac{u_2}{n-1} + \cdots + \frac{u_i}{n-i+1} \tag{3.4}$$

and since the expectation of each u_j is λ^{-1}, the expectation of $t_{(i)}$ is

$$\lambda^{-1} \sum_{j=1}^{i} (n-j+1)^{-1}. \tag{3.5}$$

This suggests a simple graphical test for exponentiality with type II censored data: A plot of $t_{(i)}$ versus $\Sigma_{j=1}^{i} (n-j+1)^{-1}$ should be approximately linear if the exponential model is suitable. This procedure is a first order approximation to that based on a plot of the log survivor function estimator as suggested in Section 2.2.1: The Kaplan–Meier estimator, for $t \in (t_{(i)}, t_{(i+1)}]$, can be written

$$\hat{F}(t) = \prod_{j=1}^{i} [1 - (n - j + 1)^{-1}]$$

so that

$$\log \hat{F}(t) = \sum_{j=1}^{i} \log[1 - (n - j + 1)^{-1}]$$
$$\doteq - \sum_{j=1}^{i} (n - j + 1)^{-1}.$$

Now consider sampling from an exponential model with an arbitrary independent censoring mechanism. The likelihood function can again be written

$$L(\lambda) = \lambda^d \exp(-\lambda V)$$

where $d = \Sigma\, \delta_i$, and $V = \Sigma\, t_i\delta_i + \Sigma\, t_i(1 - \delta_i) = \Sigma\, t_i$. Typically not only V, but also the number of failures, d, is random and the sampling distribution of the sufficient statistic (V, d) or that of the m.l.e. $\hat{\lambda} = d/V$, is complicated by the censoring mechanism.

Consider, for example, the simple case of type I censoring where the censoring times $c_1, c_2, \ldots, c_n$ are specified in advance for the n individuals on study. In this case

$$V = \sum_{i=1}^{n} [(1 - \delta_i)c_i + \delta_i t_i^0]$$

where as before t_i^0 is the (possibly unobserved) failure time of the ith individual. In this simple case, convenient expressions even for the marginal density of V are not available, and the joint distribution of d and V is quite complicated. This complication of the distribution theory leads, in most cases, to the use of asymptotic likelihood arguments for inference about the parameters.

3.4 LARGE SAMPLE LIKELIHOOD THEORY

In this section, we consider the main asymptotic results that apply to the likelihood function and the maximum likelihood estimator. Generally speaking, these results are applicable to the likelihoods derived from parametric regression models for failure time data with arbitrary censoring mechanisms. The formulation considered here, however, does not include all censoring mechanisms, but is sufficiently general to include the random censorship model of Section 3.2 and the special case of type I censoring. No derivations are given here; the reader is referred to Cox and Hinkley (1974) and the other references in the bibliographic notes for a more complete discussion.

We suppose that the data consist of n independent observations $x_1, x_2, \ldots, x_n$ of random variables whose probability laws involve $\boldsymbol{\theta}' = (\theta_1, \ldots, \theta_p)$. The likelihood function of $\boldsymbol{\theta}$ is

$$L(\boldsymbol{\theta}) = \prod_{i=1}^{n} L_i(\boldsymbol{\theta})$$

where $L_i(\boldsymbol{\theta})$ is the probability density of x_i. For the failure time application, x_i may be identified with the pair (t_i, δ_i) of the random censorship model, and the covariate vector determines the probability density $L_i(\boldsymbol{\theta})$ for the ith individual.

3.4.1 The Score Statistic

Central to asymptotic likelihood arguments are the efficient score vectors

$$\mathbf{U}_i(\boldsymbol{\theta}) = \frac{\partial}{\partial \boldsymbol{\theta}} \log L_i(\boldsymbol{\theta}) = \left[\frac{\partial}{\partial \theta_j} \log L_i(\boldsymbol{\theta})\right]_{p \times 1}, \tag{3.6}$$

$i = 1, \ldots, n$. If the operations of integration with respect to x_i and differentiation with respect to $\boldsymbol{\theta}$ can be interchanged, it can be shown that $\mathbf{U}_i(\boldsymbol{\theta})$ has expectation $\mathbf{0}$ and covariance matrix

$$\begin{aligned} \mathscr{I}_i(\boldsymbol{\theta}) &= E[\mathbf{U}_i(\boldsymbol{\theta})\mathbf{U}_i'(\boldsymbol{\theta})] \\ &= -\left[E\left(\frac{\partial^2 \log L_i}{\partial \theta_j \partial \theta_k}\right)\right]_{p \times p}. \end{aligned} \tag{3.7}$$

It should be noted that although the above conditions leading to (3.7) are quite mild, they are nonetheless sufficient to exclude threshold parameters from consideration.

Since the x_i's are independent, $\mathbf{U}_1(\boldsymbol{\theta}), \ldots, \mathbf{U}_n(\boldsymbol{\theta})$ are also independent. Thus under certain conditions, a central limit theorem will apply to the total score statistic $\mathbf{U}(\boldsymbol{\theta}) = \Sigma_{i=1}^{n} \mathbf{U}_i(\boldsymbol{\theta})$. As a consequence, $\mathbf{U}(\boldsymbol{\theta})$ is asymptotically normal with mean $\mathbf{0}$ and covariance matrix $\mathscr{I}(\boldsymbol{\theta}) = \Sigma_{i=1}^{n} \mathscr{I}_i(\boldsymbol{\theta})$. For convenience, we speak of $\mathbf{U}(\boldsymbol{\theta})$ as being asymptotically normal with mean $\mathbf{0}$ and covariance $\mathscr{I}(\boldsymbol{\theta})$, though it is of course the standardized version of $\mathbf{U}(\boldsymbol{\theta})$ that converges. For the central limit theorem to apply, the requirements are basically that the relative information $\mathscr{I}_i(\boldsymbol{\theta})\mathscr{I}(\boldsymbol{\theta})^{-1}$ approaches a zero matrix for each i as $n \to \infty$, and the total information $\mathcal{I}(\boldsymbol{\theta})$ approaches infinity at a suitable rate. Necessary and sufficient conditions for the central limit theorem for sums of independent variables were given by Lindeberg (e.g., Feller, 1971, p. 262). The sufficient conditions of Ljapunov (e.g., Feller, 1971, p. 286) are often more easily verified.

The asymptotic distribution of the score $\mathbf{U}(\boldsymbol{\theta})$ can be used for approximate inference about $\boldsymbol{\theta}$. Specifically, under a hypothesized $\boldsymbol{\theta} = \boldsymbol{\theta}_0$, the score statistic $\mathbf{U}(\boldsymbol{\theta}_0)$ is asymptotically normal with mean $\mathbf{0}$ and variance $\mathcal{I}(\boldsymbol{\theta}_0)$. If $\mathcal{I}(\boldsymbol{\theta}_0)$ is nonsingular, it follows that

$$\mathbf{U}'(\boldsymbol{\theta}_0)\mathcal{I}(\boldsymbol{\theta}_0)^{-1}\mathbf{U}(\boldsymbol{\theta}_0) \tag{3.8}$$

has an asymptotic χ^2 distribution with degrees of freedom p, the dimension of $\boldsymbol{\theta}$. The hypothesized value of $\boldsymbol{\theta}_0$ is assessed by comparing the value of (3.8) with the χ^2 tables. Alternatively, an approximate confidence region for $\boldsymbol{\theta}$ can be formed as the set of values $\boldsymbol{\theta}_0$ that yield values of (3.8) less than a specified upper level on the χ^2 distribution. It is also possible to use the score function for tests of certain composite hypotheses as illustrated in Section 3.5.2.

The limiting normal distribution of the score function is the fundamental result of asymptotic likelihood theory. It serves as the basis on which the other asymptotic results are built.

3.4.2 The Maximum Likelihood Estimator

Simpler methods of interval estimation are available than those based on the score statistic. These involve the asymptotic distribution of the maximum likelihood estimator (m.l.e.) $\hat{\boldsymbol{\theta}}$. If $\boldsymbol{\theta}$ is interior to the parameter space, and if $L(\boldsymbol{\theta})$ is thrice differentiable and certain boundedness conditions on the third derivatives are satisfied, it can be shown using the score function results that $\hat{\boldsymbol{\theta}}$ is the asymptotically unique solution to $\mathbf{U}(\boldsymbol{\theta}) = \mathbf{0}$, that $\hat{\boldsymbol{\theta}}$ is consistent for $\boldsymbol{\theta}$ and that the asymptotic distribution of $\hat{\boldsymbol{\theta}}$ is multivariate normal with mean $\boldsymbol{\theta}$ and covariance matrix $\mathcal{I}(\boldsymbol{\theta})^{-1}$. Tests of hypotheses about $\boldsymbol{\theta}$ and interval estimation can be based on this result. For example, confidence regions can be specified using the asymptotic $\chi^2_{(p)}$ distribution of

$$(\hat{\boldsymbol{\theta}} - \boldsymbol{\theta})'\mathcal{I}(\boldsymbol{\theta})(\hat{\boldsymbol{\theta}} - \boldsymbol{\theta}). \tag{3.9}$$

Similarly, if $\boldsymbol{\theta}' = (\boldsymbol{\theta}_1', \boldsymbol{\theta}_2')$ where $\boldsymbol{\theta}_1' = (\theta_1, \ldots, \theta_k)$ and

$$\mathcal{I}(\boldsymbol{\theta})^{-1} = \begin{pmatrix} \mathcal{I}^{11}(\boldsymbol{\theta}) & \mathcal{I}^{12}(\boldsymbol{\theta}) \\ \mathcal{I}_{21}(\boldsymbol{\theta}) & \mathcal{I}^{22}(\boldsymbol{\theta}) \end{pmatrix}$$

where $\mathcal{I}^{11}(\boldsymbol{\theta})$ is $k \times k$, then the asymptotic $\chi^2_{(k)}$ distribution of

$$(\hat{\boldsymbol{\theta}}_1 - \boldsymbol{\theta}_1)' \mathcal{I}^{11}(\boldsymbol{\theta})^{-1} (\hat{\boldsymbol{\theta}}_1 - \boldsymbol{\theta}_1)$$

can be used for estimation of $\boldsymbol{\theta}_1$ or testing hypotheses about $\boldsymbol{\theta}_1$. In particular, if $k = 1$, this gives a simple asymptotic standard normal variate for estimating θ_1, or equivalently θ_j, as

$$(\theta_j - \hat{\theta}_j)[i^{jj}(\boldsymbol{\theta})]^{-1/2} \tag{3.10}$$

where $i^{jj}(\boldsymbol{\theta})$ is the (j, j) element of $\mathcal{I}(\boldsymbol{\theta})^{-1}$.

All the above results can be modified by replacing the Fisher information $\mathcal{I}(\boldsymbol{\theta})$ with a consistent estimator. Thus, for example, since $\mathcal{I}(\boldsymbol{\theta})\mathcal{I}(\hat{\boldsymbol{\theta}})^{-1}$ converges in probability to a $p \times p$ identity matrix, $\mathcal{I}(\hat{\boldsymbol{\theta}})$ can replace $\mathcal{I}(\boldsymbol{\theta})$ in (3.9) and (3.10). An even simpler estimator of $\mathcal{I}(\boldsymbol{\theta})$ is provided by the observed information

$$I(\boldsymbol{\theta}) = \left(\frac{-\partial^2 \log L(\boldsymbol{\theta})}{\partial \theta_i \, \partial \theta_j} \right)_{p \times p},$$

the expected value of which is the Fisher information. Thus $\mathcal{I}(\boldsymbol{\theta})$ can be replaced in the above results with $I(\boldsymbol{\theta})$ or with $I(\hat{\boldsymbol{\theta}})$ without affecting the asymptotic distributions. It should also be noted that $\mathcal{I}(\boldsymbol{\theta}_0)$ in (3.8) may be replaced with $\mathcal{I}(\hat{\boldsymbol{\theta}})$, $I(\boldsymbol{\theta}_0)$, or $I(\hat{\boldsymbol{\theta}})$ while retaining the asymptotic χ^2 result.

3.4.3 The Likelihood Ratio Statistic

A third class of likelihood statistics is that based on the likelihood ratio

$$R(\boldsymbol{\theta}) = \frac{L(\boldsymbol{\theta})}{L(\hat{\boldsymbol{\theta}})}$$

and its asymptotic distribution. If the regularity conditions of maximum likelihood theory hold, then under the hypothesis $\boldsymbol{\theta} = \boldsymbol{\theta}_0$, the asymptotic distribution of

$$-2 \log R(\boldsymbol{\theta}_0) \tag{3.11}$$

is χ^2 with degrees of freedom p, the dimension of $\boldsymbol{\theta}$.

Similarly, if $\boldsymbol{\theta}' = (\boldsymbol{\theta}_1', \boldsymbol{\theta}_2')$ and $\hat{\boldsymbol{\theta}}_2(\boldsymbol{\theta}_1^0)$ is the m.l.e. of $\boldsymbol{\theta}_2$ given $\boldsymbol{\theta}_1 = \boldsymbol{\theta}_1^0$, then $-2 \log R[\boldsymbol{\theta}_1^0, \hat{\boldsymbol{\theta}}_2(\boldsymbol{\theta}_1^0)]$ has a χ^2 distribution with degrees of freedom the dimension of $\boldsymbol{\theta}_1$ under the hypothesis $\boldsymbol{\theta}_1 = \boldsymbol{\theta}_1^0$. This result is quite general since by reparametrization it is usually possible to test any hypothesis of interest about combinations of the θ_i's. Basically, we need to be able to reparametrize to a vector of parameters, $\boldsymbol{\lambda} = \boldsymbol{\lambda}(\boldsymbol{\theta})$, to which maximum likelihood theory applies and for which the hypothesis of interest is equivalent to specifying the values of a subset of λ_i's.

3.4.4 Discussion

One desirable feature of any inference procedure is that the conclusions drawn should be independent of the (to some extent) arbitrary parametrization used. Thus if $\boldsymbol{\theta}$ is replaced by some 1:1 function, say, $\boldsymbol{\lambda} = \boldsymbol{\lambda}(\boldsymbol{\theta})$, the conclusions should not be altered. The score test statistic (3.8) has this property as do all tests based on the likelihood ratio statistic. To see this for (3.8), we suppose that $\boldsymbol{\lambda} = \boldsymbol{\lambda}(\boldsymbol{\theta})$ is a 1:1 differentiable function of $\boldsymbol{\theta}$ and let $\boldsymbol{\lambda}_0 = \boldsymbol{\lambda}(\boldsymbol{\theta}_0)$. If $\mathbf{U}_*(\boldsymbol{\lambda})$ and $\mathcal{I}_*(\boldsymbol{\lambda})$ are the score statistic and the information for $\boldsymbol{\lambda}$, respectively, straightforward calculations show that

$$\mathbf{U}_*(\boldsymbol{\lambda}_0) = J_0\mathbf{U}(\boldsymbol{\theta}_0)$$

$$\mathcal{I}_*(\boldsymbol{\lambda}_0) = J_0\mathcal{I}(\boldsymbol{\theta}_0)J_0'$$

where J_0 is the Jacobian matrix with (i, j) element $(\partial\theta_i/\partial\lambda_j)$ evaluated at $\boldsymbol{\lambda} = \boldsymbol{\lambda}_0$. It then follows easily that

$$\mathbf{U}_*'(\boldsymbol{\lambda}_0)\mathcal{I}_*(\boldsymbol{\lambda}_0)^{-1}\mathbf{U}_*(\boldsymbol{\lambda}_0) = \mathbf{U}'(\boldsymbol{\theta}_0)\mathcal{I}(\boldsymbol{\theta}_0)^{-1}\mathbf{U}(\boldsymbol{\theta}_0).$$

The other asymptotic results discussed above do not have the property of functional invariance when applied to finite samples. Even the score function tests do not possess this property if $\mathcal{I}(\boldsymbol{\theta}_0)$ is replaced with an estimate, as is generally necessary with censored failure time data owing to the complex sample spaces over which expectations must be taken. As discussed further in Section 3.4.5, the use of the expected information can also be criticized since the inference would then depend on potential censoring times even when these have not affected the observations.

The fact that asymptotic theory applied to $\hat{\boldsymbol{\theta}}$ or $\hat{\boldsymbol{\lambda}}$ leads to different results suggests that some care need be exercised in the use of asymptotic results for the maximum likelihood estimator. Consideration should in general be given to selecting a parametrization $\boldsymbol{\gamma} = \boldsymbol{\gamma}(\boldsymbol{\theta})$ for which a normal approximation to the distribution of $\hat{\boldsymbol{\gamma}}$ is most or especially suitable. Some rough guidelines are to select $\boldsymbol{\gamma}$ so that its components do not have unnecessary range restrictions and so that its asymptotic variance matrix is reasonably stable near $\hat{\boldsymbol{\gamma}}$. It is also possible, in some instances, to choose $\boldsymbol{\gamma}$ to make the likelihood nearly symmetric in shape by making the third derivative of the log likelihood, evaluated at $\hat{\boldsymbol{\gamma}}$, small. The asymptotic distribution of $\hat{\boldsymbol{\theta}}$ is obtained from that of $\mathbf{U}(\boldsymbol{\theta})$ by approximating the log likelihood with a quadratic function. This suggests that the normal shape of the likelihood is important in large sample theory. It should be noted that if the likelihood can be made very close to normal in shape, the likelihood ratio statistic and the maximum likelihood estimates yield nearly identical inferences.

The above discussion would suggest superiority of the likelihood ratio results, and these same general conclusions have been reached by many authors.

3.4.5 Exponential Sampling Illustration

In this section, we consider application of the methods of Section 3.4 to a sample from the exponential distribution with type I censoring as defined at the end of Section 3.3. We suppose, as there, that failure times T_i^0 arise as independent exponential variates with failure rate λ and the ith individual has a censoring time c_i fixed in advance of experimentation. Note that the same analysis would apply also to a random censorship model where independent censoring times C_i, $i = 1, \ldots, n$, are determined according to some distribution free of λ. In this latter case, the analysis is conditioned upon $C_i = c_i$, the censoring times being ancillary statistics for the estimation of λ.

As noted before, the contribution of the ith individual to the likelihood is

$$\begin{aligned} L_i(\lambda) &= (\lambda e^{-\lambda t_i})^{\delta_i}(e^{-\lambda t_i})^{1-\delta_i} \\ &= \lambda^{\delta_i} e^{-\lambda t_i}, \qquad i = 1, \ldots, n, \end{aligned}$$

so that the efficient scores are

$$U_i(\lambda) = \frac{\delta_i}{\lambda} - t_i$$

and $-\partial^2 \log L_i(\lambda)/\partial\lambda^2 = \delta_i/\lambda^2$. It is easily seen that

$$\begin{aligned} E[U_i(\lambda)] &= \frac{E(\delta_i)}{\lambda} - E(t_i) \\ &= 0 \end{aligned}$$

while

$$\mathscr{I}_i(\lambda) = \frac{E(\delta_i)}{\lambda^2} = \frac{1 - p_i}{\lambda^2}$$

where $p_i = e^{-\lambda c_i}$ is the probability that the ith item is censored.

If, as before, $d = \Sigma\, \delta_i$ represents the total number of failures and $V = \Sigma\, t_i$, the total accumulated survival, it follows from Section 3.4.1 that the score function,

$$U(\lambda) = \frac{d}{\lambda} - V,$$

is asymptotically normal with mean 0 and variance

$$\mathscr{I}(\lambda) = \frac{E(d)}{\lambda^2} = \frac{\Sigma(1-p_i)}{\lambda^2}. \tag{3.12}$$

The requirement for asymptotic normality of $U(\lambda)$ is simply that the expected number of deaths approaches infinity at a sufficient rate as $n \to \infty$. This in turn places some very mild restriction on the censoring times to the effect that the c_i's must not converge too rapidly to zero.

Now we have a common situation in the application of asymptotic methods with censored data: (3.12) involves the potential censoring times c_i for all individuals under study whereas these would often not be known for individuals that fail. Further, it is not clear that potential, but unobserved, censoring times should affect the inference even if they are available. For these reasons, the observed information,

$$I(\lambda) = \frac{-\partial^2 \log L(\lambda)}{\partial \lambda^2},$$

or $I(\hat{\lambda})$, is commonly substituted for $\mathscr{I}(\lambda)$ as was discussed earlier. Replacing $\mathscr{I}(\lambda)$ with $I(\lambda) = d/\lambda^2$, we find that $\lambda d^{-1/2} U(\lambda)$ has an asymptotic standard normal distribution.

For example, suppose failure time in excess of 100 days in the group 1 data of Section 1.1.1 is exponentially distributed with failure rate λ. Then $d = 17$ and $V = (43 + 64 + \cdots + 144) = 2195$. The above results give an approximate 95% confidence interval for the failure rate, λ, as the set of λ values for which

$$|\lambda d^{-1/2} U(\lambda)| = |d^{1/2} - \lambda d^{-1/2} V| < 1.96,$$

since ± 1.96 bracket the central 95% probability from a standard normal distribution. Direct solution gives (.00406, .01143) as the approximate confidence interval. Maximum likelihood results applied to $\hat{\lambda} = d/V$, with the information estimated by $I(\hat{\lambda}) = V^2/d$, give

$$(\hat{\lambda} - \lambda) I_0(\hat{\lambda})^{1/2} = \left(\frac{d}{V} - \lambda\right) d^{-1/2} V$$
$$= d^{1/2} - \lambda d^{-1/2} V$$

as an approximate standard normal variate which, in this case, is precisely the same as the score procedure. Alternatively, we may prefer to apply asymptotic m.l.e. results to $\gamma = \log \lambda$ since γ has unrestricted range and the asymptotic variance of $\hat{\gamma} = \log \hat{\lambda}$ is a constant, d^{-1}. This latter statement follows from direct manipulation of the likelihood for γ, $L_*(\gamma) = L(e^\gamma) = \exp(\gamma d - e^\gamma V)$. It follows that $d^{1/2} \log \hat{\lambda}$ has an asymptotic

ratio statistic will usually require iterative calculation. This is discussed further in Section 3.7.

3.5.2 Comparisons of Two Exponential Samples

A special case in which $\hat{\boldsymbol{\beta}}$ can be explicitly calculated is that of comparing two exponential samples. Suppose failure times from two groups of individuals are exponential with failure rates λ_1 and λ_2, respectively. Comparison of the two groups then involves comparison of λ_1 and λ_2. Such data can be placed in the above regression framework by defining for the ith individual a regression vector $\mathbf{z}_i = (z_{1i}, z_{2i})$, where $z_{1i} = 1$, as before, and $z_{2i} = 0$ if individual i is in group 1 and $z_{2i} = 1$ if individual i is in group 2. Then $e^{\beta_1} = \lambda_1$ and $e^{\beta_1+\beta_2} = \lambda_2$ and equality of λ_1 and λ_2 is equivalent to $\beta_2 = 0$. The score statistic (3.16) can be written

$$U_1(\boldsymbol{\beta}) = d_1 + d_2 - e^{\beta_1}V_1 - e^{\beta_1+\beta_2}V_2$$

$$U_2(\boldsymbol{\beta}) = \quad d_2 \quad - e^{\beta_1+\beta_2}V_2,$$

where (d_j, V_j) is the number of failures and total survival time in sample j, $j = 1, 2$. This gives $e^{\hat{\beta}_1} = d_1/V_1$ and $e^{\hat{\beta}_2} = d_2V_1/d_1V_2$. The observed Fisher information matrix is

$$I(\boldsymbol{\beta}) = \begin{pmatrix} e^{\beta_1}V_1 + e^{\beta_1+\beta_2}V_2 & e^{\beta_1+\beta_2}V_2 \\ e^{\beta_1+\beta_2}V_2 & e^{\beta_1+\beta_2}V_2 \end{pmatrix}$$

so that

$$I(\hat{\boldsymbol{\beta}}) = \begin{pmatrix} d_1 + d_2 & d_2 \\ d_2 & d_2 \end{pmatrix} \quad \text{and} \quad I(\hat{\boldsymbol{\beta}})^{-1} = \frac{1}{d_1d_2}\begin{pmatrix} d_2 & -d_2 \\ -d_2 & d_1 + d_2 \end{pmatrix}.$$

Consider now a test of the hypothesis $\beta_2 = 0$ which corresponds to equality of the failure time distributions for the two groups. The asymptotic distribution of $\hat{\beta}_2$ is most convenient for this purpose. From the (2, 2) element of $I(\hat{\boldsymbol{\beta}})^{-1}$ we note that $\hat{\beta}_2$ has an asymptotic normal distribution with mean β_2 and variance estimated by $(d_1 + d_2)/d_1d_2$. The corresponding test for $\beta_2 = 0$ then involves a comparison of

$$\hat{\beta}_2\sqrt{\frac{d_1d_2}{d_1 + d_2}} \tag{3.18}$$

with standard normal tables.

For illustration we suppose that survival times in excess of 100 days are exponential for both groups of rats in Section 1.1.1. Then $d_1 = 17$, $V_1 = 2195$, as before, and $d_2 = 19$, $V_2 = 2923$ so that $\hat{\beta}_2 = \log(d_2/V_2) - \log(d_1/V_1) = -.1752$ and (3.18) has value $-.5248$. This value is central to the standard normal distribution. A normal approximation to the distribution of (3.18) may be expected to be suitable with only moderate values of d_1, d_2 because of the stable variance estimate and the absence of range restrictions on $\hat{\beta}_2$.

The asymptotic distribution of the score statistic can also be used to test $\beta_2 = \beta_2^0$ upon maximizing out β_1. Let $\hat{\beta}_1(\beta_2^0)$ represent the maximum likelihood estimate of β_1, assuming $\beta_2 = \beta_2^0$. The asymptotic conditional distribution of $U_2(\boldsymbol{\beta})$ given $U_1(\boldsymbol{\beta}) = 0$ [that is, given $\beta_1 = \hat{\beta}_1(\beta_2^0)$] is, from the asymptotic distribution of $\mathbf{U}(\boldsymbol{\beta})$ and multivariate normal theory, normal with mean zero and variance estimated by

$$e^{\beta_1+\beta_2^0}V_2 - \frac{(e^{\beta_1+\beta_2^0}V_2)^2}{e^{\beta_1}V_1 + e^{\beta_1+\beta_2^0}V_2}, \tag{3.19}$$

evaluated at $\beta_1 = \hat{\beta}_1(\beta_2^0)$. Under the hypothesis $\beta_2 = 0$, we have from $U_1(\boldsymbol{\beta})$, $\exp[\hat{\beta}_1(0)] = (d_1 + d_2)/(V_1 + V_2)$ so that $U_2(\boldsymbol{\beta})$ evaluated at $(\hat{\beta}_1(0), 0)$ becomes

$$d_2 - \frac{(d_1 + d_2)V_2}{V_1 + V_2} = \frac{d_2V_1 - d_1V_2}{V_1 + V_2}$$

with variance estimate, from (3.19), of $(d_1 + d_2)V_1V_2/(V_1 + V_2)^2$. This gives an asymptotic standard normal statistic

$$\frac{d_2V_1 - d_1V_2}{[(d_1 + d_2)V_1V_2]^{1/2}} \tag{3.20}$$

for testing $\beta_2 = 0$. For the carcinogenesis data, (3.20) has value $-.5255$ in excellent agreement with (3.18). Either the m.l.e. or score procedure could be used to form an approximate confidence interval for β_2, though the asymptotic distribution of $\hat{\beta}_2$ is more convenient for this purpose.

The likelihood ratio method could also be used to test $\beta_2 = 0$. The log likelihood is written

$$\log L(\boldsymbol{\beta}) = d_1\beta_1 - e^{\beta_1}V_1 + d_2(\beta_1 + \beta_2) - e^{\beta_1+\beta_2}V_2$$

so that

$$\begin{aligned}\log R(\beta) &= \log L(\beta) - \log L(\hat{\beta})\\ &= \log L(\boldsymbol{\beta}) - d_1 \log\left(\frac{d_1}{V_1}\right) + d_1 - d_2 \log\left(\frac{d_2}{V_2}\right) + d_2.\end{aligned}$$

For the hypothesis $\beta_2 = 0$, we have $\exp[\hat{\beta}_1(0)] = (d_1 + d_2)/(V_1 + V_2)$ so that $-2 \log R(\boldsymbol{\beta})$ evaluated at $\beta_1 = \hat{\beta}_1(0)$, $\beta_2 = 0$ has value

$$2\left[d_1 \log\left(\frac{d_1}{V_1}\right) + d_2 \log\left(\frac{d_2}{V_2}\right) - (d_1 + d_2) \log\left(\frac{d_1 + d_2}{V_1 + V_2}\right)\right] \tag{3.21}$$

which will have an asymptotic $\chi^2_{(1)}$ distribution under the hypothesis $\beta_2 = 0$. Again with the carcinogenesis data, (3.21) has value .274 which is in excellent agreement with the approximate $\chi^2_{(1)}$ statistics $(-.5248)^2 = .275$ and $(-.5255)^2 = .276$ from the m.l.e. and score procedures. None of the three procedures suggests any survival difference between the two groups of rats. Of course, all the tests are based on an assumed exponential model for time in excess of 100 days which is suspect on the basis of the survival curves of Figure 1.2. The more general parametric regression models of Chapter 2 would be expected to provide an improved fit.

3.6 ESTIMATION IN LOG-LINEAR REGRESSION MODELS

The likelihood methods of Section 3.5 are easily generalized to any specific log-linear regression model, such as those arising from the parametric models of Section 2.2.

Suppose the p.d.f. for $y = \log t$ can be written

$$\sigma^{-1}f(w),$$

where $w = (y - \mathbf{z}\boldsymbol{\beta})/\sigma$ and $\mathbf{z} = (z_1, \ldots, z_s)$ is a regression vector corresponding to failure time t. Note that σ is a scale constant and that, if $z_1 = 1$ identically, the first component of $\boldsymbol{\beta}' = (\beta_1, \ldots, \beta_s)$ represents the general location of y. As simple examples, a Weibull regression model is given by $f(w) = \exp(w - e^w)$; the exponential model of Section 3.5 requires, in addition, that $\sigma = 1$.

Again for an independent censoring mechanism where δ_i and y_i represent the censoring indicator and the logarithm of the observed survival time for the ith individual, the likelihood function may be written

$$L(\boldsymbol{\beta}, \sigma) = \prod_1^n [\sigma^{-1}f(w_i)]^{\delta_i} F(w_i)^{1-\delta_i}$$

where $w_i = (y_i - \mathbf{z}_i\boldsymbol{\beta})/\sigma$ and $F(w) = \int_w^\infty f(u)\,du$. The score statistic can be written

$$\begin{aligned} U_j(\boldsymbol{\beta}, \sigma) &= \frac{\partial \log L}{\partial \beta_j} = \sigma^{-1} \sum z_{ji} a_i, \qquad j = 1, \ldots, s \\ U_{s+1}(\boldsymbol{\beta}, \sigma) &= \frac{\partial \log L}{\partial \sigma} = \sigma^{-1} \sum (w_i a_i - \delta_i), \end{aligned} \tag{3.22}$$

where

$$\begin{aligned} a_i &= -\left(\delta_i \frac{d \log f(w_i)}{dw_i} + (1 - \delta_i) \frac{d \log F(w_i)}{dw_i}\right) \\ &= -\delta_i \frac{d \log f(w_i)}{dw_i} + (1 - \delta_i)\lambda(w_i) \end{aligned}$$

and $\lambda(w_i)=f(w_i)/F(w_i)$. The observed information matrix, $I(\boldsymbol{\beta},\sigma)$ has entries

$$\frac{-\partial^2 \log L}{\partial\beta_j\partial\beta_k}=\sigma^{-2}\sum z_{ji}z_{ki}A_i, \qquad j=1,\ldots,s, \quad k=1,\ldots,s$$

$$\frac{-\partial^2 \log L}{\partial\beta_j\partial\sigma}=\sigma^{-2}\sum z_{ji}w_iA_i+\sigma^{-1}U_j(\boldsymbol{\beta},\sigma), \qquad j=1,\ldots,s$$

$$\frac{-\partial^2 \log L}{\partial\sigma^2}=\sigma^{-2}\sum (w_i^2A_i+\delta_i)+2\sigma^{-1}U_{s+1}(\boldsymbol{\beta},\sigma), \tag{3.23}$$

where

$$\begin{aligned} A_i &= \frac{da_i}{dw_i} \\ &= -\delta_i\frac{d^2\log f(w_i)}{dw_i^2}+(1-\delta_i)\left[\lambda(w_i)\frac{d\log f(w_i)}{dw_i}+\lambda^2(w_i)\right]. \end{aligned}$$

$I(\hat{\boldsymbol{\beta}},\hat{\sigma})$ is somewhat simpler by virtue of the fact that $\mathbf{U}(\hat{\boldsymbol{\beta}},\hat{\sigma})=\mathbf{0}$. The same criteria as in Section 3.5 will be associated with the suitability of a normal approximation to the distribution of $\mathbf{U}(\boldsymbol{\beta},\sigma)$ and $(\hat{\boldsymbol{\beta}},\hat{\sigma})$. Of course, convergence to normality would usually be more rapid for error distributions close to normal.

The likelihood derivatives (3.22) and (3.23) depend on (a_i, A_i), $i=1,\ldots,n$, which are straightforward to obtain provided $\lambda(w_i)$ can be conveniently calculated. For instance, the Weibull regression model has $f(w_i)=\exp(w_i-e^{w_i})$ so that $\lambda(w_i)=e^{w_i}$, $a_i=e^{w_i}-\delta_i$, and $A_i=da_i/dw_i=e^{w_i}$. Similarly, the log-logistic regression model has $f(w_i)=e^{w_i}(1+e^{w_i})^{-2}$, so that $F(w_i)=(1+e^{w_i})^{-1}$, $\lambda(w_i)=e^{w_i}(1+e^{w_i})^{-1}$, $a_i=-\delta_i+(1+\delta_i)\times e^{w_i}(1+e^{w_i})^{-1}$, and $A_i=(1+\delta_i)e^{w_i}(1+e^{w_i})^{-2}$. The log-normal regression model is given by $f(w_i)=(2\pi)^{-1/2}\exp(-w_i^2/2)$ so that $a_i=\delta_iw_i+(1-\delta_i)\lambda(w_i)$ and $A_i=\delta_i+(1-\delta_i)\lambda(w_i)[\lambda(w_i)-w_i]$. The likelihood derivatives involve, through $\lambda(w_i)$, the incomplete normal integral. An approximation, such as that given in Abramowitz and Stegun (1965, p. 932, 26.2.19) gives rise to a straightforward application of the log-normal model.

Fitting such models almost always requires an iterative solution to the likelihood equations. The next section describes a standard numerical procedure for likelihood maximization.

3.7 THE NEWTON–RAPHSON TECHNIQUE

3.7.1 Methodology

Consider a general likelihood function $L(\boldsymbol{\theta})$ for a column vector $\boldsymbol{\theta}$. The Newton–Raphson technique is based on the first order Taylor series expansion of $\mathbf{U}(\boldsymbol{\theta}) = \partial \log L(\boldsymbol{\theta})/\partial\boldsymbol{\theta}$. Given a trial value $\boldsymbol{\theta}_0$ the score statistic at $\hat{\boldsymbol{\theta}}$ can be written

$$\mathbf{U}(\hat{\boldsymbol{\theta}}) = \mathbf{U}(\boldsymbol{\theta}_0) - \boldsymbol{I}(\boldsymbol{\theta}^*)(\hat{\boldsymbol{\theta}} - \boldsymbol{\theta}_0)$$

where $\boldsymbol{\theta}^*$ is "between" $\boldsymbol{\theta}_0$ and $\hat{\boldsymbol{\theta}}$. For $\boldsymbol{\theta}_0$ in the vicinity of $\hat{\boldsymbol{\theta}}$, $I_0(\boldsymbol{\theta}^*)$ approximately equals $\boldsymbol{I}(\boldsymbol{\theta}_0)$, so that setting $\mathbf{U}(\hat{\boldsymbol{\theta}}) = \mathbf{0}$ and solving gives

$$\hat{\boldsymbol{\theta}} = \boldsymbol{\theta}_0 + \boldsymbol{I}(\boldsymbol{\theta}_0)^{-1}\mathbf{U}(\boldsymbol{\theta}_0). \tag{3.24}$$

The right side of (3.24) gives a new trial value for $\boldsymbol{\theta}$ with which the process is repeated until successive $\boldsymbol{\theta}$ estimates agree to a specified extent, and, of course, $\mathbf{U}(\boldsymbol{\theta}) = \mathbf{0}$ at convergence. The method works well if $\boldsymbol{I}(\boldsymbol{\theta})$ is reasonably stable over a range of values near $\hat{\boldsymbol{\theta}}$ (i.e., if the likelihood function is close to normal in shape). Asymptotic likelihood theory guarantees the normal shape for large samples, in which case this method is expected to be efficient. The method typically fails if the likelihood is multimodal. In this instance, however, if the secondary modes are nontrivial in relation to the mode at $\hat{\boldsymbol{\theta}}$ giving rise to the largest likelihood, then application of the asymptotic normal theory is improper; in fact, $\hat{\boldsymbol{\theta}}$ or any other single statistic with a measure of precision would provide a seriously incomplete data summary.

Note that the Newton–Raphson procedure produces as a by-product the observed information matrix evaluated at the m.l.e., $\boldsymbol{I}(\hat{\boldsymbol{\theta}})$.

3.7.2 Illustration

As an extremely simple example consider again the comparison of two exponential samples discussed in Section 3.5. For survival time beyond 100 days and the carcinogenesis data of Section 1.1.1, the m.l.e.'s $\hat{\beta}_1 = \log(d_1/V_1) = -4.8607$ and $\hat{\beta}_2 = \log(d_2/V_2) - \hat{\beta}_1 = -.1752$ can be calculated explicitly. Alternatively $\hat{\beta}_1$, $\hat{\beta}_2$ could have been computed using a Newton–Raphson procedure, perhaps with starting values $\beta_{10} = \log[(d_1 + d_2)/(V_1 + V_2)] = -4.9570$ and $\beta_{20} = 0$. With these starting values the score statistic has value $U_1(\boldsymbol{\beta}_0) = 0$, $U_2(\boldsymbol{\beta}_0) = -1.5604$. The inverse observed information matrix is

$$\begin{pmatrix} 36 & 20.5604 \\ 20.5604 & 20.5604 \end{pmatrix}^{-1} = \begin{pmatrix} .0648 & -.0648 \\ -.0648 & .1134 \end{pmatrix},$$

so that updated values of $\boldsymbol{\beta}$ are

$$\begin{pmatrix} -4.9570 \\ 0 \end{pmatrix} + \begin{pmatrix} .0648 & -.0648 \\ -.0648 & .1134 \end{pmatrix}\begin{pmatrix} 0 \\ -1.5604 \end{pmatrix} = \begin{pmatrix} -4.8559 \\ -.1770 \end{pmatrix}.$$

Use of (3.24) through two more iterations gives the maximum likelihood estimates to 4 decimal places as $\hat{\beta}_1 = -4.8607$ and $\hat{\beta}_2 = -.1752$. A variance estimate for $\hat{\boldsymbol{\beta}}$ is

$$\begin{pmatrix} .0588 & -.0588 \\ -.0588 & .1115 \end{pmatrix}$$

and the maximized log likelihood has value -42.999.

Some of the more general parametric models discussed in Section 3.7 may provide an improved fit to these data. A Newton–Raphson procedure using a Weibull regression model gives

$$\begin{pmatrix} \hat{\beta}_1 \\ \hat{\beta}_2 \\ \hat{\sigma} \end{pmatrix} = \begin{pmatrix} 4.87 \\ .213 \\ .323 \end{pmatrix} \qquad \text{with} \qquad I(\hat{\boldsymbol{\beta}}, \hat{\sigma})^{-1} = 10^{-3} \begin{pmatrix} 6.24 & -6.09 & -0.45 \\ -6.09 & 11.62 & 0.14 \\ -0.45 & 0.14 & 1.76 \end{pmatrix}$$

and a maximized log likelihood of -20.618. Note that the parametrization is such that $\boldsymbol{\beta}$ here corresponds to $-\boldsymbol{\beta}$ from the exponential model of Section 3.5. The hypothesis $\sigma = 1$ is that of exponentiality within the Weibull model. A likelihood ratio test for $\sigma = 1$ gives a $\chi^2_{(1)}$ statistic of $-2(20.618 - 42.999) = 44.76$ which is very highly significant. In comparison, the asymptotic distribution of $\hat{\sigma}$ gives rise to a standard normal value of $(.323 - 1)/(.00176)^{1/2} = -16.57$ or a $\chi^2_{(1)}$ value of 274.5. The poor agreement of this test with the likelihood ratio test is indicative of a lack of symmetry in the likelihood in the σ direction. It may be checked that asymptotic theory applied to $\log \hat{\sigma}$ gives much better agreement. In any case there is very strong evidence that $\sigma < 1$; or equivalently that the Weibull shape parameter $p = \sigma^{-1} > 1$; that is, the failure rate is clearly increasing with time on study. The fact that the exponential model assumes a highly inflated value of σ leads to grossly overestimated variances for $\hat{\boldsymbol{\beta}}$, even though $\hat{\boldsymbol{\beta}}$ itself appears to be in the proper range. The test for $\beta_2 = 0$, based on the asymptotic distribution of $\hat{\beta}_2$, gives rise to an approximate standard normal statistic with value

$$\frac{.213}{(.01162)^{1/2}} = 1.98,$$

which indicates that there is evidence that survival of the group 2 rats exceeds that of group 1. This is quite different from the tests of Section 3.5 which were by no means suggestive of a survival difference.

A log-normal regression model applied to the same data (survival time beyond 100 days in Section 1.1.1) gives

$$\begin{pmatrix} \hat{\beta}_1 \\ \hat{\beta}_2 \\ \hat{\sigma} \end{pmatrix} = \begin{pmatrix} 4.73 \\ .154 \\ .416 \end{pmatrix} \qquad \text{with} \qquad I(\hat{\boldsymbol{\beta}}, \hat{\sigma})^{-1} = 10^{-3} \begin{pmatrix} 9.44 & -9.43 & 0.21 \\ -9.43 & 17.95 & -0.04 \\ 0.21 & -0.04 & 2.45 \end{pmatrix}$$

and a maximized log likelihood of −22.890. This time a test for $\beta_2 = 0$ gives an asymptotic standard normal statistic of

$$\frac{.154}{(.01795)^{1/2}} = 1.15$$

which, in contrast to the Weibull model, does not provide even moderately strong evidence of a survival difference. This raises the question of which of the two models fits the data better and, more importantly, whether either model gives an adequate representation of the data.

3.8 ILLUSTRATIONS IN MORE COMPLEX DATA SETS

3.8.1 Accelerated Life Testing

Consider the accelerated life test data of Nelson and Hahn (1972) as given in Table 1.2 (Section 1.1.4). Hours to failure of motorettes are given as a function of operating temperatures of 150°C, 170°C, 190°C, or 220°C. The primary purpose of the experiment was to estimate certain percentiles of the failure time distribution at a design temperature of 130°C. Nelson and Hahn applied a log-normal model to these data with the single regressor variable, $z = 1000/(273.2 + °\text{C})$. They used a weighted least squares method of estimation which required at least two failures at each test condition so that the 150°C data had to be excluded. By this method, they obtained an estimate of 10.454 for the log-median lifetime at 130°C (they use base 10 logarithms rather than base e used here) with an associated 90% confidence interval $10.454 \pm 1.645(.417)$. This gives an estimated median life time of $\exp(10.454) = 34{,}700$ hours, and approximate 90% confidence interval (17,500, 68,900).

For comparative purposes the models of Section 3.6 are fitted to only the 30 observations at test temperatures of 170°C or greater, though elimination of the 150°C data is unnecessary for the maximum likelihood methods. A log-normal regression model and Newton–Raphson iterative technique yield

$$\hat{\alpha} = -10.471, \qquad \hat{\beta} = 8.322, \qquad \text{and} \qquad \hat{\sigma} = 0.6040,$$

at which parameter values the maximum log likelihood is −24.474.

The estimated covariance matrix of $(\hat{\alpha}, \hat{\beta}, \hat{\sigma})$ is

$$\begin{pmatrix} 7.684 & -3.556 & .0327 \\ -3.556 & 1.649 & -.0128 \\ .0327 & -.0128 & .0123 \end{pmatrix}.$$

A comparison of $|\hat{\beta}|$ with its estimated standard error verifies the important effect of temperature on failure time.

The maximum likelihood estimate of the 100 pth percentile of the distribution of y (log-failure time) at $z = z_0$ is simply

$$\hat{y}_p = \hat{\alpha} + z_0\hat{\beta} + \hat{\sigma}w_p \tag{3.25}$$

where w_p is the 100 pth percentile of the error distribution. Also, if Σ represents the estimated covariance matrix of $(\hat{\alpha}, \hat{\beta}, \hat{\sigma})$, then the estimated variance of $\hat{y}_p$ is

$$(1, z_0, w_p) \sum (1, z_0, w_p)'.$$

At 130°C, $z_0 = 1000/(273.2 + 130) = 2.480$, so that the estimated log-median lifetime is

$$\hat{y}_{.5} = -10.471 + 8.322(2.480) + .6040(0) = 10.170$$

while var $\hat{y}_{.5} = (.433)^2$. This gives an approximate 90% confidence interval for $\hat{y}_{.5}$ of $10.170 \pm 1.645(.433)$. The estimate of the median lifetime is $\exp(10.170) = 26100$ with an associate approximate 90% confidence interval (12800, 53200).

A Weibull model may equally well be taken for failure time. In fact, if the motorettes are such that failure occurs when the first of any of several essentially independent components fails, there would be some theoretical reason for considering a Weibull model. A Newton–Raphson iteration gives

$$\hat{\alpha} = -11.891, \qquad \hat{\beta} = 9.038, \qquad \hat{\sigma} = .3613$$

and a maximized log likelihood of -22.952. Since the median of an extreme value minimum distribution (2.1) is log(log 2), the maximum likelihood estimate of the log-median lifetime at 130°C is $\hat{y}_{.5} = 10.391$ from (3.25). The standard error of $\hat{y}_{.5}$ is estimated as .303. The estimated median lifetime is then $\exp(10.391) = 32600$ with approximate 90% confidence interval (19800, 53600). Note the increase in precision of the Weibull analysis over the log-normal and weighted least squares procedures. The Weibull model is to some extent preferable to the log-normal model on account of the larger maximized log likelihood. Further work with these models could for example, include additional or alternative functions of temperature in the regression vector. Because of the small number of failures involved further work should also be carried out, perhaps by simulation, to validate the use of asymptotic methods.

3.8.2 Clinical Trial Data

As a further example, consider the Veteran's Administration lung cancer data of Appendix 1. In this trial, males with advanced inoperable lung cancer were randomized to either a standard or test chemotherapy. The primary end point for therapy comparison was time to death. Only 9 of the 137 survival times were censored. As is common in such studies, there was much heterogeneity between patients in, for example, disease extent and pathology, previous treatment of the disease, demographic background, and initial health status. The data in the appendix include information on a number of covariables measuring this heterogeneity:

1. A measure, at randomization, of the patients' performance status (Karnofsky rating); 10–30 completely hospitalized, 40–60 partial confinement, 70–90 able to care for self.
2. Time in months from diagnosis to randomization.
3. Age in years.
4. Prior therapy: 0 = no, 10 = yes.
5. Histological type of tumor: squamous, small cell, adeno, large cell.
6. Treatment: 0 = standard, 1 = test.

After preliminary investigations described below, a Weibull regression model was fitted to these data with eight regressor variables; the results are summarized in Table 3.1. Single indicator variables distinguish treatment and prior therapy groups, and three indicator variables for

Table 3.1 ASYMPTOTIC LIKELIHOOD INFERENCE ON LUNG CANCER DATA USING A WEIBULL REGRESSION MODEL

Regressor Variable	Regression Coefficient ($\hat{\beta}$)	χ^2 Statistic
Performance status (Karnofsky scale)	.0301	38.79
Disease duration (months)	−.0005	0.00
Age (years)	.0061	0.51
Prior therapy (0 no, 10 yes)	−.0041	0.04
Cell type		
Squamous vs. large	.3977	
Small cell vs. large	−.4285	22.03
Adeno vs. large	−.7350	
Treatment (standard 0, test 1)	.2061	1.30

squamous, small cell, and adenocarcinoma permit arbitrary log-linear location effects for the four cell-type classes. The other factors enter as indicated in Table 3.1. The asymptotic $\chi^2_{(1)}$ statistics given in the table are formed for the ith component as $\{\hat{\beta}_i/(\text{estimated standard error of } \hat{\beta}_i)\}^2$. The asymptotic $\chi^2_{(3)}$ statistic corresponding to cell type differences is calculated as $\mathbf{b}'\Sigma^{-1}\mathbf{b}$ where $\mathbf{b}' = (\hat{\beta}_5, \hat{\beta}_6, \hat{\beta}_7)$ is the vector of maximum likelihood estimates and Σ is the estimated covariance matrix for $\mathbf{b}$. This statistic, of course, does not depend on which three cell types are used to define the indicator variables.

From Table 3.1, it is clear that a strong prognostic effect of initial performance status is indicated and also that survival times in the different cell type groups differ significantly. This analysis would indicate, however, that patient survival does not differ significantly between treatment groups after taking account of the prognostic effect of other variables. There is as well no apparent dependence of survival time on age or disease duration. It should be stressed that, even in a randomized study such as this, it is important to take account of prognostic factors in the analysis. Otherwise, imbalances in the prognostic factors between the treatment groups may bias the estimates of treatment differences.

Weibull and log-normal analyses of these data with only performance status and cell type as factors yielded maximized log likelihoods of -197.10 and -196.75, respectively. It is of some interest to test the adequacy of the exponential regression model relative to the Weibull model. The Weibull model reduces to the exponential at $\sigma = p^{-1} = 1$. The maximum likelihood estimate of σ under the Weibull model is $\hat{\sigma} = .983$ with estimated standard error of .116. Clearly a test of the hypothesis $\sigma = 1$ provides no evidence against the exponential model relative to the encompassing Weibull model. Further results of fitting various models to these data are given in Sections 3.9.2 and 4.5.

Graphical methods can be very useful with such data both in preliminary data exploration and in checking the validity of fitted models. For example, if regression variables do not severely dominate, a plot of $\log[-\log \hat{F}(t)]$ versus $\log t$ ($\hat{F}$ is the product limit estimator) may be used to give an overall impression of adequacy of the Weibull model. Such a plot should be approximately linear with slope $\sigma^{-1} = p$ if the data were homogeneous and Weibull and so, in this case, yields also an informal estimate p of the Weibull shape parameter. Similar plots may be constructed in strata defined by components of the regression vector, for example, by low, medium, and high initial performance status groups in the lung cancer study. The corresponding plots of $\log[-\log \hat{F}(t)]$ versus $\log t$ should each be roughly linear with approximately common slope if the Weibull regression model holds. Further, the distance between these plots should be roughly proportional to the difference between z values used in forming the strata. Once a Weibull regression model has been fitted, the same type of plot may be used as a check on the model. For such a plot the original survival times t_i are replaced with $t_i' = t_i \exp(-\mathbf{z}_i\hat{\boldsymbol{\beta}})$ and the product limit estimator computed on the basis of these.

An exploratory tool that can be useful if the distribution of the data is nearly exponential is to compute hazard rate estimators (d/V in the notation of Sections 3.3 and 3.5) in various subsets of the data. For example, in Figure 3.1 exponential failure rate estimators are plotted on a logarithmic scale versus performance status for the lung cancer data. A straight line relationship agrees with a linear modeling of performance status on log t. Often tabulations of exponential failure rates, taking regressor variables one or two at a time, point out the most important prognostic factors and suggest a form for a log-linear modeling of regression variables. A graphical estimator, $\tilde{p}$, of the Weibull shape parameter may be used to bring to bear this procedure when a Weibull, but not an exponential, model is appropriate. Each censored or uncensored failure time t is simply replaced by $\tilde{t} = t^{\tilde{p}}$ before computing hazard rate estimators (d/V). If a Weibull model is appropriate, the $\tilde{t}$ values will have a distribution closely approximated by an exponential regression model. An alternative and more generally applicable approach can be based on the log-rank test of Sections 1.4 and 4.2.5. This is illustrated in the example of Section 8.5.

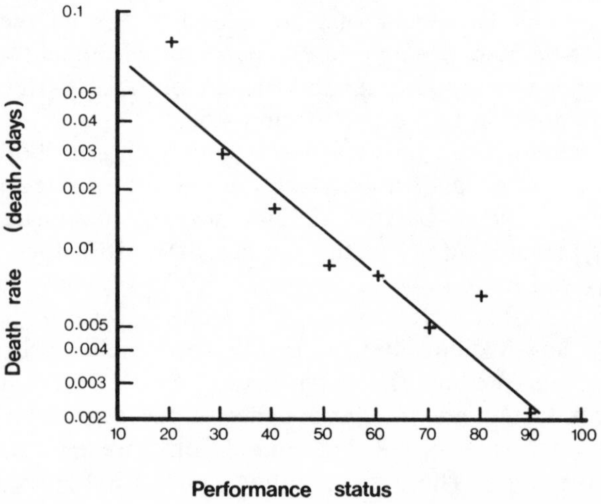

Figure 3.1 Log death rates estimated from an exponential model for the nine performance status groups (VA lung cancer data).

3.9 DISCRIMINATION AMONG PARAMETRIC MODELS

3.9.1 Methods

There are many formal as well as informal methods of assessing the goodness of fit of data to a specified probability model or of selecting a best model among several competitors. One approach, for a log-linear model $y = \mathbf{z}\boldsymbol{\beta} + \sigma w$ where w has an error distribution of specified form, is to compute the residuals $\hat{w}_i = (y_i - \mathbf{z}_i\hat{\boldsymbol{\beta}})/\hat{\sigma}$ $(i = 1, \ldots, n)$ which should resemble to some extent a (censored) sample from the specified error distribution. As suggested above for the Weibull model, graphical methods based on the Kaplan–Meier estimator computed from these residuals can then provide the tool for an informal assessment of fit. More formally, however, the generalized F model of Section 2.2.7 includes the other parametric models of Chapter 2 as special cases and thus permits their evaluation relative to each other and to a more general model.

Recall that the generalized F is a log-linear model $y = \alpha + \mathbf{z}\boldsymbol{\beta} + \sigma w$ for $y = \log t$ where the error density $f(w)$ is that of the logarithm of an F variate on $2m_1$ and $2m_2$ degrees of freedom. Its special forms were discussed in Section 2.2.7 and we review these briefly here. The distribution of t is log-logistic for $m_1 = m_2 = 1$, Weibull for $m_1 = 1$, $m_2 \to \infty$, (degenerate) log-normal for $m_1 = m_2 = \infty$, and generalized gamma as $m_2 \to \infty$; that is,

$$\lim_{m_2 \to \infty} f(w) = \frac{m_1^{m_1} \exp(m_1 w - m_1 e^w)}{\Gamma(m_1)}. \tag{3.26}$$

For specified m_1 and m_2, the results of Section 3.6 can be used to fit this model to data. If m_1 and m_2 are both finite, then from (2.5)

$$\frac{d \log f(w_i)}{dw_i} = m_1 - \frac{(m_1 + m_2)k_i}{1 + k_i}$$

$$\frac{d^2 \log f(w_i)}{dw_i^2} = -\frac{(m_1 + m_2)k_i}{(1 + k_i)^2}$$

where $k_i = m_1 e^{w_i}/m_2$ and $F(w_i) = I(s_i; m_2, m_1)$. Here $s_i = (1 + k_i)^{-1}$ and I represents the incomplete beta ratio that can be calculated using results of Osborn and Madley (1968). Similarly from (3.26) at finite m_1 and $m_2 = \infty$, $d \log f(w_i)/dw_i = m_1 - m_1 e^{w_i}$; $d^2 \log f(w_i)/dw_i^2 = -m_1 e^{w_i}$ and $F(w_i) = 1 - P(s_i; m_1)$, where $s_i = e^{m_1 w_i}$ and P is the incomplete gamma ratio that may be calculated using Abramowitz and Stegun (1965, p. 262, 6.5.29). The model at $(m_1 = \infty, m_2)$ can be fit by replacing each w_i by w_i^{-1} and using the method just indicated with m_2 replacing m_1. The log-normal model $(m_1 = \infty, m_2 = \infty)$ can be applied as indicated in Section 3.6. As

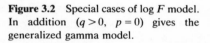

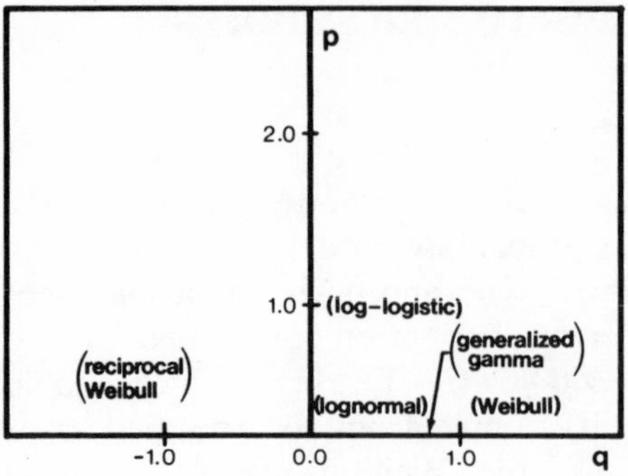

Figure 3.2 Special cases of log F model. In addition $(q>0,\ p=0)$ gives the generalized gamma model.

shown in Prentice (1975) a reparametrization from (m_1, m_2) to $(q, p \geq 0)$ where

$$q = (m_1^{-1} - m_2^{-1})(m_1^{-1} + m_2^{-1})^{-1/2}$$
$$p = 2(m_1 + m_2)^{-1}$$

will lead to a regular maximized log likelihood function (finite, not identically zero likelihood derivatives) everywhere on the boundary $m_1 = \infty$ or $m_2 = \infty$ $(p = 0)$. In the new parametrization (Fig. 3.2) the log-normal, Weibull, log-logistic, reciprocal Weibull, and generalized gamma model occur at respective (q, p) values of $(0, 0)$, $(1, 0)$, $(0, 1)$, $(-1, 0)$, and $(q>0, 0)$. For inference on (q, p) or equivalently on (m_1, m_2) we may calculate the maximized log likelihood over a grid of (q, p) values and, because of the regularity of the log likelihood, we may use the asymptotic distribution of the likelihood ratio statistic to form approximate confidence regions for (q, p) and for evaluating the specific models relative to the generalized F model.

3.9.2 Illustrations

Consider again the two sample carcinogenesis data of Section 1.1.1. As in Farewell and Prentice (1977), the generalized F model was applied to the variables $t - 100$ at a range of boundary values $(p = 0)$. Note that at $p = 0$

$$q = \begin{cases} m_1^{-1/2} & \text{at } m_2 = \infty \\ -m_2^{1/2} & \text{at } m_1 = \infty, \end{cases}$$

and that the log-normal and Weibull models previously applied occur at $q = 0$ and $q = 1$, respectively. Figure 3.3 presents a plot of the maximized

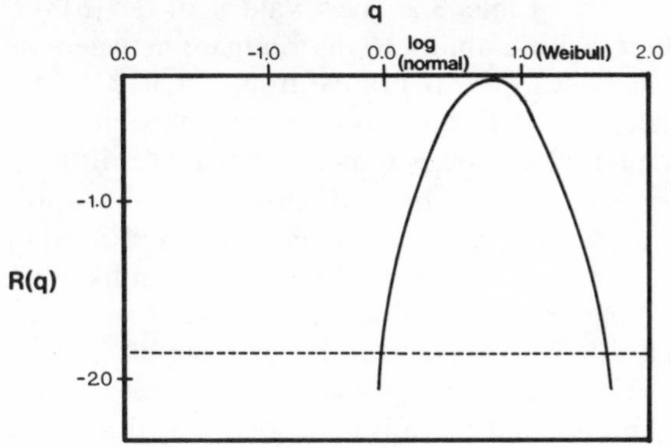

Figure 3.3 Maximized log likelihoods assuming generalized F model with $p = 0$ and based on data of Section 1.1.1.

log relative likelihood, $R(q)$ (maximized log likelihood standardized to have maximal value zero). Note that the m.l.e. $\hat{q}$ (at $p=0$) has value .87 and that $R(1)=-.05$ while $R(0)=-2.32$. The asymptotic $\chi^2_{(1)}$ distribution for $-2R(q)$ gives an approximate 95% confidence interval for q as those values of q for which $R(q)>-1.92$. There is then evidence against the log-normal model relative to this more general class but not against the Weibull model, which, as noted above, indicated an improved survival for the group 2 rats. In fact the previous calculations of Section 3.7.2 alone, giving maximized log likelihood of -20.618 and -22.890 for Weibull and log-normal models, respectively, are sufficient to provide evidence against the log-normal model. The fact that the difference between log likelihoods $-20.618+22.890=2.27$ exceeds $1.92=3.84/2$ indicates that the log-normal model will be excluded from an approximate 95% confidence interval based on $R(q)$. Note that $P(\chi^2_{(1)}>3.84)=.05$.

The above analysis may be extended to estimate the duration, δ, of the initial failure-free period, rather than specify it as 100 days. Pike (1966) calculates $\hat{\delta}=98.9$ assuming a Weibull model. Since δ is a threshold parameter, the likelihood function does not possess the required regularity to permit the use of standard asymptotic likelihood results for the estimation of δ. To examine whether inclusion of δ gives rise to a significant improvement in the generalized F, the model with $p=0$ was applied as above to $y_\delta = \log(t-\delta)$, where t is the time from insult to diagnosis, for several values of δ between 0 and the smallest observed time of 142 days. In each case the log likelihood was maximized over ($\boldsymbol{\beta}$, σ, q). Note that the Jacobian factor $(\exp y-\delta)/\exp y$ needs to be introduced into the maximized log likelihood, $l^*(\delta)$, to describe the change in scale from y to y_δ. Table 3.2 gives values of $R^*(\delta) = l^*(\delta) - l^*(125)$. Apparently there is little ability to discriminate between values of δ with these data. The values $\hat{q} = \hat{q}(\delta)$ range from .36 at $\delta = 0$ to about 1.75 at $\delta = 140$. The generalized F model is sufficiently flexible in this case that the inclusion of a guarantee type parameter contributes little.

This section ends with a brief discussion of the application of the generalized F regression model to the illustration of Section 3.8. With the accelerated life test data of Table 1.2 the maximum likelihood estimate of the "skewness" parameter q (subject to a value of zero for the "kurtosis" parameter p) is 1.6. The maximized log relative likelihood yields $R(0) = -1.73$ and $R(1) = -.21$ so that some doubt is cast on the suitability of the log-normal model, but there is no evidence against the Weibull model relative to this more general model. At $\hat{q} = 1.6$ one obtains an estimated log-median failure time estimate and standard error at 130°C of 10.499 ± .252. The corresponding median failure time estimate and approximate 90% confidence interval are then 36300 and (24000, 54900), respectively. The standard error estimate for $\hat{y}_{.5}$ is appropriate assuming $q = 1.6$ but does not take account of the correlation between $\hat{y}_{.5}$ and $\hat{q}$.

A similar application to the lung cancer data (Appendix 1) with performance status and three cell type indicators as regressor variables gives $\hat{q} = .47$. There is evidence against both Weibull and log-normal models as $R(0) = -2.59$ and $R(1) = -2.94$. Further, the shape of the failure time distribution was found to depend on whether or not the patient had received prior therapy. Data on the 40 patients who had received therapy prior to the start of the study give $\hat{q} = 1.05$ whereas data on the 97 without prior therapy yield $\hat{q} = .43$. A likelihood ratio test for equality of the q's gives $\chi^2_{(1)} = 9.0$ which is significant at the 1% level. Separate analyses for the two prior therapy groups are therefore indicated. Table 3.3 gives some results from such analyses. Note the interaction between prior therapy and performance status. Performance status is an important prognostic factor for both groups of patients but is particularly dominating among patients who have received prior therapy.

Table 3.2 MAXIMIZED LOG RELATIVE LIKELIHOOD FOR A GUARANTEE TIME δ BASED ON THE DATA OF SECTION 1.1.1

δ	0	25	50	75	100	125	140
$R^*(\delta)$	−.55	−.49	−.40	−.32	−.19	.00	−.16

Table 3.3 ANALYSIS OF VETERAN'S ADMINISTRATION LUNG CANCER DATA USING A GENERALIZED F REGRESSION MODEL

	Log-Normal		Weibull		Log F	
Regression Variable	Coeff.	Std. Error	Coeff.	Std. Error	Coeff.	Std. Error
No Prior Therapy—97 Patients					($q = .43$, $p = 0$)	
Performance status	.0297	.006	.0219	.006	.0264	.006
Squamous vs. large	−.085	.34	.175	.31	.086	.32
Small vs. large	−.762	.31	−.521	.28	−.669	.29
Adeno vs. large	−.804	.34	−.840	.30	−.795	.32
Prior Therapy—40 Patients					($q = 1.05$, $p = 0$)	
Performance status	.0587	.010	.0537	.009	.0532	.009
Squamous vs. large	−.199	.46	.428	.38	.450	.38
Small vs. large	−.388	.49	−.044	.42	−.033	.41
Adeno vs. large	−.694	.61	−.787	.51	−.794	.50

3.10 DISCUSSION

Parametric regression models such as the Weibull, log-normal, and log-logistic may involve stronger distributional assumptions than it is suitable to make and the inference procedures mentioned may not be sufficiently robust to departures from these assumptions. This seems particularly to be the case in medical applications in which only limited experimentation in similar situations may have preceded the study in question, or in which data were recorded by a number of individuals. When the primary interest is in the effects of regression variables, a variety of approaches may be considered to achieve greater robustness. The more general parametric models just discussed represent one such approach, though the method indicated may be computationally too complicated to achieve wide use. A natural approach would be to extend the M estimate procedures developed for linear regression (Huber, 1972, 1973; Andrews et al., 1972; Hampel, 1974; Andrews, 1974) to include censoring; that is, to develop specific (pseudo) score functions for which the corresponding regression estimates have both good efficiency and robustness properties. Such studies, however, do not seem to have been made yet. A third and related approach is to consider more general models that are nonparametric, or only partially parametric, such as the proportional hazards and accelerated failure time models of Section 2.3. The objective with such models is to develop inference procedures that will be efficient regardless of which member of the class obtains. The next chapter describes the remarkable work of Cox (1972) on regression estimation within the proportional hazards model.

BIBLIOGRAPHIC NOTES

Inference from type II (order statistic) censored samples has been much discussed in the literature. For example, Epstein and Sobel (1953) gave some basic exponential sampling results. General reference books on order statistic properties include those by Sarhan and Greenberg (1962) and David (1970). Johnson and Kotz (1970a, b) give a comprehensive account of estimation for homogeneous populations assumed to have the parametric models considered in this chapter (exponential, Weibull, log-normal, gamma, log-logistic). Mann et al. (1974) summarize, from an industrial life testing point of view, estimation procedures for these as well as other distributions, both for single sample and two sample problems, with censoring. Gross and Clark (1975) give similar results from the biomedical point of view. In general the methods given are based on asymptotic maximum likelihood procedures with some exact results for order statistic sampling. Bartholomew (1957) provides an early example of a discussion of statistical properties of the m.l.e. with type I (time) censoring. Cox (1953) indicates that under some circumstances a suitable approximation involves treating type I censored data as if it were type II censored. Some authors (e.g., Gilbert, 1962; Efron, 1967; Breslow, 1970; Breslow and Crowley, 1974) postulate a probability distribution for censoring times in an attempt to derive exact properties or to more easily develop asymptotic distribution theory for estimators. Exact conditional confidence interval estimates have been obtained for type II censored data from the Weibull (or) extreme value distribution by Lawless (1973, 1978). He compares these exact results with those obtained from asymptotic theory. Similar arguments could be applied to type II censored data from other models in the accelerated failure time class.

The literature on parametric regression with censored data does not appear to have been collected. The following papers have been concerned with exponential, Weibull, or log-normal regression: Cox (1964), Feigl and Zelen (1965), Zippin and Armitage (1966), Pike (1966), Glasser (1967), Cox and Snell (1968), Sprott and Kalbfleisch (1969), Nelson (1970), Mantel and Myers (1971), Nelson and Hahn (1972), Peto and Lee (1973), Prentice (1973), Myers et al. (1973), Breslow (1974), Byar et al. (1974), Kalbfleisch (1974), and Prentice and Shillington (1975). Farewell and Prentice (1977) consider regression in the generalized gamma distribution.

The accelerated failure time application of Sections 3.8 and 3.9.2 considered the use of log-linear regression models for extrapolation in an industrial setting. Even more extreme extrapolations beyond dosages (regression variable values) actually considered is necessary in carcinogenesis testing. Typically, experimental animals are tested at highly accelerated dosages of a suspected carcinogen and times to tumor incidence are recorded. Downward extrapolation is required to dosages that yield some "safe" level of risk or to dosages comparable to those found in the environment. The regression models considered in this chapter form the basis for proposed methods of extrapolation. The associated literature gives some interesting further insights into biological mechanisms that can give rise, for example, to Weibull or log-normal regression models. For example, "multi-stage" or "multi-hit" carcinogenesis theories can lead to the Weibull model (e.g., Armitage and Doll, 1954, 1961; Crump et al., 1976). Rather different assumptions concerning the existence of thresholds leads to log-normal (probit) type extrapolations (e.g., Mantel and Bryan, 1961; Mantel et al., 1975). Differences in tail shape for these models generally lead to completely different low dose risk estimates. Some recent work in this area (e.g., Hartley and Sielken, 1977) utilizes models more general than the Weibull, but still of a proportional hazards (Section 2.3.2) type.

A comprehensive summary of asymptotic likelihood theory has recently been given by Cox and Hinkley (1974). Such procedures were proposed by Fisher (1922, 1925) and important contributions were made by Neyman and Pearson, Wilks, Wald, Cramér, Bartlett, and Le Cam, among others. In particular, the reader is referred to Le Cam (1970) for a technical discussion of asymptotic normality. Moran (1971) discusses some properties of maximum likelihood estimators of parameters on the boundary of the parameter space. Methods of improving the asymptotic approximation to the distribution of maximum likelihood estimators have been considered by many authors. Variance stabilizing transformation are well-known and discussed in several standard texts (e.g., Cox and Hinkley, 1974, p. 275). Transformations to improve symmetry in the log likelihood (by eliminating the cubic term in the Taylor series expansion) have been considered by Anscombe (1964) and by Sprott (1975b).

Cox (1961, 1962a) and Atkinson (1970) consider general procedures for discriminating among several families of hypotheses. Cox's work involves approximating the distribution of the likelihood ratio. Assuming uncensored data Dumonceaux and Antle (1973) simulate the distribution of the likelihood ratio statistic in order to discriminate between Weibull and log-normal hypotheses. Hagar and Bain (1970) consider the problem of testing for a Weibull model within the generalized gamma family.

CHAPTER 4

The Proportional Hazards Model

4.1 INTRODUCTION

In this chapter, attention is focused on methods of estimation and testing based on data arising from the proportional hazards model (2.10) or from related discrete models. In the parametric case discussed in Chapter 3, the failure time distribution is assumed known except for a few scalar parameters. The proportional hazards model, however, is non-parametric in the sense that it involves an unspecified function in the form of an arbitrary base-line hazard function. In consequence, this model is more flexible, but different approaches are required for estimation and testing.

As before, $\mathbf{z}$ is a row vector of s measured covariates, $\boldsymbol{\beta}$ is a column vector of s regression parameters, and T is the associated failure time. The proportional hazards model is specified by the hazard relationship

$$\lambda(t; \mathbf{z}) = \lambda_0(t) \exp(\mathbf{z}\boldsymbol{\beta}) \tag{4.1}$$

as in (2.10), where $\lambda_0(t)$ is an arbitrary and unspecified base-line hazard function. For ease of reference we record also here the survivor function

$$F(t; \mathbf{z}) = \exp\left[-\int_0^t \lambda_0(u) e^{\mathbf{z}\boldsymbol{\beta}}\, du\right] \tag{4.2}$$

and the density function

$$f(t; \mathbf{z}) = \lambda(t; \mathbf{z}) F(t; \mathbf{z}) \tag{4.3}$$

The main problems addressed in this chapter are those related to the estimation of $\boldsymbol{\beta}$ and $\lambda_0(\cdot)$ in (4.1). Estimation of $\boldsymbol{\beta}$ is considered in Section 4.2, and Section 4.3 deals with the estimation of $\lambda_0(\cdot)$, or equivalently of $F_0(\cdot)$, where

$$F_0(t) = \exp\left[-\int_0^t \lambda_0(u)\, du\right].$$

Other topics include the extension of (4.1) to include strata, the relationship between the log-rank test and the model (4.1), and the asymptotic properties of the estimation procedures including a discussion of efficiency. The analysis of the related discrete model (2.21) is discussed in Section 4.6.

Throughout this chapter, failure times are presumed to be subject to arbitrary independent censoring (see Sections 3.2 and 5.2), unless otherwise specified.

4.2 METHODS OF ESTIMATION

4.2.1 Background

In this section a number of approaches to the problem of estimating the parameters $\boldsymbol{\beta}$ in the model (4.1) are discussed. In Section 4.2.2, the method of marginal likelihood is explored, Section 4.2.3 exemplifies the partial likelihood approach, and Section 4.2.4 gives a brief examination of other methods. A simple numerical example is considered in Section 4.2.6. Historically, the method of partial likelihood was the first applied to the model (4.1) and in many ways is the most general of the methods outlined here. The method of marginal likelihood is discussed on account of its close relationship to the rank tests discussed in Chapter 6 and also because of the directness of the marginal likelihood solution.

4.2.2 Marginal Likelihood

Suppose that n individuals are observed to fail at $t_1, \ldots, t_n$ with corresponding covariates $\mathbf{z}_1, \ldots, \mathbf{z}_n$. For the moment we assume that all failures are distinct and that no censoring is present in the data. Central to our discussion will be the order statistic $\mathbf{O(t)} = [t_{(1)}, \ldots, t_{(n)}]$ and the rank statistic $\mathbf{r(t)} = [(1), \ldots, (n)]$. The order statistic refers to the t_i's ordered from smallest to largest (i.e., $t_{(1)} < t_{(2)} < \cdots < t_{(n)}$) and the notation (i) in the rank statistic refers to the label attached to the ith order statistic. Thus for example, if $n = 4$ and $t_1 = 5$, $t_2 = 17$, $t_3 = 12$, $t_4 = 15$ are observed, then $\mathbf{O} = [5, 12, 15, 17]$ and $\mathbf{r} = [1, 3, 4, 2]$.

Consider the model (4.1) and define $u = g^{-1}(t)$ where $g \in G$, the group of strictly increasing and differentiable transformations of $(0, \infty)$ onto $(0, \infty)$. The conditional distribution of u given $\mathbf{z}$ has the hazard

$$\lambda_1(u)e^{\mathbf{z}\boldsymbol{\beta}}$$

where $\lambda_1(u) = \lambda_0(g(u))g'(u)$. Thus if the data were presented in the form $u_1, \ldots, u_n$ and $\mathbf{z}_1, \ldots, \mathbf{z}_n$ where $g(u_i) = t_i$, the inference problem about $\boldsymbol{\beta}$

would be the same provided $\lambda_0(\cdot)$ were completely unknown. In effect, the estimation problem for $\boldsymbol{\beta}$ is invariant under the group G of transformations on the survival time, t.

On the other hand, if we consider the action of G on the sample space, the order statistic $\mathbf{O(t)}$ can be mapped to any specified order statistic while the rank statistic $\mathbf{r(t)}$ is left unchanged. For example, if the transformation $u = t^2$ is applied to the sample above we obtain $\mathbf{O(u)} = [25, 144, 225, 289]$ and $\mathbf{r(u)} = [1, 3, 4, 2] = \mathbf{r(t)}$. Further, any specified order statistic can clearly be obtained for $\mathbf{u}$ by a judicious choice of $g \in G$. Since the estimation problem for $\boldsymbol{\beta}$ is the same under any such transformation and since the order statistic can be made arbitrary by such a transformation, only the ranks $\mathbf{r}$ can carry information about $\boldsymbol{\beta}$ when $\lambda_0(t)$ is completely unknown. Barnard (1963) has discussed such situations and has extended the definition of sufficiency: the rank statistic $\mathbf{r(t)}$ is said to be marginally sufficient for the estimation of $\boldsymbol{\beta}$. [Barnard used the expression "sufficient for $\boldsymbol{\beta}$ in the absence of knowledge of $\lambda_0(\cdot)$"].

For inference about $\boldsymbol{\beta}$, the marginal distribution of the ranks is available and the marginal likelihood (c.f. Fraser, 1968, and Kalbfleisch and Sprott, 1970) is proportional to the probability that the rank vector should be that observed. That is, the marginal likelihood is proportional to

$$
\begin{aligned}
P(\mathbf{r};\boldsymbol{\beta}) &= P\{\mathbf{r} = [(1), \ldots, (n)]; \boldsymbol{\beta}\} \\
&= \int_0^\infty \int_{t_{(1)}}^\infty \cdots \int_{t_{(n-1)}}^\infty \prod_1^n f(t_{(i)}; \mathbf{z}_{(i)})\, dt_{(n)} \ldots dt_{(1)} \\
&= \frac{\exp(\Sigma_1^n \mathbf{z}_i\boldsymbol{\beta})}{\Pi_1^n \left[\Sigma_{l \in R(t_{(i)})} \exp(\mathbf{z}_l\boldsymbol{\beta})\right]} \qquad (4.4)
\end{aligned}
$$

where $R(t_{(i)})$ is the set of labels attached to the individuals at risk just prior to $t_{(i)}$; that is, $R(t_{(i)}) = \{(i), (i+1), \ldots, (n)\}$.

To handle censored data, some modification of this argument is required. If all items are simultaneously put on test and followed to the kth failure time (type II censoring), a marginal likelihood is again easily obtained. In this case, the group acts transitively on the censoring time and the invariant in the sample space is the first k rank variables $(1), \ldots, (k)$. The argument could be extended to progressive type II censoring patterns where items are withdrawn from test with each failure.

More general independent censoring cannot be handled directly by this approach since the censored model will not in general possess the group invariance properties. We can note, however, that had the entire sample been observed, the rank statistic would be marginally sufficient for $\boldsymbol{\beta}$. When a censored sample is obtained, only partial information is observed on the ranks. For example, if the observed survival times of four tested items were

114, 90*, 63, 108*, where the asterisks indicate censoring, the rank statistic is known to be one of the following six possibilities:

$$[3,2,4,1]; \quad [3,4,2,1]; \quad [3,2,1,4];$$
$$[3,4,1,2]; \quad [3,1,2,4]; \quad [3,1,4,2].$$

In order to make an inference about $\boldsymbol{\beta}$, the marginal probability that the rank statistic should be one of those possible can be used. The observed part of the marginally sufficient statistic $\mathbf{r}$ is generating the likelihood. Note that the exact time of censoring is ignored, but the invariance of the uncensored model suggests that the lengths of the intervals between successive failures is irrelevant for inference about $\boldsymbol{\beta}$. Consequently it would seem reasonable to suppose that the exact time of censoring, relative to adjacent uncensored times, should not contribute to the inference about $\boldsymbol{\beta}$.

Suppose that k items labeled $(1), \ldots, (k)$ give rise to observed failure times $t_{(1)} < t_{(2)} < \cdots < t_{(k)}$ with corresponding covariates $\mathbf{z}_{(1)}, \ldots, \mathbf{z}_{(k)}$ and suppose further that m_i items with covariates $\mathbf{z}_{i1}, \ldots, \mathbf{z}_{im_i}$ are censored in the ith interval $[t_{(i)}, t_{(i+1)})$, $i = 0, 1, \ldots, k$, where $t_{(0)} = 0$ and $t_{(k+1)} = \infty$. The marginal likelihood of $\boldsymbol{\beta}$ is computed as the probability that the rank statistic should be one of those possible on the sample and is, therefore, the sum of a large number of terms like (4.4). The set of possible rank vectors can be characterized, however, by

$$\begin{gathered} t_{(1)} < \cdots < t_{(k)} \\ t_{(i)} < t_{i1}, \ldots, t_{im_i} \qquad (i = 0, 1, \ldots, k) \end{gathered} \tag{4.5}$$

where $t_{i1}, \ldots, t_{im_i}$ are the unobserved failure times associated with individuals censored in $[t_{(i)}, t_{(i+1)})$. Writing the event as (4.5) allows simple computation of the marginal likelihood since, given $t_{(i)}$, the event $t_{(i)} < t_{i1}, \ldots, t_{im_i}$ has the conditional probability

$$h(t_{(i)}) = \exp\left[-\sum_{j=1}^{m_i} \exp(\mathbf{z}_{ij}\boldsymbol{\beta}) \int_0^{t_{(i)}} \lambda_0(u)\, du\right] \qquad i = 0, 1, \ldots, k.$$

The marginal likelihood is then proportional to the probability of the event (4.5). This probability is

$$\int_0^\infty \int_{t_{(1)}}^\infty \cdots \int_{t_{(k-1)}}^\infty \prod_{i=1}^k f(t_{(i)}; \mathbf{z}_{(i)}) h(t_{(i)})\, dt_{(k)} \cdots dt_{(1)}$$

$$= \frac{\exp(\Sigma_1^k \mathbf{z}_{(i)}\boldsymbol{\beta})}{\Pi_{i=1}^k [\Sigma_{l \in R(t_{(i)})} \exp(\mathbf{z}_l \boldsymbol{\beta})]} \tag{4.6}$$

where $R(t_{(i)}) = \{[(j), j1, \ldots, jm_j],\ j = i, \ldots, k\}$ is the risk set at $t_{(i)} - 0$. It should be noted that (4.6) is not the probability of observing the event (4.5) in the censored experiment. This probability would depend on the censoring mechanism and in general also on $\lambda_0(t)$. The expression (4.6) is the probability that, in the underlying uncensored version of the experiment, the event (4.5) would occur.

Although survival time can, in most practical situations, be thought of as a continuous variable, the recording of survival data always involves some measurement error, and ties may result. To include ties, the same approach can be adopted as was used to incorporate censoring into the model. Suppose that, of n individuals under test, individuals $i1, \ldots, id_i$ are observed to fail at $t_{(i)}$, $i = 1, \ldots, k$ where $t_{(1)} < \cdots < t_{(k)}$ and $\Sigma\, d_i = n$. Here again (considering the ties to result from grouping the continuous model) only partial information is available on the rank vector. Although the ranks of individuals failing at $t_{(i)}$ are known to be less than those failing at $t_{(j)}$ $(i < j)$, the arrangement of the ranks of the d_i individuals failing at $t_{(i)}$ is otherwise unknown. The probability that the rank vector should be one of those possible given the sample is therefore the sum of $\Pi_1^k\, d_i!$ terms like (4.4).

The calculation of this is considerably simplified by noting that the ranks assigned to the d_i individuals who fail at $t_{(i)}$ is unaffected by the ranks assigned to the d_j individuals who fail at $t_{(j)}$. The sum then reduces to the product of k sums, one for each failure time. Let Q_i be the set of permutations of the symbols $i1, \ldots, id_i$ and $\mathbf{P} = (p_1, \ldots, p_{d_i})$ be an element in Q_i. As before, $R(t_{(i)})$ is the risk set at $t_{(i)} - 0$ and let $R(t_{(i)}, p_r)$ be the set difference $R(t_{(i)}) - \{p_1, \ldots, p_{r-1}\}$. The marginal likelihood for $\boldsymbol{\beta}$ is then

$$\prod_{i=1}^{k} \left\{ \exp(\mathbf{s}_i\boldsymbol{\beta}) \sum_{\mathbf{P}\in Q_i} \prod_{r=1}^{d_i} \left[\sum_{l \in R(t_{(i)}, p_r)} \exp(\mathbf{z}_l\boldsymbol{\beta}) \right]^{-1} \right\} \tag{4.7}$$

where $\mathbf{s}_i = \Sigma\, \mathbf{z}_{ij}$ is the sum of the covariates of individuals observed to fail at $t_{(i)}$. The notation used in (4.7) is sufficiently general to include also the case where censored data are present. It is easily seen that the likelihoods (4.4) and (4.6) are special cases of (4.7).

Computationally, the result (4.7) is extremely awkward if the number of ties is large at any failure time. If the number of individuals failing, d_i, at each failure point is small compared to the number of items in the risk set, the expression (4.7) will be well approximated (Peto, 1972b, and Breslow, 1974) by

$$L = \prod_{i=1}^{k} \frac{\exp(\mathbf{s}_i\boldsymbol{\beta})}{[\Sigma_{l\in R(t_{(i)})} \exp(\mathbf{z}_l\boldsymbol{\beta})]^{d_i}}\,. \tag{4.8}$$

Note that (4.4) and (4.6) are special cases of (4.8). Efron (1977) suggested the alternative approximation

$$L = \prod_{i=1}^{k} \left(\frac{\exp(\mathbf{s}_i \boldsymbol{\beta})}{\prod_{r=1}^{d_i} [\sum_{l \in R(t_{(i)})} \exp(\mathbf{z}_l \boldsymbol{\beta}) - (r-1) d_i^{-1} \sum_{l \in D(t_{(i)})} \exp(\mathbf{z}_l \boldsymbol{\beta})]} \right)$$

where $D(t_{(i)})$ indexes the set of individuals failing at $t_{(i)}$.

Numerical investigation indicates that if the fraction d_i/n_i of ties at any failure time is large, (4.7) still gives resonably good estimates whereas (4.8) may exhibit a severe bias. Efron's approximation is likely to yield some improvement over (4.8), but we have not examined this numerically. In a situation with frequent ties, a discrete model can be used and is preferable. Estimation based on the discrete model formed by grouping failure times from (4.1) is discussed in Section 4.6. In the meantime we assume that, when ties exist, the ratios d_i/n_i are sufficiently small that the approximation (4.8) is satisfactory. It should be kept in mind, however, that the results are "exact" if no ties are present in the data set.

The maximum likelihood estimate $\hat{\boldsymbol{\beta}}$, from (4.8), can be obtained as a solution to the system of equations

$$U_j(\boldsymbol{\beta}) = \frac{\partial \log L}{\partial \beta_j} = \sum_{i=1}^{k} [s_{ji} - d_i A_{ji}(\boldsymbol{\beta})] = 0 \qquad (j = 1, \ldots, s), \tag{4.9}$$

where s_{ji} is the jth element in the vector $\mathbf{s}_i$ and

$$A_{ji}(\boldsymbol{\beta}) = \frac{\sum_{l \in R(t_{(i)})} z_{jl} \exp(\mathbf{z}_l \boldsymbol{\beta})}{\sum_{l \in R(t_{(i)})} \exp(\mathbf{z}_l \boldsymbol{\beta})}.$$

Similarly,

$$I_{hj}(\boldsymbol{\beta}) = -\frac{\partial^2 \log L}{\partial \beta_h \partial \beta_j} = \sum_{i=1}^{k} d_i C_{hji} \tag{4.10}$$

where

$$C_{hji} = \frac{\sum_{l \in R(t_{(i)})} z_{hl} z_{jl} e^{\mathbf{z}_l \boldsymbol{\beta}}}{\sum_{l \in R(t_{(i)})} e^{\mathbf{z}_l \boldsymbol{\beta}}} - A_{hi}(\boldsymbol{\beta}) A_{ji}(\boldsymbol{\beta}) \qquad (h, j = 1, \ldots, s). \tag{4.11}$$

The value $\hat{\boldsymbol{\beta}}$ that maximizes (4.8) can typically be obtained by a Newton–Raphson iteration utilizing (4.9) and (4.10). It appears that only mild conditions on the covariates and censoring are required to ensure the asymptotic normality of $\hat{\boldsymbol{\beta}}$. In the absence of ties, this asymptotic normal distribution has mean $\boldsymbol{\beta}$ and estimated covariance matrix $I(\hat{\boldsymbol{\beta}})^{-1} = [I_{hj}(\hat{\boldsymbol{\beta}})]^{-1}$. Sufficient conditions for this asymptotic result are reviewed in Section 4.8. The same asymptotic results hold with tied failure times except that there is some asymptotic bias in both the estimation of $\hat{\boldsymbol{\beta}}$ and in the estimate of its covariance matrix owing to the approximation used in arriving at (4.8).

The score statistic $\mathbf{U}(\boldsymbol{\beta}_0)$ from (4.9) can be used to test $\boldsymbol{\beta} = \boldsymbol{\beta}_0$. Under the hypothesis $\boldsymbol{\beta} = \boldsymbol{\beta}_0$, and under additional mild conditions, $\mathbf{U}(\boldsymbol{\beta}_0)$ will be asymptotically normal with mean $\mathbf{0}$ and variance matrix estimated by $\boldsymbol{I}(\boldsymbol{\beta}_0)$ from (4.10) provided there are no tied failure times. Once again, some bias in both the mean (for $\boldsymbol{\beta}_0 \neq \mathbf{0}$) and the estimate of variance of $\mathbf{U}(\boldsymbol{\beta}_0)$ can be expected in the presence of ties. Score function tests are discussed in more detail in Section 4.2.5.

The numerical illustrations in this chapter assume that the conditions for asymptotic normality of the score statistic $\mathbf{U}(\boldsymbol{\beta})$ and $\hat{\boldsymbol{\beta}}$ are met and that the asymptotic bias in $\hat{\boldsymbol{\beta}}$ estimation can be ignored. As in Chapter 3, however, the question of the adequacy with which the asymptotic form of the distribution approximates the actual sampling distribution must be kept in mind in any particular application. There has not been sufficient numerical investigation to give explicit guidelines on this matter. The normal approximation to the distribution of the score statistic (4.9) for the special case of comparing several uncensored samples without ties, however, seems adequate for surprisingly small samples of size 10 or possibly even fewer in each group. Somewhat larger, but probably not appreciably larger, sample sizes are likely to be necessary for an adequate approximation to the $\hat{\boldsymbol{\beta}}$ distribution provided the regression variables and censoring mechanism are not too "extreme." As with the parametric models extreme and isolated $\mathbf{z}$ values or very severe censoring increase the total sample size necessary to ensure the adequacy of normal approximations. In situations where, for example, the results of significance tests are equivocal some numerical work may be necessary to develop an improved approximation for, or to produce an estimate of, the actual sampling distributions for $\mathbf{U}(\boldsymbol{\beta})$ or $\hat{\boldsymbol{\beta}}$.

4.2.3 Partial Likelihood

The method of partial likelihood proposed by Cox (1975) gives essentially equivalent results for an analysis based on (4.1) to those given above. In this section, the partial likelihood argument is given for the proportional hazards model whereas in Section 5.4 the argument is discussed in a more general setting. In Section 5.5, it is shown that the results in this section hold when the covariates $\mathbf{z}$ are allowed to vary with time.

Consider the set $R(t_{(i)})$ of individuals at risk at $t_{(i)} - 0$. The conditional probability that item (i) fails at $t_{(i)}$ given that the items $R(t_{(i)})$ are at risk and that exactly one failure occurs at $t_{(i)}$ is

$$\frac{\lambda(t_{(i)}; \mathbf{z}_{(i)})}{\sum_{l \in R(t_{(i)})} \lambda(t_{(i)}; \mathbf{z}_l)} = \frac{\exp(\mathbf{z}_{(i)}\boldsymbol{\beta})}{\sum_{l \in R(t_{(i)})} \exp(\mathbf{z}_l\boldsymbol{\beta})} \tag{4.12}$$

for $i = 1, \ldots, k$.

The argument proceeds by noting that, since $\lambda_0(t)$ is completely unspecified, no additional information about $\boldsymbol{\beta}$ is obtained from the observation that no failure occurs in $(t_{(i-1)}, t_{(i)})$ $(i = 1, \ldots, k)$. This is intuitively the case since we can account for this observation merely by taking $\lambda_0(t)$ to be very close to zero over this interval. If one had additional information on $\lambda_0(t)$, for example, a parametric form, there would of course be contributions to the inference about $\boldsymbol{\beta}$ from the intervals with no failures.

The partial likelihood for $\boldsymbol{\beta}$ is now formed by taking the product over all failure points $t_{(i)}$ of (4.12) to give

$$L(\boldsymbol{\beta}) = \prod_{i=1}^{k} \left(\frac{\exp(\mathbf{z}_{(i)}\boldsymbol{\beta})}{\sum_{l \in R(t_{(i)})} \exp(\mathbf{z}_l\boldsymbol{\beta})} \right),$$

identical to the marginal likelihood (4.6). The partial likelihood, it should be noted, is not a likelihood in the usual sense in that the general construction does not give a result that is proportional to the conditional or marginal probability of any observed event. Nonetheless, it has been shown somewhat informally by Cox that the method used to construct this likelihood gives maximum "partial" likelihood estimates that are consistent and asymptotically normally distributed with asymptotic covariance matrix estimated consistently by the inverse of the matrix of second partial derivatives of the log likelihood function. These arguments are discussed further in Section 5.4. It then follows that the same asymptotic results for $\boldsymbol{\beta}$ hold for estimation from partial likelihood as for the usual likelihood function.

If ties are present in the data, the partial likelihood can be obtained by applying a similar argument to the discrete logistic model of Section 2.4.3. For this model, the hazard relationship is given by

$$\frac{\lambda(t; \mathbf{z})\, dt}{1 - \lambda(t; \mathbf{z})\, dt} = \frac{\lambda_d(t)\, dt \exp(\mathbf{z}\boldsymbol{\beta})}{1 - \lambda_d(t)\, dt} \tag{4.13}$$

where, as outlined in Section 2.4.3, $\lambda_d(t)$ is an unspecified discrete hazard giving positive contributions at the observed failure times $t_{(1)}, t_{(2)}, \ldots, t_{(k)}$. A direct generalization of the above argument can then be used to compute, at each failure time, the probability that the d_i failures should be those observed given the risk set and the multiplicity d_i. A simple computation gives the conditional probability as the ith term in the product,

$$\prod_{i=1}^{k}\left(\frac{\exp(\mathbf{s}_i\boldsymbol{\beta})}{\sum_{l\in R_{d_i}(t_{(i)})}\exp(\mathbf{s}_l\boldsymbol{\beta})}\right). \tag{4.14}$$

In this expression, $\mathbf{s}_i$ is the sum of the covariates associated with the d_i failures at $t_{(i)}$, $\mathbf{s}_l = \sum_{j=1}^{d_i}\mathbf{z}_{l_j}$ and $l = (l_1, \ldots, l_{d_i})$; $R_{d_i}(t_{(i)})$ is the set of all subsets of d_i items chosen from the risk set $R(t_{(i)})$ without replacement. The partial likelihood (4.14) is very difficult computationally, even more difficult than (4.7), but an approximation to it is again afforded by the expression (4.8). It is of some interest to note that (4.14) can also be obtained by applying the same conditional argument directly to the continuous model (4.1).

The partial likelihood (4.14) does not give rise to a consistent estimator of the parameter $\boldsymbol{\beta}$ in (4.1) if the ties arise by the grouping of continuous failure times. This inconsistency in the partial likelihood occurs since (4.14) must be thought of as arising from the discrete model (4.13) and so estimates the odds ratio parameter $\boldsymbol{\beta}$ in that model. Since (4.13) does not arise as a grouping of the continuous model, the two parameters do not have identical interpretations. These two parameters agree if the failure time grouping is fine; they will become more disparate, however, as the failure time grouping becomes more severe. Kalbfleisch and Prentice (1973) give some numerical results on this topic.

4.2.4 Other Methods

An alternative derivation of the likelihoods (4.4), (4.6), and (4.8) is given by a maximum likelihood technique suggested by Breslow (1974). The hazard function is approximated by a step function with discontinuities at each observed failure time; that is,

$$\lambda_0(t) = \lambda_i,\ t_{(i-1)} < t \le t_{(i)} \qquad (i = 1, \ldots, k+1), \tag{4.15}$$

where $t_{(0)} = 0$ and $t_{(k+1)} = \infty$.

If an individual censored in the interval $[t_{(i-1)}, t_{(i)})$ is taken to have been censored at $t_{(i-1)}$, the likelihood on the data is

$$\prod_{i=1}^{k}\left\{\lambda_0(t_{(i)})^{d_i}e^{\mathbf{s}_i\boldsymbol{\beta}}\exp\left[-\int_0^{t_{(i)}}\lambda_0(u)\,du\sum_{l\in H(t_{(i)})}e^{\mathbf{z}_l\boldsymbol{\beta}}\right]\right\} \tag{4.16}$$

where $H(t_{(i)})$ is the set of labels attached to the individuals either failing or censored at $t_{(i)}$. Using (4.15) and reorganizing terms, (4.16) reduces to

$$\prod_{i=1}^{k}\left\{\lambda_i^{d_i}e^{\mathbf{s}_i\boldsymbol{\beta}}\exp\left[-\lambda_i(t_{(i)} - t_{(i-1)})\sum_{l\in R(t_{(i)})}e^{\mathbf{z}_l\boldsymbol{\beta}}\right]\right\} \tag{4.17}$$

as the likelihood of $\lambda_1, \ldots, \lambda_k$ and $\boldsymbol{\beta}$ jointly. No information is contained in the data about λ_{k+1}. For any fixed $\boldsymbol{\beta}$, (4.17) is easily maximized with respect to λ_i, $i = 1, \ldots, k$ at

$$\tilde{\lambda}_i = \frac{d_i}{(t_{(i)} - t_{(i-1)}) \sum_{l \in R(t_{(i)})} \exp(\mathbf{z}_l \boldsymbol{\beta})} \qquad (i = 1, \ldots, k),$$

and substitution in (4.17) gives the maximum likelihood function of $\boldsymbol{\beta}$ as proportional to (4.8).

Note that (4.8) is obtained by this approach as the appropriate result when ties are present in the data, whereas in the marginal or partial likelihood approach, (4.8) is obtained as an approximation to an exact result.

This approach can be criticized on the grounds that the data are, in effect, determining the model which is specified given the failure times $t_{(1)}, \ldots, t_{(k)}$. It is also well-known that maximizing a likelihood over a large number of nuisance parameters can lead to misleading and biased results.

Similar results can be obtained by a Bayesian argument once the model (4.15) is selected provided the distributions for $\log \lambda_i$, $i = 1, \ldots, k$ are taken to be independent uniform priors on $(-\infty, \infty)$ independently of the proper prior $p(\boldsymbol{\beta})$. A more satisfactory Bayesian approach to the proportional hazards model is discussed in Section 8.4.

4.2.5 The Log-Rank Test

Before proceeding to look at an application of these results, it is worth studying the score function tests that arise from the marginal and partial likelihoods given above, thereby deriving the important log-rank test of Section 1.4.

First, if there are no ties or censoring in the data, the score test statistic for the hypothesis $\boldsymbol{\beta} = \mathbf{0}$ that arises from (4.4) or (4.8) can be written as follows:

$$\mathbf{U}(\mathbf{0}) = \sum_{i=1}^{n} \mathbf{z}'_{(i)}\{1 - [n^{-1} + (n-1)^{-1} + \cdots + (n-i+1)^{-1}]\} \qquad (4.18)$$

obtained by substitution in (4.9).

This is of the form of a linear rank statistic (see Chapter 6) with the score corresponding to the ith ordered failure time being one minus the expected ith order statistic from a unit exponential distribution (see Section 3.3). This is the Savage (1956) or exponential scores test.

Generalizations of the Savage test for tied or censored data can be obtained from the score function tests of the appropriate marginal and partial likelihoods. From the approximate likelihood (4.8) the score test statistic for the global null hypothesis $\boldsymbol{\beta} = \mathbf{0}$ can be written

$$\mathbf{U}(\mathbf{0}) = \sum_{i=1}^{k} \left(\mathbf{s}'_i - d_i n_i^{-1} \sum_{l \in R(t_{(i)})} \mathbf{z}'_l \right) \tag{4.19}$$

which reduces to (4.18) if there are no ties ($d_i = 1$) and no censoring. The special case of the comparison of $s+1$ survival curves labeled 0, 1, 2, . . . , s arises upon defining $\mathbf{z}_i = (z_{1i}, \ldots, z_{si})$, where z_{ji} equals one or zero according to whether or not the ith study subject is in the jth sample. It is then easy to see that (4.19) is precisely the log-rank or generalized Savage statistic (1.16) introduced in Section 1.4. This statistic can be written

$$\mathbf{U}(\mathbf{0}) = \mathbf{O} - \mathbf{E}$$

where $\mathbf{O} = \mathbf{s}'_1 + \cdots + \mathbf{s}'_k$ gives the observed number of failures in each of the samples $1, 2, \ldots, s$ and

$$\mathbf{E} = \sum_{i=1}^{k} d_i n_i^{-1} \sum_{l \in R(t_{(i)})} \mathbf{z}'_l$$

is a corresponding vector of "expected" failures. Actually, $\mathbf{E}$ is not an overall expected number of failures but is rather the sum over failure times of the conditional expected number of failures in each sample, the expectation being under the null hypothesis and, at each time, being conditional upon the total number of failures at that time (see Section 1.4). The fact that $\mathbf{E}$ can be thought of as a vector of expected failures only in an informal sense is clear since elements of $\mathbf{E}$ are themselves random variables.

It is important to note that this same score statistic (4.19) arises also out of the partial likelihood (4.14) as an exact result corresponding to the discrete logistic model (4.13). In fact, the construction of the partial likelihood relates very closely to the $\mathbf{O}-\mathbf{E}$ interpretation of the log-rank statistic. Further, the asymptotic results for partial likelihoods as discussed in Chapter 5 show that the log-rank statistic, $\mathbf{O}-\mathbf{E}$, is asymptotically normal with estimated covariance matrix given by V, where

$$V_{hj} = \sum_{i=1}^{k} [d_i(n_{ji}n_i^{-1}\delta_{hj} - n_{hi}n_{ji}n_i^{-2})(n_i - d_i)(n_i - 1)^{-1}]$$

is obtained as the second partial derivative $\partial^2/(\partial\beta_h\partial\beta_j)$ of the logarithm of (4.14). In this expression, n_{ji} is the size of the risk set in sample j just prior to $t_{(i)}$ and $\delta_{hj} = 1$ or 0 according to whether or not $h = j$. The appropriate test statistic for testing $\boldsymbol{\beta} = \mathbf{0}$ is then

$$\mathbf{U}(\mathbf{0})' V^{-1} \mathbf{U}(\mathbf{0})$$

which will, under the hypothesis, have a $\chi^2_{(s)}$ distribution. This is the log-rank test exemplified in Sections 1.4 and 8.5.

The covariance matrix of $\mathbf{U}(\mathbf{0})$ might also be estimated from the approximate likelihood (4.8). This gives $I(\mathbf{0})$ with h, j element

$$I_{hj}(\mathbf{0}) = \sum_{i=1}^{k} d_i(n_{ji}n_i^{-1}\delta_{hj} - n_{hi}n_{ji}n_i^{-2}).$$

It is of interest to note that $I(\mathbf{0})$ agrees with V if there are no ties ($\delta_i = 1$) which corresponds to the fact that (4.8) is then an exact result. If there are ties, however, the elements of $I(\mathbf{0})$ will tend to overestimate the score statistic variance. This overestimate will be severe if the d_i's are appreciable compared to the n_i's. The corresponding statistic

$$\mathbf{U}(\mathbf{0})'I(\mathbf{0})^{-1}\mathbf{U}(\mathbf{0})$$

is less than or equal to the log-rank statistic and so provides a conservative test.

Peto and Pike (1973) consider the statistic $X = \Sigma_0^s (O_i - E_i)^2/E_i$ and show that this statistic is also less than or equal to $\mathbf{U}(\mathbf{0})'V^{-1}\mathbf{U}(\mathbf{0})$. A comparison of X with a $\chi^2_{(s)}$ distribution gives, as well, a conservative test. The work of Crowley and Breslow (1975) indicates that the conservatism is likely to be mild unless the censoring patterns differ markedly between samples. There seems, however, to be little reason for using such a procedure since the computational advantage is small in spite of the avoidance of a matrix inversion.

Other generalizations of the Savage exponential scores test (4.18) can be generated from the exact marginal likelihood (4.7) or from the discrete model that arises from grouping the continuous model (4.1). This last approach is considered further in Section 4.6.

4.2.6 Application to the Carcinogenesis Data of Section 1.1.1

The first step in applying the results of this section to data is to order the survival times from smallest to largest with the additional convention that failure times precede censored times in the case of ties. This ordering is presented in Table 4.1 for the carcinogenesis data. Also recorded in the table are the numbers of failures d_i occurring at each distinct failure time $t_{(i)}$, the covariates of the failures, and the covariates of censored times in $[t_{(i)}, t_{(i+1)})$. An efficient computer solution to the problem would require essentially the same organization of the data set. In general, there is advantage to begin the calculation at the last failure time $t_{(k)}$ since the risk set at $t_{(i)}$ can then be formed by adding, to that at $t_{(i+1)}$, the labels of items failing or censored in $[t_{(i)}, t_{(i+1)})$. The contributions to the likelihood (4.8) at a specified β value are then easily computed. With these data, the approximation in (4.8) is not necessary since all items failing at any given

Table 4.1 PROPORTIONAL HAZARDS MODEL APPLIED TO CARCINOGENESIS DATA

i	$t_{(i)}$	d_i	Failures z_l	Censored z_l	Contribution to Likelihood
1	142	1	1		$e^{\beta}/(19+21e^{\beta})$
2	143	1	0		$1/(19+20e^{\beta})$
3	156	1	1		$e^{\beta}/(18+20e^{\beta})$
4	163	1	1		$e^{\beta}/(18+19e^{\beta})$
5	164	1	0		$1/(18+18e^{\beta})$
6	188	2	0, 0		$1/(17+18e^{\beta})^2$
7	190	1	0		$1/(15+18e^{\beta})$
8	192	1	0		$1/(14+18e^{\beta})$
9	198	1	1	1(204)	$e^{\beta}/(13+18e^{\beta})$
10	205	1	1		$e^{\beta}/(13+16e^{\beta})$
11	206	1	0		$1/(13+15e^{\beta})$
12	209	1	0		$1/(12+15e^{\beta})$
13	213	1	0		$1/(11+15e^{\beta})$
14	216	1	0	0(216)	$1/(10+15e^{\beta})$
15	220	1	0		$1/(8+15e^{\beta})$
16	227	1	0		$1/(7+15e^{\beta})$
17	230	1	0		$1/(6+15e^{\beta})$
18	232	2	1, 1		$e^{2\beta}/(5+15e^{\beta})^2$
19	233	4	1, 1, 1, 1		$e^{4\beta}/(4+13e^{\beta})^4$
20	234	1	0		$1/(5+9e^{\beta})$
21	239	1	1		$e^{\beta}/(4+9e^{\beta})$
22	240	1	1	0(244)	$e^{\beta}/(4+8e^{\beta})$
23	246	1	0		$1/(3+7e^{\beta})$
24	261	1	1		$e^{\beta}/(2+7e^{\beta})$
25	265	1	0		$1/(2+6e^{\beta})$
26	280	2	1, 1		$e^{2\beta}/(1+6e^{\beta})^2$
27	296	2	1, 1		$e^{2\beta}/(1+4e^{\beta})^2$
28	304	1	0		$1/(1+2e^{\beta})$
29	323	1	1	1(344)	$e^{\beta}/2e^{\beta}$

time have the same covariates. The ties may thus be broken in any way at all and the likelihood (4.4) used. The approximation (4.8) has been used here for illustration. Only a small difference arises through breaking the ties and using the exact result.

Three iterations of the Newton–Raphson procedure (Section 3.7), with an initial estimate of $\beta = 0$, gives the maximum likelihood estimate to three figure accuracy, as $\hat{\beta} = -.596$. The observed information is $I(\hat{\beta}) =$

8.237. Thus $\hat{\beta}\sqrt{I(\hat{\beta})} = -1.71$ is, under assumption that $\beta = 0$, an observation from a $N(0, 1)$ distribution leading to a significance level of .087. There is some indication of a treatment effect though the evidence is not strong. The log relative likelihood

$$R(\beta) = \log L(\beta) - \log L(\hat{\beta}).$$

is plotted for these data in Figure 4.1. The close agreement of this plot with the normal log likelihood $-\frac{1}{2}I(\hat{\beta})(\hat{\beta} - \beta)^2$ suggests that the large sample procedures are reasonably accurate.

An alternative test of $\beta = 0$ is based on the likelihood ratio statistic. We find $-2R(0) = 2.86$, and comparison with the $\chi^2_{(1)}$ distribution gives a significance level of .091 in close agreement with the test based on the asymptotic distribution of $\hat{\beta}$. The latter gives a $\chi^2_{(1)}$ statistic of $(1.71)^2 = 2.92$. An approximate 95% confidence interval for β is obtainable from the likelihood ratio test as $\{\beta \mid -2R(\beta) \leq 3.84\}$. Note that $P(\chi^2_{(1)} > 3.84) = .05$. This yields the interval $(-1.27, .11)$ compared to $(-1.28, .09)$ as the interval based on the normal pivotal $(\hat{\beta} - \beta)\sqrt{I(\hat{\beta})}$.

A third procedure for testing $\beta = 0$ is provided by the score test based on $U(0)$. We find that $U(0) = 4.763$ and $I(0) = 7.560$ which gives the test statistic $U(0)^2 I(0)^{-1} = 3.00$. Recall that $U(0) = O - E$ is the log-rank statistic which was also considered in Section 1.4. The variance estimate $I(0)$ is based on the approximate likelihood (4.8); as discussed above, a better estimate of the asymptotic variance is $V = 7.263$, which arises as the observed information from the likelihood (4.14). The log-

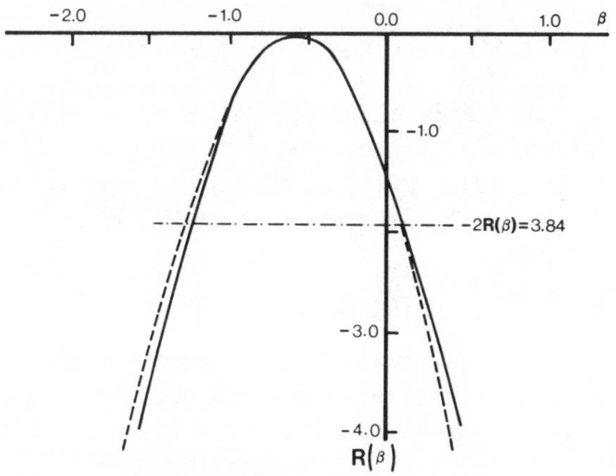

Figure 4.1 Marginal log likelihood function of β (solid line) and the approximating normal likelihood arising from the carcinogenesis data.

rank test then gives a $\chi^2_{(1)}$ statistic $U(0)^2V^{-1} = 3.12$. The test based on the variance estimate $I(0)$ is therefore conservative though the differences are too small to be of practical importance. The log-rank test may also be approximated by the approximate $\chi^2_{(1)}$ statistic

$$X = \sum_{j=0}^{1} \frac{(O_j - E_j)^2}{E_j} = (4.763)^2\left(\frac{1}{12.237} + \frac{1}{23.584}\right)$$
$$= 2.82.$$

which again gives a test that is somewhat conservative, though the differences are not large.

As was noted earlier, since all items failing at any given time have the same covariates, the ties may be broken and the analysis could be based on the log-rank statistic with no ties. This gives $U(0) = 4.584$ and $I = V = 7.653$. The log-rank $\chi^2_{(1)}$ statistic is then $(4.584)^2/7.653 = 2.75$.

One advantage of this analysis of the carcinogenesis data is that the extended initial period with no observed failures is easily handled. Except for the generalized F, all other models discussed have required the insertion of a guarantee time. This was usually done by arbitrarily specifying the origin of measurement to be 100 days, though the threshold parameter could be included and estimated from the data. In Section 3.10, we found the generalized F to be sufficiently flexible to be able to account for this failure-free period with no guarantee parameter. The proportional hazards approach, however, is conceptually simpler and involves much less computation.

4.3 ESTIMATION OF THE SURVIVOR FUNCTION

The proportional hazards model (4.1) was generalized in Section 2.4.2 to include both discrete and continuous survival distributions. It was noted that the general model can be described as follows: let $F_0(t)$ be an arbitrary survivor function (discrete, continuous or mixed). The survivor function of T given $\mathbf{z}$ is then taken as

$$F(t; \mathbf{z}) = F_0(t)^{\exp(\mathbf{z}\boldsymbol{\beta})}. \tag{4.20}$$

Suppose now that data are available from the extended model (4.20) and consider the calculation of the nonparametric maximum likelihood estimate of $F_0(t)$. In doing this, an argument analogous to that used in obtaining the Kaplan–Meier estimate (Section 1.3) is employed.

As before let $t_{(1)}, \ldots, t_{(k)}$ be the distinct failure times, let D_i be the set of labels associated with individuals failing at $t_{(i)}$, and C_i be the set of labels associated with individuals censored in $[t_{(i)}, t_{(i+1)})$ $(i = 0, \ldots, k)$, where $t_{(0)} = 0$ and $t_{(k+1)} = \infty$. The censoring times in the interval $[t_{(i)}, t_{(i+1)})$ are t_l where l ranges over C_i. The contribution to the likelihood of an individual with covariates $\mathbf{z}$ who fails at $t_{(i)}$ is, under 'independent' censorship,

$$F_0(t_{(i)})^{\exp(\mathbf{z}\boldsymbol{\beta})} - F_0(t_{(i)} + 0)^{\exp(\mathbf{z}\boldsymbol{\beta})}$$

and the contribution of a censored observation at time t is

$$F_0(t + 0)^{\exp(\mathbf{z}\boldsymbol{\beta})}.$$

The likelihood function can then be written

$$\mathscr{L} = \prod_{i=0}^{k} \left\{ \prod_{l \in D_i} [F_0(t_{(i)})^{\exp(\mathbf{z}_l\boldsymbol{\beta})} - F_0(t_{(i)} + 0)^{\exp(\mathbf{z}_l\boldsymbol{\beta})}] \prod_{l \in C_i} F_0(t_l + 0)^{\exp(\mathbf{z}_l\boldsymbol{\beta})} \right\} \tag{4.21}$$

where D_0 is empty.

As with the Kaplan–Meier estimate, it is clear that $\mathscr{L}$ is maximized by taking $F_0(t) = F_0(t_{(i)} + 0)$ for $t_{(i)} < t \leq t_{(i+1)}$ and allowing probability mass to fall only at the observed failure times $t_{(1)}, \ldots, t_{(k)}$. These observations lead to the consideration of a discrete model with hazard contribution $1 - \alpha_j$ at $t_{(j)}$ $(j = 1, \ldots, k)$. Thus we take

$$F_0(t_{(i)}) = F_0(t_{(i-1)} + 0) = \prod_{j=0}^{i-1} \alpha_j, \qquad i = 1, \ldots, k,$$

where $\alpha_0 = 1$. On substitution in (4.20) and rearranging terms, we obtain

$$\prod_{i=1}^{k} \left[\prod_{j \in D_i} (1 - \alpha_i^{\exp(\mathbf{z}_j\boldsymbol{\beta})}) \prod_{l \in R(t_{(i)}) - D_i} \alpha_i^{\exp(\mathbf{z}_l\boldsymbol{\beta})} \right] \tag{4.22}$$

as the likelihood function to be maximized.

The estimation of the survivor function can be carried out by joint estimation of the α's and $\boldsymbol{\beta}$ in (4.22) (Meier, 1978, personal communication). More simply, however, we might take $\boldsymbol{\beta} = \hat{\boldsymbol{\beta}}$ as estimated from the marginal likelihood function and then maximize (4.22) with respect to $\alpha_1, \ldots, \alpha_k$. Differentiating the logarithm of (4.22) with respect to α_i gives the maximum likelihood estimate of α_i as a solution to

$$\sum_{j \in D_i} \frac{\exp(\mathbf{z}_j\hat{\boldsymbol{\beta}})}{1 - \hat{\alpha}_i^{\exp(\mathbf{z}_j\hat{\boldsymbol{\beta}})}} = \sum_{l \in R(t_{(i)})} \exp(\mathbf{z}_l\hat{\boldsymbol{\beta}}). \tag{4.23}$$

If only a single failure occurs at $t_{(i)}$, (4.23) can be solved directly for $\hat{\alpha}_i$ to give

$$\hat{\alpha}_i = \left(1 - \frac{\exp(\mathbf{z}_{(i)}\hat{\boldsymbol{\beta}})}{\sum_{l \in R(t_{(i)})} \exp(\mathbf{z}_l\hat{\boldsymbol{\beta}})}\right)^{\exp(-\mathbf{z}_{(i)}\hat{\boldsymbol{\beta}})}$$

Otherwise an iterative solution is required; a suitable initial value for the iteration is α_{i_0} where

$$\log \alpha_{i_0} = \frac{-d_i}{\Sigma_{l \in R(t_{(i)})} \exp(\mathbf{z}_l \hat{\boldsymbol{\beta}})}$$

which is obtained by substituting

$$\begin{aligned} \hat{\alpha}_i^{\exp(\mathbf{z}_j \hat{\boldsymbol{\beta}})} &= \exp(e^{\mathbf{z}_j \hat{\boldsymbol{\beta}}} \log \hat{\alpha}_i) \\ &\simeq 1 + \exp(\mathbf{z}_j \hat{\boldsymbol{\beta}}) \log \hat{\alpha}_i \end{aligned}$$

in (4.23). Note that the $\hat{\alpha}_i$'s can be separately calculated.

The maximum likelihood estimate of the base-line survivor function is then

$$\hat{F}_0(t) = \prod_{i \mid t_{(i)} < t} \hat{\alpha}_i, \tag{4.24}$$

which is a step function, like the Kaplan–Meier estimate, with discontinuities at each observed failure time $t_{(i)}$. When the covariate $\mathbf{z} = \mathbf{0}$ for all individuals sampled, (4.24) reduces to the Kaplan–Meier estimate. Clearly, the estimated survivor function for covariate value $\tilde{\mathbf{z}}$ is

$$\hat{F}(t; \tilde{\mathbf{z}}) = \prod_{i \mid t_{(i)} < t} \hat{\alpha}_i^{\exp(\tilde{\mathbf{z}} \hat{\boldsymbol{\beta}})}. \tag{4.25}$$

Many other estimates of the survivor function have been proposed. One such, a modified life table estimate, involves partitioning the time axis into intervals $I_1, \ldots, I_k$ and supposing the base-line hazard $\lambda_0(t)$ to be constant within each interval. A simple estimation of the hazard function and consequently of $F_0(t)$ is then available. This approach is analogous to that giving (1.11) in the single sample problem of Section 1.3 and is outlined in Problem 31 of Appendix 2. The maximum likelihood approach of Section 4.2.3 gives yet another estimator. All these estimators are in reasonable agreement and, since the use of such estimators is largely descriptive, it probably does not matter much which is used. Large sample properties of the estimates (4.24) and (4.25) are discussed in Section 4.7.2.

Figure 4.2 gives the estimated survivor functions from (4.25) for the carcinogenesis data for each of the two samples ($z = 0, 1$). To carry out these calculations, the data are again ordered as in Table 4.1 and the hazard contribution at each observed failure time $t_{(i)}$ calculated from (4.23). Note that the assumed model constrains the estimates so that one survivor function dominates the other. Such graphs can give a misleading impression that one of the treatments is consistently preferable and suggest significant differences even when they are not present. In this example, a better description is given by the Kaplan–Meier estimates in Figure 1.2.

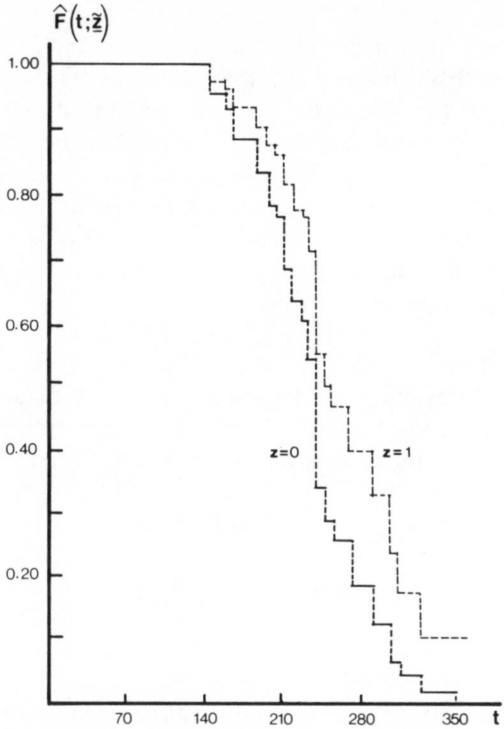

Figure 4.2 Survivor function estimates (4.25) for the carcinogenesis data.

4.4 THE INCLUSION OF STRATA

The proportional hazards model (4.1) requires that, for any two covariate sets $\mathbf{z}_1$ and $\mathbf{z}_2$, the hazard functions are related by

$$\lambda(t;\mathbf{z}_1) \propto \lambda(t;\mathbf{z}_2), \qquad 0<t<\infty. \tag{4.26}$$

Although this relation is descriptive of many situations, sometimes there are important factors, the different levels of which produce hazard functions which differ markedly from proportionality. To accommodate such factors, a simple extension of the model (4.1) is available.

Suppose there is a factor that occurs on q levels and for which (4.26) may be violated. We define the hazard function for an individual in the jth stratum (or level) of this factor as

$$\lambda_j(t;\mathbf{z}) = \lambda_{0j}(t)\exp(\mathbf{z}\boldsymbol{\beta}) \tag{4.27}$$

for $j = 1, 2, \ldots, q$ where $\mathbf{z}$ is the vector of covariates for which (4.26) is descriptive. The base-line hazard functions, $\lambda_{01}(\cdot), \ldots, \lambda_{0q}(\cdot)$, for the q strata are allowed to be arbitrary and are completely unrelated.

Let $t_{j1}, \ldots, t_{jn_j}$ be the failure times of the n_j items in the jth stratum and $\mathbf{z}_{j1}, \ldots, \mathbf{z}_{jn_j}$ be the corresponding covariates. We let $\mathbf{r}_j$ be the rank vector for the jth stratum. The model (4.27) is invariant under the direct product of groups of differentiable monotone increasing transformations acting on the time scale in each stratum. The marginal likelihood of $\boldsymbol{\beta}$ is then proportional to the marginal probability of $\mathbf{r}_1, \ldots, \mathbf{r}_q$;

$$L(\boldsymbol{\beta}) \propto \prod_{j=1}^{q} f_j(\mathbf{r}_j; \boldsymbol{\beta}),$$

where $f_j(\mathbf{r}_j; \boldsymbol{\beta})$ is the marginal likelihood (4.4) which arises out of the distribution of the ranks for the jth stratum. The results are again easily generalized to take account of right censoring and ties in the data set. In this more general situation, the marginal likelihood of $\boldsymbol{\beta}$ is approximately the product of terms like (4.8), one arising from each stratum. In general,

$$L(\boldsymbol{\beta}) = \prod_{j=1}^{q} L_j(\boldsymbol{\beta}) \tag{4.28}$$

where $L_j(\boldsymbol{\beta})$ is the marginal likelihood of $\boldsymbol{\beta}$ arising from the jth stratum alone.

The division into strata tends to reduce the computational complexities of marginal likelihood inference on $\boldsymbol{\beta}$ by reducing the number of ties. The maximization of the likelihood (4.28) is easily accomplished on a computer; the first and second derivatives are merely sums over strata of those computed earlier [c.f. (4.9) and (4.10)] and a Newton–Raphson technique generally leads to quick convergence to the estimate $\hat{\boldsymbol{\beta}}$. Although some loss of efficiency is encountered in the estimate of $\boldsymbol{\beta}$ when stratification is used unnecessarily, it is shown in Section 4.7 that this loss is generally not severe.

Once an estimate of $\boldsymbol{\beta}$ is obtained, the methods of Section 4.3 can be used to give estimates of the survivor functions in each of the q strata separately. This provides a graphical check of the appropriateness of a proportional hazards modeling for those factors used in defining strata. If $\hat{F}_{0j}(t)$ is the estimate of the survivor function for the jth level of a factor, a check to determine whether the factor can be modeled in the proportional hazards way is afforded by plotting $\log[-\log \hat{F}_{0j}(t)]$, $j = 1, \ldots, q$, versus $\log t$. Such plots for any two values of j should exhibit approximately constant differences over time. In addition, the plots should be approximately linear if a Weibull regression model provides a good fit.

If the hypothesis $\boldsymbol{\beta} = \mathbf{0}$ is of interest, the score function test from (4.28) can be used for inference. The score statistic is $\Sigma_1^q \mathbf{v}_j$ where $\mathbf{v}_j = \mathbf{U}_j(\mathbf{0})$ is the contribution arising from L_j and is thus a log-rank statistic (4.19) computed on items in the jth stratum. The resulting test is the stratified log-rank test discussed in Section 1.4. Briefly, if V_j is the covariance matrix corresponding to $\mathbf{v}_j$, and if ΣV_j is nonsingular then

$$\left(\sum \mathbf{v}_j\right)\left(\sum V_j\right)^{-1}\left(\sum \mathbf{v}_j\right)$$

is, under the hypothesis, an approximate $\chi^2_{(s)}$ variate.

4.5 ILLUSTRATIONS

We first look at some results of applying the proportional hazards model to the lung cancer data discussed in Section 3.8 and listed in Appendix 1 for comparison with the parametric methods of Sections 3.8 and 3.9.2.

Table 4.2 summarizes the maximum likelihood estimates and asymptotic χ^2 statistics based on the marginal likelihood (4.8) for the proportional hazards model (4.1) and for the Weibull analysis of Section 3.8. The Weibull model is a special case of the proportional hazards model and the extremely good agreement between the χ^2 statistics even in the presence of strong prognostic factors suggests that little efficiency is lost in using the partially parametric model (4.1) relative to the fully parametric regression model. Theoretical efficiency comparisons are considered in Section 4.7. Table 4.2 also shows good agreement between the absolute values of the regression coefficient estimates. If a Weibull model with regression coefficient $\mathbf{b}$ and shape parameter $p = \sigma^{-1}$ holds, then the proportional hazards regression parameter is $\boldsymbol{\beta} = -\mathbf{b}/\sigma$. The fact that $\hat{\sigma} = .938$ is close to unity accounts for the close correspondence between $\hat{\boldsymbol{\beta}}$ and $-\hat{\mathbf{b}}$ for these data.

As a second example, consider the clinical trial discussed in Section 1.1.2. In that study patients with primary tumors at any of four sites in the head and neck were randomly assigned to a test or standard treatment policy. The data for one of the sites are given in Appendix 1. Each treatment policy dictated the treatment to be administered during a 90 day period. After this, each patient received medical care as deemed prudent by the participating institution. No restrictions, except a prohibition of the study treatment, were placed on the post 90 day care. The primary purpose of the study was to compare patient survival for the two treatment policies in the four primary disease sites. The data considered here are those collected by the eight institutions with the largest patient accession. There were 438 patients entered by these institutions, 217 assigned to the standard, and 221 to the test treatment groups.

As noted in Section 1.1.2, and exemplified in the data of Appendix 1, there

Table 4.2 ASYMPTOTIC LIKELIHOOD INFERENCE ON LUNG CANCER DATA

	Weibull Model		Proportional Hazards Model	
Regressor Variable	Coefficient ($-\hat{b}$)	χ^2 Statistic	Coefficient ($\hat{\beta}$)	χ^2 Statistic
Performance status	−.0301	38.79	−.0328	35.08
Disease duration (months)	.0005	0.00	−.0002	0.00
Age (years)	−.0061	0.51	−.0086	0.85
Prior therapy	.0041	0.04	.0090	0.15
Cell type				
Squamous	−.3977		−.4087	
Small	.4285	22.03	.4597	18.15
Adeno	.7350		.7879	
Treatment	−.2061	1.30	−.2979	2.07

are many covariates measured on individuals under study and available for consideration. Institution is one such covariate and is of particular importance here since there was considerable variability in patient treatment following the 90 day study period. In addition, the TN staging classification, the grade or degree of differentiation of the primary tumor, the site of the primary tumor, age, sex, and general condition are covariates considered here. A discussion of these covariates is contained in Section 1.2.3.

In the analysis of such an extensive set of failure time data, the first step is exploratory, its purpose being the identification of which covariates correlate with subsequent survival. One approach makes extensive use of the log-rank test to check for a dependence of failure time on each covariate taken one at a time. The log-rank test with stratification allows an additional check for possible interactions with other covariates. In the present case, the failure time distribution is reasonably close to an exponential distribution and, as is discussed in Section 3.8, estimated failure rates based on assumed exponential distributions were computed in two way tables with covariates being examined in pairs. This examination identified four covariates (namely, sex, general condition, and TN staging) as being highly related to subsequent survival.

As a preliminary model, these four factors were included as covariates in (4.24) and the eight institutions were allowed to define the strata with base-line hazards $\lambda_{0j}(t)$. In order to check whether institution might reasonably be incorporated in the regression portion of (4.24), the corresponding survivor function estimates $\hat{F}_{0j}(t)$ were obtained and $\log[-\log \hat{F}_{0j}(t)]$ was plotted against time for each j. Although any of the estimates of the survivor function discussed in Section 4.3 would be adequate, we have used the modified life table estimate outlined in Problem 31 of Appendix 2. Figure 4.3 gives the resulting plots and the curves are seen to have approximately constant differences over time. This suggests that institution might be incorporated as a covariate in (4.24). Modeling of the other covariates can be checked in similar ways. Figure 4.4 gives the estimated survivor functions when general condition $\times$ sex forms the strata and the survivals are adjusted for T, N, and institution differences. Again there is close correspondence to constant separation. It is further apparent, on examining the differences between males and females within the levels of general condition that the factors of sex and general condition operate approximately additively on the log($-$log) survivor function. This suggests that the linear modeling in the exponential factor of (4.23) is appropriate for these variables.

In investigating the factors of treatment and region, the covariates in (4.23) were sex, general condition, TN staging, and institution, while region $\times$ treatment formed the strata. Figures 4.5 and 4.6 give the survivor

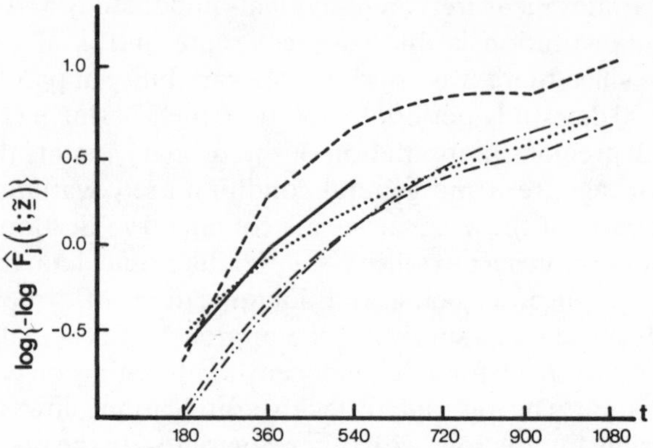

Figure 4.3 Log minus log plot for five largest institutions. Survival curve estimates standardized to Male, General Condition 1, T classification 3, N classification 2.

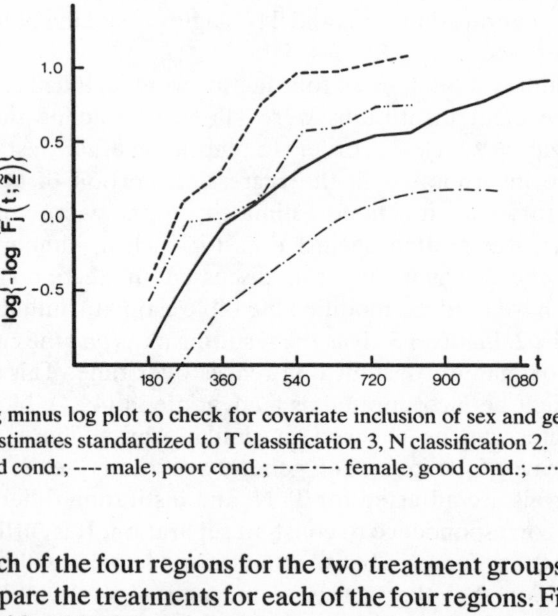

Figure 4.4 Log minus log plot to check for covariate inclusion of sex and general condition. Survival curve estimates standardized to T classification 3, N classification 2. (—— male, good cond.; ---- male, poor cond.; —·—·— female, good cond.; —···—·· female, poor cond.)

curves for each of the four regions for the two treatment groups, and Figures 4.7–4.10 compare the treatments for each of the four regions. Figures 4.5 and 4.6 show considerable departures from constant separation and suggest that site of primary tumor is probably best not included in the regression portion of (4.24). Figures 4.7–4.10 suggest that, with the possible exception of region 3, the proportional hazards specification for treatment is reasonable within region. In what follows, treatment is included in the regression portion for all

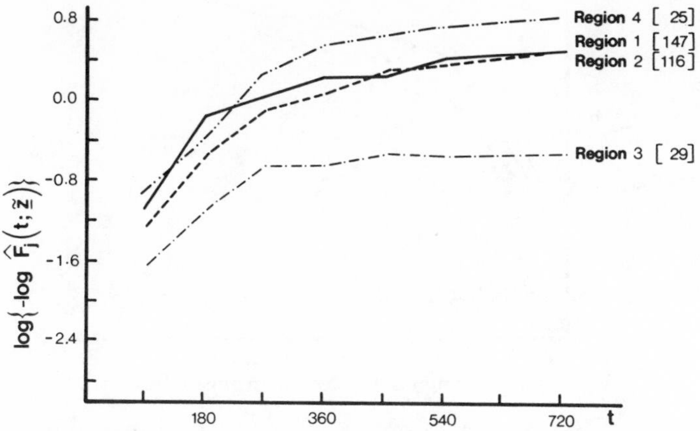

Figure 4.5 Log minus log plot to check for covariate inclusion of region (standard treatment group). Bracketed numbers are sample sizes.

regions. However some further investigation of region 3 would be useful. This could be done by partitioning the time axis into two parts and investigating separately the effect of treatment on the hazard for early and late times. These figures (4.7–4.10) give adjusted comparisons of the treatments within each site. It should be noted that these comparisons are not constrained by the proportional hazards relationship since treatments are here determining the strata. It would be possible, once treatment is

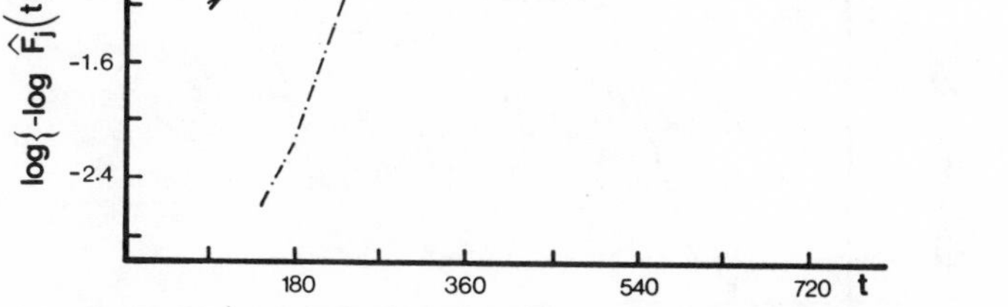

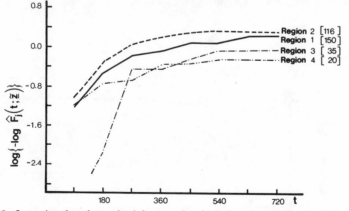

Figure 4.6 Log minus log plot to check for covariate inclusion of region (test treatment group). Bracketed numbers are sample sizes.

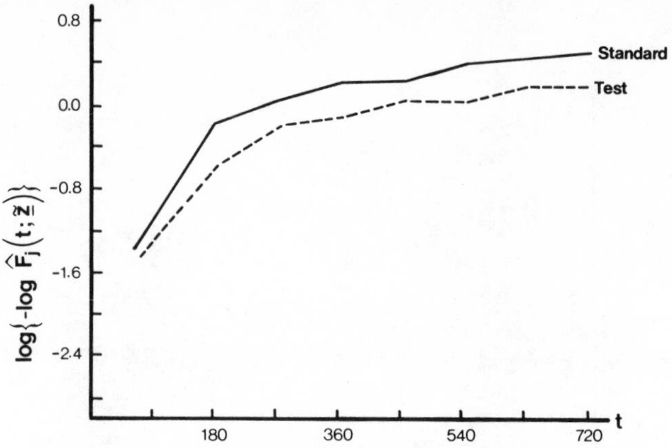

Figure 4.7 Log minus log plot for treatment, region 1.

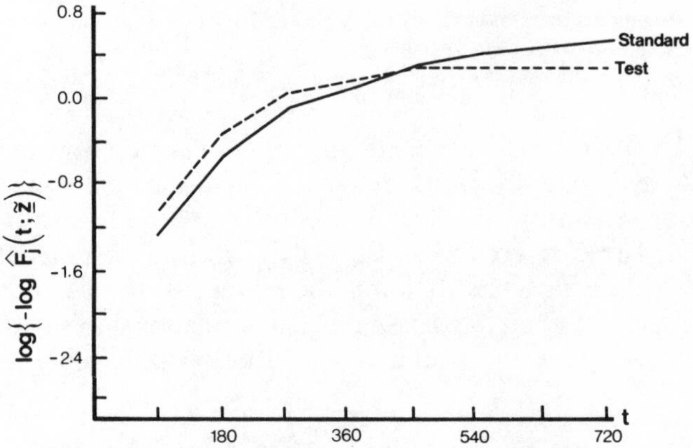

Figure 4.8 Log minus log plot for treatment, region 2.

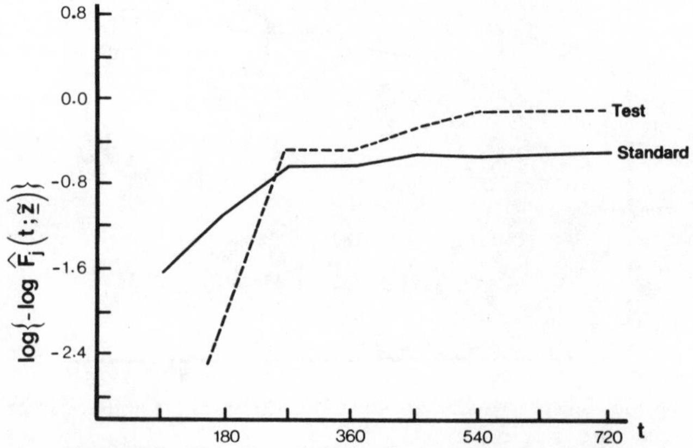

Figure 4.9 Log minus log plot for treatment, region 3.

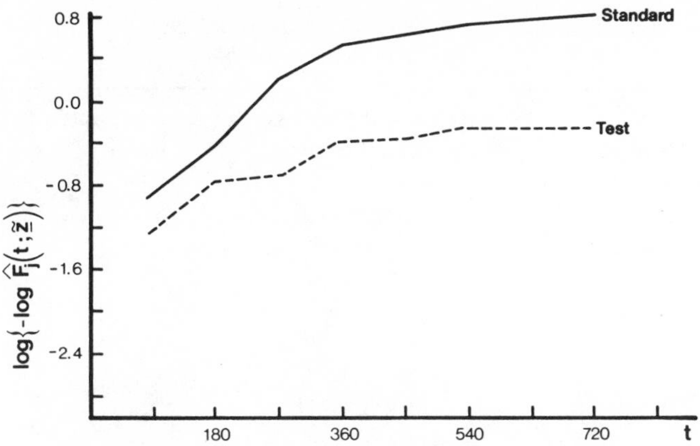

Figure 4.10 Log minus log plot for treatment, region 4.

incorporated as a covariate, to produce estimates of the survivor function that are constrained in this way. Such plots would, however, tend to accentuate treatment differences and are subject to the same criticism as was Figure 4.2 in the carcinogenesis example. The model incorporating treatment as a regression variable is useful, however, in that simple tests for treatment differences are then available.

These considerations lead to the tentative model for these data with hazard function

$$\lambda_j(t; \mathbf{z}, i, x) = \lambda_{0j}(t) \exp(\mathbf{z}\boldsymbol{\beta} + \gamma_i + \alpha_j x), \qquad j = 1, \ldots, 4, \qquad (4.29)$$

where $\mathbf{z}$ is the vector giving sex, general condition, the T and the N classifications, j denotes the region of the primary tumor, i is the institution number (with $\gamma_1 = 0$), and x takes values 0 or 1 for the standard and test treatments, respectively. The coefficient α_j gives a measure of treatment differences within the jth region. Table 4.3 gives the estimates for the regression parameters (except the $\hat{\gamma}_i$'s corresponding to institution) and the estimated standard errors of the estimates. The calculations were done using a Newton–Raphson routine with initial values 0 for all parameters.

From Table 4.3 it is easily seen that sex, general condition, and the TN classification are all important in evaluating survival prognosis. The treatment effects, however, are not significant. Only in region 4 is there any evidence of dependence on treatment, and there the significance level is marginal. In this case, four independent tests for treatment differences have been made. Thus the nominal significance level of about 7% for the treatment effect in region 4 must, to some extent, be discounted owing to these multiple comparisons.

Table 4.3 REGRESSION COEFFICIENTS AND ESTIMATED VARIANCES FOR THE MODEL (4.26)

Variable	Estimated Regression Coefficients	Estimated Standard Error of the Estimate
Sex	$\hat{\beta}_1 = -.446$	.154
General condition	$\hat{\beta}_2 = .483$	.103
T classification	$\hat{\beta} = .358$	.098
N classification	$\hat{\beta}_4 = .267$	.055
Trt. rg. 1	$\hat{\alpha}_1 = -.313$	.267
Trt. rg. 2	$\hat{\alpha}_2 = .102$	.162
Trt. rg. 3	$\hat{\alpha}_3 = -.101$	.349
Trt. rg. 4	$\hat{\alpha}_4 = -.656$	.369

Once the model (4.29) is fitted, it is possible to check its appropriateness by forming residuals and carrying out residual plots such as in ordinary linear regression. In the model (4.26) let t_{ji} be the survival time of the ith individual in the jth stratum and define

$$e_{ji} = \Lambda_{0j}(t_{ji})e^{\mathbf{z}_{ji}\boldsymbol{\beta}}$$

where $\Lambda_{0j}(t) = \int_0^t \lambda_{0j}(u)\,du$. The e_{ji}'s are a censored sample from the exponential distribution with failure rate 1. If Λ_{0j} and $\boldsymbol{\beta}$ are replaced with estimates, we obtain estimates of the e_{ji}'s or residuals

$$\hat{e}_{ji} = \hat{\Lambda}_{0j}(t_{ji})e^{\mathbf{z}_{ji}\hat{\boldsymbol{\beta}}}.$$

In forming these residuals in the present case, the continuous estimate $\hat{\Lambda}_{0j}(t)$ obtained above and the marginal likelihood estimate $\hat{\boldsymbol{\beta}}$ have been used. If the model is appropriate, the $\hat{e}_{ji}$'s should be similar, at least to some extent, to a censored exponential sample (an $\hat{e}_{ji}$ is taken as censored if the corresponding t_{ji} is censored). Survival curve estimates based on the residuals should, when plotted on a log scale, yield approximately a straight line with slope -1.

The overall adequacy of the model can be partially checked by plotting the survivor curve estimate arising from the residuals, as is illustrated in Figure 4.11. The correspondence with the anticipated line is extremely good. A check that the adjustment for covariates has been adequate in the four regions under study is provided by plotting the residual survivor curve for each of the regions (Figure 4.12). Again close correspondence to the expected line is observed. The modeling of the covariates can be checked in a similar way. Figure 4.13 gives the residual survivor curves for sex and Figure 4.14 for general condition. Again, the fit seems adequate.

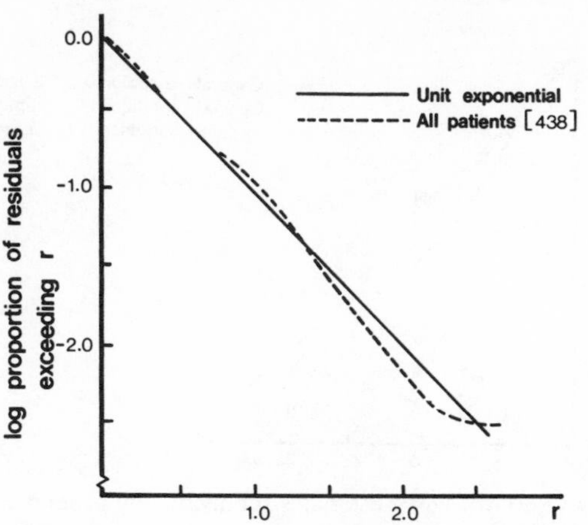

Figure 4.11 Survivor curve estimate from residuals.

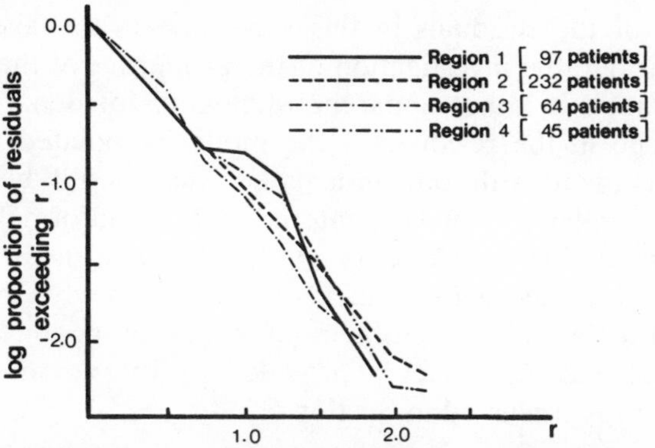

Figure 4.12 Survivor curve estimates from residuals subdivided by region.

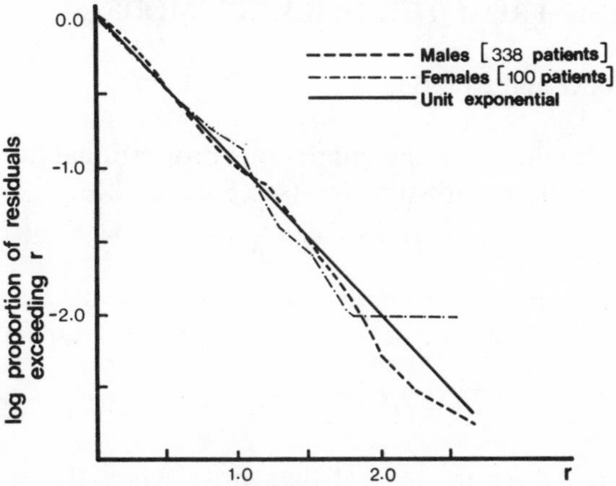

Figure 4.13 Survivor curve estimates from residuals subdivided by sex.

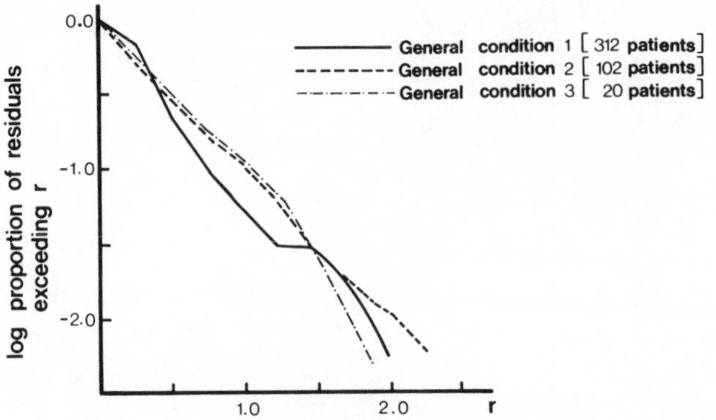

Figure 4.14 Survivor curve estimates subdivided by general condition.

The use of the residuals in this way appears to have some promise, though considerable investigation of the properties of the residuals would seem desirable. It is not apparent what kinds of departures one would expect to see in the residuals if the model is violated or even to what extent agreement with the anticipated line should be expected. This situation is similar to that encountered with the informal use of residuals in the normal linear regression model. Further investigation of these procedures is needed before their use can be strongly advocated. A more formal approach to the problem of checking goodness of fit in the proportional hazards model is provided by the use of time dependent covariates, as is discussed in Section 5.5.

4.6 SAMPLING FROM THE DISCRETE MODEL

4.6.1 The Model and Notation

The discrete analogue to the continuous proportional hazards model is characterized by the hazard function (see Section 2.4.2)

$$\lambda(t; \mathbf{z})\, dt = 1 - [1 - \lambda_d(t)\, dt]^{\exp(\mathbf{z}\boldsymbol{\beta})} \tag{4.30}$$

where $\lambda_d(t)\,dt$ is the function that gives positive hazard contributions $\lambda_1, \lambda_2, \ldots$ to the discrete set of points $x_1 < x_2 < \cdots$, respectively. That is,

$$\lambda_d(t) = \sum \lambda_i \delta(t - x_i). \tag{4.31}$$

Most commonly, discrete survival data arise when the survival time is subject to interval grouping. Thus it is only recorded into which of a number of disjoint intervals, $A_j = [a_{j-1}, a_j)$, $j = 1, \ldots, k+1$, the survival time actually falls ($a_0 = 0$, $a_{k+1} = \infty$). If the underlying survival variable has the continuous hazard (4.1), then as remarked in Section 2.4.2, if x_i represents failure in the ith interval, the hazard contribution at x_i for covariate $\mathbf{z}$ is

$$1 - (1 - \lambda_i)^{\exp(\mathbf{z}\boldsymbol{\beta})}$$

as in (4.30), where $1 - \lambda_i = \exp[-\int_{a_{i-1}}^{a_i} \lambda_0(u)\, du]$. In other instances, the time measurement may truly be discrete, as, for example, when T represents the number of attempts required to successfully perform a certain task.

Suppose that data are available from the discrete model (4.30) and let D_i represent the set of labels attached to individuals failing at x_i and R_i the set of labels attached to individuals censored at x_i or observed to survive past x_i. We suppose that an item censored at x_i contributes the information that its underlying survival time is observed to *exceed* x_i but nothing further is known. In effect, censored observations at x_i are being supposed to follow failures at x_i. In grouping continuous data, this is equivalent to allowing censoring to occur only just prior to the end of an interval. It is possible, of course, that censoring may be observed within an interval, but such a case requires essentially ad hoc solutions.

4.6.2 Maximum Likelihood Estimation

Maximum likelihood calculations are simplified and Newton–Raphson convergence is improved by taking $\gamma_i = \log[-\log(1 - \lambda_i)]$, $i = 1, \ldots, k$ ($\lambda_{k+1} = 1$). This transformation removes range restrictions on the parameters.

The log likelihood of $\boldsymbol{\gamma} = (\gamma_1, \gamma_2, \ldots, \gamma_k)$, $\boldsymbol{\beta}$ can be written

$$\log L(\boldsymbol{\gamma}, \boldsymbol{\beta}) = \sum_{i=1}^{k} \Big(\sum_{l \in D_i} \log\{1 - \exp[-\exp(\gamma_i + \mathbf{z}_l\boldsymbol{\beta})]\} - \sum_{l \in R_i} \exp(\gamma_i + \mathbf{z}_l\boldsymbol{\beta}) \Big). \tag{4.32}$$

The argument leading to (4.32) is basically the same as that leading to (4.21); the relationships (1.5) and (1.6) have been used to obtain the survivor and probability functions associated with (4.30).

The score statistic can be written

$$\frac{\partial \log L}{\partial \gamma_i} = \sum_{l \in D_i} b_{il} - \sum_{l \in R_i} h_{il} \tag{4.33}$$

and

$$\frac{\partial \log L}{\partial \beta_j} = \sum_{i=1}^{k} \left(\sum_{l \in D_i} z_{jl} b_{il} - \sum_{l \in R_i} z_{jl} h_{il} \right)$$

where $h_{il} = \exp(\gamma_i + \mathbf{z}_l \boldsymbol{\beta})$ and $b_{il} = h_{il} e^{-h_{il}}(1 - e^{-h_{il}})^{-1}$.

Maximum likelihood estimators $(\hat{\boldsymbol{\gamma}}, \hat{\boldsymbol{\beta}})$ are solutions to

$$\mathbf{c}' = \left(\frac{\partial \log L}{\partial \gamma_1}, \ldots, \frac{\partial \log L}{\partial \gamma_k}, \frac{\partial \log L}{\partial \beta_1}, \ldots, \frac{\partial \log L}{\partial \beta_s} \right) = \mathbf{0}'.$$

A Newton–Raphson approach to the calculation of $(\hat{\boldsymbol{\gamma}}, \hat{\boldsymbol{\beta}})$ requires second derivatives of $\log L$. The "observed" Fisher information can be written

$$H = \begin{pmatrix} H_{11} & H_{12} \\ H_{21} & H_{22} \end{pmatrix} = \begin{pmatrix} \dfrac{-\partial^2 \log L}{\partial \boldsymbol{\gamma}\, \partial \boldsymbol{\gamma}} & \dfrac{-\partial^2 \log L}{\partial \boldsymbol{\gamma}\, \partial \boldsymbol{\beta}} \\ \dfrac{-\partial^2 \log L}{\partial \boldsymbol{\beta}\, \partial \boldsymbol{\gamma}} & \dfrac{-\partial^2 \log L}{\partial \boldsymbol{\beta}\, \partial \boldsymbol{\beta}} \end{pmatrix}$$

where H_{11} is diagonal with ith element

$$\frac{-\partial^2 \log L}{\partial \gamma_i^2} = \sum_{l \in D_i} g_{il} + \sum_{l \in R_i} h_{il}$$

and

$$\frac{-\partial^2 \log L}{\partial \gamma_i\, \partial \beta_j} = \sum_{l \in D_i} z_{jl} g_{il} + \sum_{l \in R_j} z_{jl} h_{il}$$

$$\frac{-\partial^2 \log L}{\partial \beta_j\, \partial \beta_m} = \sum_{i=1}^{k} \left(\sum_{l \in D_i} z_{jl} z_{ml} g_{il} + \sum_{l \in R_i} z_{jl} z_{ml} \right)$$

where $g_{il} = b_{il}(e^{-h_{il}} + h_{il} - 1)(1 - e^{-h_{il}})^{-1}$.

A Newton–Raphson iteration (Section 3.7) to compute $(\hat{\boldsymbol{\gamma}}, \hat{\boldsymbol{\beta}})$ involves updating current values $(\boldsymbol{\gamma}_0, \boldsymbol{\beta}_0)$ to $(\boldsymbol{\gamma}_1, \boldsymbol{\beta}_1)$ until convergence is reached using the formula

$$\begin{pmatrix} \boldsymbol{\gamma}_1 \\ \boldsymbol{\beta}_1 \end{pmatrix} = \begin{pmatrix} \boldsymbol{\gamma}_0 \\ \boldsymbol{\beta}_0 \end{pmatrix} + H_0^{-1} \mathbf{c}_0,$$

where H_0 and $\mathbf{c}_0$ represent H and $\mathbf{c}$ evaluated at $(\boldsymbol{\gamma}_0, \boldsymbol{\beta}_0)$. Since $(k + s)$, the dimension of H, may be large, direct numerical inversion of H may be time consuming or inaccurate. The fact that the first $(k \times k)$ block of H is diagonal can be exploited, since

$$H^{-1} = \begin{pmatrix} H_{11} & H_{12} \\ H_{21} & H_{22} \end{pmatrix}^{-1} = \left(\begin{array}{c|c} H_{11}^{-1} + FJ^{-1}F' & -FJ^{-1} \\ \hline -J^{-1}F' & J^{-1} \end{array} \right)$$

where $F = H_{11}^{-1} H_{12}$ and $J = H_{22} - H_{21} F$. Consequently only a matrix J of dimension s need be inverted numerically.

A simple starting value system takes $\boldsymbol{\beta}_0 = \mathbf{0}$ and $\boldsymbol{\gamma}_0 = \hat{\boldsymbol{\gamma}}(\mathbf{0})$, the maximum likelihood estimate at $\boldsymbol{\beta} = \mathbf{0}$. Suppose n_i individuals are at risk just prior to x_i (n_i is the total number of study subjects in D_i and R_i), of which d_i (the number of subjects in D_i) fail at x_i. The ith component of $\hat{\boldsymbol{\gamma}}(\mathbf{0})$ is then

$$\hat{\gamma}_i(0) = \log\left[-\log\left(1 - \frac{d_i}{n_i}\right)\right].$$

Experience thus far indicates that instability in the Newton–Raphson procedure may occur if the numbers of failures in specific time intervals are small. Such situations would usually correspond to a rather fine grouping of failure times in which case estimation based on the marginal likelihood (4.7), the partial likelihood (4.14), or the approximation (4.8) may provide a more attractive alternative. With course grouping or even with a relatively fine grouping and large sample sizes, many ties will occur in the failure time data and (4.7) or (4.14) will be computationally impractical. With broad grouping the approximation attendant to (4.8) would be poor. The methods of this section provide a useful estimation procedure in such circumstances, and one that involves precisely the same "relative risk" parameter $\exp(\mathbf{z}\boldsymbol{\beta})$ as in the continuous model (4.1). Further, since the discrete model involves only a finite number $(k + s)$ of parameters, ordinary asymptotic likelihood theory can be applied. This theory leads, of course, to an asymptotic normal distribution for $(\hat{\boldsymbol{\gamma}}, \hat{\boldsymbol{\beta}})$ with mean vector $(\boldsymbol{\gamma}, \boldsymbol{\beta})$ and variance matrix estimated by H^{-1}, at least under mild restrictions on the $\mathbf{z}$ vectors and the censoring.

Consider now estimation of the survivor function at a specified covariate vector $\mathbf{z}$. The maximum likelihood estimator at x_i is

$$\hat{F}(x_i; \mathbf{z}) = \prod_{j=1}^{i-1} \exp[-\exp(\hat{\gamma}_j + \mathbf{z}\hat{\boldsymbol{\beta}})]. \tag{4.34}$$

Because of the fixed number of parameters the asymptotic distribution of $\hat{F}$ can easily be written. In order to improve the adequacy of the asymptotic approximations in moderate sized samples, one may apply asymptotic results to $\hat{X} = \log[-\log \hat{F}(x_i; \mathbf{z})]$ rather than to $\hat{F}(x_i; \mathbf{z})$ itself, since $\hat{X}$ is devoid of range restrictions. The asymptotic distribution of $\hat{X}$ is normal with mean $X = \log[-\log F(x_i; \mathbf{z})]$ and variance $\sigma_X^2 = \mathbf{w}'H^{-1}\mathbf{w}$, where $\mathbf{w}' = (\partial X/\partial \boldsymbol{\gamma}', \partial X/\partial \boldsymbol{\beta}')$ has components

$$\frac{\partial X}{\partial \gamma_k} = \begin{cases} \dfrac{h_k}{\Sigma_1^{i-1} h_j} & k < i \\ 0 & k \geq i \end{cases}$$

$$\frac{\partial X}{\partial \beta_j} = \frac{\Sigma_{j=1}^{i-1} z_j h_j}{\Sigma_{j=1}^{i-1} h_j}$$

where $h_j = \exp(\gamma_j + \mathbf{z}\boldsymbol{\beta})$.

Note that at $\boldsymbol{\beta} = \mathbf{0}$ the survival curve estimator (4.34) reduces to the Kaplan–Meier estimator

$$\hat{F}(x_i; \mathbf{z}) = \prod_{j=1}^{i-1} \left(1 - \frac{d_j}{n_j}\right)$$

by virtue of $\hat{\boldsymbol{\gamma}}(\mathbf{0})$.

The methods of this section as well as an illustration to survival data on a large set (11,442) of breast cancer patients are given in Prentice and Gloeckler (1978).

4.6.3 Score Test for $\boldsymbol{\beta} = \mathbf{0}$

A test for $\boldsymbol{\beta} = \mathbf{0}$ is of interest, for example, in the comparison of $s+1$ survival curves. A test for equality of the $s+1$ curves corresponds to testing $\boldsymbol{\beta} = \mathbf{0}$ in the above regression notation provided $\mathbf{z}$ consists of indicator variables for any s of the groups. This may be viewed as an alternative procedure to the log-rank procedure of Sections 4.2.5 and 1.4.

A fully efficient score test can be based on $\partial \log L/\partial\boldsymbol{\beta}$ evaluated at $\boldsymbol{\beta} = \mathbf{0}$ and $\boldsymbol{\gamma} = \hat{\boldsymbol{\gamma}}(\mathbf{0})$. On substitution in (4.33) this statistic can be written

$$\mathbf{c} = -\sum_{i=1}^{k} n_i d_i^{-1} \log\left(1 - \frac{d_i}{n_i}\right)\left(\sum_{l \in D_i} \mathbf{z}_l - d_i n_i^{-1} \sum_{l \in R_i \cup D_i} \mathbf{z}_l\right) \qquad (4.35)$$

The variance matrix for $\mathbf{c}$, under the hypothesis $\boldsymbol{\beta} = \mathbf{0}$, is estimated by the matrix J evaluated at $\boldsymbol{\beta} = \mathbf{0}$ and $\boldsymbol{\gamma} = \hat{\boldsymbol{\gamma}}(\mathbf{0})$. Some simplification occurs if expectations conditional on (n_i, d_i), $i = 1, \ldots, k$ are taken over elements of H before the matrix J is calculated. The desired expectations, evaluated at $[\mathbf{0}, \hat{\boldsymbol{\gamma}}(\mathbf{0})]$, are as follows:

$$E\left(\frac{-\partial^2 \log L}{\partial \gamma_i^2}\right) = q_i, \qquad E\left(\frac{-\partial^2 \log L}{\partial \gamma_i\, \partial\boldsymbol{\beta}}\right) = \bar{\mathbf{z}}_i \boldsymbol{\beta}$$

and

$$E\left(\frac{-\partial^2 \log L}{\partial\boldsymbol{\beta}\, \partial\boldsymbol{\beta}}\right) = \sum_{i=1}^{k} q_i n_i^{-1} \sum_{l \in D_i \cup R_i} \mathbf{z}_l' \mathbf{z}_l$$

where

$$q_i = \frac{n_i(n_i - d_i)}{d_i}\left[\log\left(1 - \frac{d_i}{n_i}\right)\right]^2.$$

It follows easily that an estimator of asymptotic variance for $\mathbf{c}$ is given by

$$C = \sum_{i=1}^{k} q_i \left[n_i^{-1} \sum_{l \in D_i \cup R_i} (\mathbf{z}_l - \bar{\mathbf{z}}_i)'(\mathbf{z}_l - \bar{\mathbf{z}}_i)\right],$$

a weighted average of the covariance matrices of the finite population of $\mathbf{z}$ values for individuals at risk within each time interval.

The hypothesis $\boldsymbol{\beta} = \mathbf{0}$ may be tested by comparing

$$\mathbf{c}'C^{-1}\mathbf{c} \tag{4.36}$$

with χ^2 tables on s degrees of freedom (if C is nonsingular). Specialization of (4.33) to the comparison of two samples gives

$$\frac{[\Sigma_1^k (d_{1i}n_{2i} - d_{2i}n_{1i})d_i^{-1}\log(1 - d_in_i^{-1})]^2}{\Sigma_1^k q_in_{1i}n_{2i}n_i^{-2}}$$

as an approximate $\chi^2_{(1)}$ statistic, where subscripts 1 and 2 index the two samples.

A straightforward generalization of (4.33) leads to a test for $\boldsymbol{\beta} = \mathbf{0}$ when data have been divided into strata. The approximate $\chi^2_{(s)}$ test statistic is given by

$$\left(\Sigma\, \mathbf{c}\right)'\left(\Sigma\, C\right)^{-1}\left(\Sigma\, \mathbf{c}\right)$$

where the summation is over strata. Note that both the nuisance function $\{\lambda_j\}$ and the time intervals themselves can vary over strata.

The statistic (4.35) can be compared to the log-rank statistic (4.19) which arises out of the partial likelihood for data from the discrete model (4.13). In fact, the last term in parentheses in expression (4.35) is exactly the contribution of the ith failure point to the log-rank statistic (4.19). The test based on $\mathbf{c}$ will therefore be similar to the log-rank statistic provided d_i/n_i is small, in which case the term $n_id_i^{-1}\log(1 - d_in_i^{-1}) \simeq -1$. In fact, an alternative estimate of the covariance matrix of $\mathbf{c}$ can be obtained by exploiting this fact. See Problem 36 in Appendix 2. It should be noted that the test based on (4.35) is best for alternatives with the same relative risk parameter e^β as in (4.1) or (4.30). The log-rank test, however, is most efficient for alternatives in which sample differences are specified by the odds ratio parameter in (4.19).

4.7 EFFICIENCY OF THE REGRESSION ESTIMATOR

Consider now the efficiency of estimators of $\boldsymbol{\beta}$ that maximize the likelihood functions developed in Section 4.2. We restrict attention to situations in which tied failure times cannot occur so that the marginal likelihood (4.7), the partial likelihood from (4.13), and the approximate likelihood (4.8) all reduce to (4.6). In this section $\boldsymbol{\beta}$ values that maximize (4.6) are denoted $\hat{\boldsymbol{\beta}}_r$, the r indicating that it is based on failure time ranks.

Two important questions arise concerning the efficiency of $\hat{\boldsymbol{\beta}}_r$. First, with the hazard function $\lambda_0(\cdot)$ unspecified, can $\hat{\boldsymbol{\beta}}_r$ be improved upon in regard to (asymptotic) efficiency? Second, what is the efficiency of $\hat{\boldsymbol{\beta}}_r$ with respect to that of a maximum likelihood estimator, $\hat{\boldsymbol{\beta}}_p$, based on a special case of (4.1) in which $\lambda_0(\cdot)$ is specified up to certain unknown parameters?

The development of (4.6) as a likelihood based on a "marginally sufficient statistic" suggests that no more efficient estimator than $\hat{\boldsymbol{\beta}}_r$ can be anticipated "in the absence of knowledge of $\lambda_0(\cdot)$," at least if there is no censoring. Efron (1977) and, somewhat less explicitly, Oakes (1977) argue that the asymptotic variance matrix for $\hat{\boldsymbol{\beta}}_r$ will be "close" to that for $\hat{\boldsymbol{\beta}}_p$ provided the parametric family giving rise to $\hat{\boldsymbol{\beta}}_p$ is reasonably rich.

The second question concerns the efficiency of $\hat{\boldsymbol{\beta}}_r$ relative to estimators from parametric models that may be highly specialized. Kalbfleisch (1974) examined the efficiency of $\hat{\boldsymbol{\beta}}_r$ relative to $\hat{\boldsymbol{\beta}}_p$ from the exponential special case $\{\lambda_0(\cdot) \equiv \lambda\}$. The results of Oakes and Efron provide general efficiency expressions. The material presented below makes use of some aspects of each of these works, but makes particular use of the results of Efron and of Oakes. Emphasis is placed on the interpretation of the efficiency expressions. The second question is addressed first, though the same approach leads to an answer to the first question also.

The most extreme parametric specification of the $\lambda_0(\cdot)$ function in (4.1) that one can make is to suppose that it is completely known. Such a specification, however, involves an asymmetry in that the hazard function at $\mathbf{z} = \mathbf{0}$ is completely known whereas, at other $\mathbf{z}$ values, it is known only up to the factor $\exp(\mathbf{z}\boldsymbol{\beta})$. To avoid this situation we consider a model in which $\lambda_0(\cdot)$ is specified up to a scale parameter; that is, $\lambda_0(t) = \lambda h_0(t)$ where $h_0(t)$ is assumed known.

In the notation of Section 4.2, the likelihood function from this parametric model can be written

$$L_1 = \prod_{i=1}^{k} [\lambda_0(t_{(i)}) \exp(\mathbf{z}_{(i)}\boldsymbol{\beta})] \exp\left[-\int_0^\infty N(t)\lambda_0(t)\,dt\right] \tag{4.37}$$

where $N(t) = N(t, \boldsymbol{\beta}) = \Sigma_{l \in R(t)} \exp(\mathbf{z}_l\boldsymbol{\beta})$. The likelihood (4.37) is valid for any independent censoring mechanism as is discussed further in Section 5.2, and the results obtained here are valid for this same class. In order to investigate the efficiency of the regression estimator from the proportional hazards model, we compare the information content of (4.37) with the information from the marginal or partial likelihood (4.6) given by

$$L_2 = \prod_{i=1}^{k} \left(\frac{\exp(\mathbf{z}_{(i)}\boldsymbol{\beta})}{N(t_{(i)})} \right). \tag{4.38}$$

We are here assuming, as above, that the usual asymptotic results apply to the marginal or partial likelihood (4.38). Sufficient conditions for this to hold are discussed in Section 4.8.

The observed information matrix from (4.37) has (u, v) element

$$\frac{-\partial^2 \log L_1}{\partial \beta_u \, \partial \beta_v} = \int \sum_{l \in R(t)} z_{ul} z_{vl} \exp(\mathbf{z}_l \boldsymbol{\beta}) \lambda_0(t) \, dt \tag{4.39}$$

where $u, v = 0, 1, \ldots, s$ and we have defined $\beta_0 = \log \lambda$ and $z_{0l} = 1$ for all l. The expected information is then

$$\mathcal{I}_1 = \begin{pmatrix} \int \eta \lambda_0 \, dt & (\int \boldsymbol{\eta}_1 \lambda_0 \, dt)' \\ \int \boldsymbol{\eta}_1 \lambda_0 \, dt & \int \boldsymbol{\eta}_2 \lambda_0 \, dt \end{pmatrix}$$

where $\eta = E[N(t)]$, $\boldsymbol{\eta}_1 = (\eta_{1u}) = (E[\partial N(t)/\partial \beta_u])$ is an $s \times 1$ vector and $\boldsymbol{\eta}_2 = (\eta_{2uv}) = (E[\partial^2 N(t)/\partial \beta_u \, \partial \beta_v])$ is an $s \times s$ matrix. Although η, $\boldsymbol{\eta}_1$, and η_2 are in fact functions of t, $\boldsymbol{\beta}$, λ this dependence is suppressed for ease of notation. Note that if $A = (a_{ij}(t))$ is an $n \times m$ matrix valued function of t, then $\int A \, dt$ represents the $n \times m$ matrix with (i, j) element $\int a_{ij}(t) \, dt$. The "marginal" information matrix for $\boldsymbol{\beta} = (\beta_1, \ldots, \beta_s)$, that is, the variance matrix for $\partial \log L / \partial \boldsymbol{\beta}$ after maximizing out λ, is then

$$\begin{aligned} \mathcal{I}_p &= \int \boldsymbol{\eta}_2 \lambda_0 \, dt - \left(\int \boldsymbol{\eta}_1 \lambda_0 \, dt \right) \left(\int \eta \lambda_0 \, dt \right)^{-1} \left(\int \boldsymbol{\eta}_1 \lambda_0 \, dt \right)' \\ &= \int (\boldsymbol{\eta}_2 - \eta^{-1} \boldsymbol{\eta}_1 \boldsymbol{\eta}_1') \lambda_0 \, dt + \int \eta (\eta^{-1} \boldsymbol{\eta}_1 - \mathbf{c})(\eta^{-1} \boldsymbol{\eta}_1 - \mathbf{c})' \lambda_0 \, dt \end{aligned} \tag{4.40}$$

where

$$\mathbf{c} = \left(\int \eta \lambda_0 \, dt \right)^{-1} \int \boldsymbol{\eta}_1 \lambda_0 \, dt.$$

The information computation for the partial likelihood (4.38) is not so straightforward. The information matrix is

$$\begin{aligned} \mathcal{I}_r &= E\left(\frac{-\partial^2 \log L_2}{\partial \boldsymbol{\beta} \, \partial \boldsymbol{\beta}} \right) \\ &= E\left(\sum_{i=1}^{k} \frac{\partial^2 \log N(t_{(i)})}{\partial \boldsymbol{\beta} \, \partial \boldsymbol{\beta}} \right) \\ &= E\left[\int_0^\infty \frac{\partial^2 \log N(t)}{\partial \boldsymbol{\beta} \, \partial \boldsymbol{\beta}} \, dk(t) \right] \end{aligned} \tag{4.41}$$

where $dk(t)$ is the counting measure which has value 1 at $t = t_{(i)}$, $i = 1, \ldots, k$ (that is, $dk(t) = 1$ if there is a failure in time $[t, t + dt)$) and has value zero otherwise. Since

$$P[dk(t) = 1 | R(t)] = N(t) \lambda_0(t) \, dt,$$

the expectation in (4.41) can be evaluated by writing (4.41) as a Reimann sum and conditioning at each t on $R(t)$. Since, given $R(t)$, $N(t)$ is fixed and $dk(t)$ has expectation $N(t) \lambda_0(t) \, dt$, it follows that

$$\begin{aligned}\mathscr{I}_r &= \int_0^\infty E\left[\frac{\partial^2 \log N(t)}{\partial \boldsymbol{\beta}\, \partial \boldsymbol{\beta}}\, dk(t)\right] \\ &= \int_0^\infty E\left[\frac{\partial^2 \log N(t)}{\partial \boldsymbol{\beta}\, \partial \boldsymbol{\beta}}\, N(t)\right]\lambda_0(t)\, dt \\ &= \int (\boldsymbol{\eta}_2 - \eta^{-1}\boldsymbol{\eta}_1\boldsymbol{\eta}_1')\lambda_0\, dt + \int (\eta^{-1}\boldsymbol{\eta}_1\boldsymbol{\eta}_1' - B)\lambda_0\, dt \end{aligned} \tag{4.42}$$

where

$$B = E\left(\frac{(\partial N(t)/\partial \boldsymbol{\beta})(\partial N(t)/\partial \boldsymbol{\beta})'}{N(t)}\right).$$

The second term in (4.42) is typically small. For example, Efron (1977, p. 560, remark I) shows that, in the case of a random censorship model, the integrand of this term is $O(\eta^{-1})$ and so can be ignored for purposes of evaluating asymptotic efficiency.

The rank based estimator $\hat{\boldsymbol{\beta}}_r$ will have full asymptotic efficiency (Fisher, 1956, p. 147) if $\mathscr{I}_r\mathscr{I}_p^{-1}$ converges to an identity matrix, or equivalently if $(\mathscr{I}_p - \mathscr{I}_r)\mathscr{I}_p^{-1}$ converges to a zero matrix. A comparison of (4.42) with (4.40) shows that full asymptotic efficiency will be achieved, for example, if in the limit

$$\eta^{-1}\boldsymbol{\eta}_1 = \frac{E(\sum_{i \in R(t)} \mathbf{z}_i' e^{\mathbf{z}_i\boldsymbol{\beta}})}{E(\sum_{i \in R(t)} e^{\mathbf{z}_i\boldsymbol{\beta}})} = \text{constant} \tag{4.43}$$

in which case the second term in (4.40) is zero. If $\boldsymbol{\beta} = \mathbf{0}$ and censorship does not depend on $\mathbf{z}$, all risk sets $R(t)$ of the same size are equally probable at time t and (4.43) is satisfied. The rank estimator is then fully efficient. This also shows that the log-rank test for $\boldsymbol{\beta} = \mathbf{0}$ will have full Pitman efficiency (see, for example, Cox and Hinkley, 1974, p. 338, for definitions) under these circumstances. Censoring rates that depend on $\mathbf{z}$ (e.g., differ among samples in the s-sample problem) give rise to a distribution of $\mathbf{z}$ values with expectation that varies over time and thereby to some loss in efficiency since (4.43) is not satisfied. Crowley and Thomas (1975) give some calculations for the efficiency of the log-rank test under random censorship but with differing censoring distributions in the samples. Even in situations with rather extreme differences in the censoring rates, the log rank has respectable efficiency, usually greater than .90.

Perhaps of more concern is the typical decline of the relative efficiency of $\hat{\boldsymbol{\beta}}_r$ as $\boldsymbol{\beta}$ departs from zero. If, for example, $s = 1$ and $\beta > 0$ and there is no censoring, then as t increases $R(t)$ with high probability contains more small values of z than large and $\eta^{-1}\eta_1$ decreases with increasing t. The dependence of efficiency on $\boldsymbol{\beta}$ is a situation unfamiliar to ordinary linear regression.

If $c_i(t)$ is an indicator variable with value 1 for $i \in R(t)$ and 0 otherwise, then

$$N(t) = \sum_{i=1}^{n} c_i(t)e^{\mathbf{z}_i\boldsymbol{\beta}}, \qquad \frac{\partial N(t)}{\partial \boldsymbol{\beta}} = \sum_{i=1}^{n} c_i(t)\mathbf{z}_i' e^{\mathbf{z}_i\boldsymbol{\beta}}$$

and if $p_i(t) = E[c_i(t)] = P[i \in R(t)]$, then

$$\eta(t) = \sum_{i=1}^{n} p_i(t)e^{\mathbf{z}_i\boldsymbol{\beta}}, \qquad \boldsymbol{\eta}_1(t) = \sum_{i=1}^{n} p_i(t)\mathbf{z}_i' e^{\mathbf{z}_i\boldsymbol{\beta}}.$$

Under reasonably weak conditions involving the long run stability of the process so that sample means converge to their corresponding expectations, the ratio $\eta(t)^{-1}\boldsymbol{\eta}_1(t)$ is consistently estimated at each t by

$$e(t; \boldsymbol{\beta}) = \frac{\sum_{i\in R(t)} \mathbf{z}_i e^{\mathbf{z}_i\boldsymbol{\beta}}}{\sum_{i\in R(t)} e^{\mathbf{z}_i\boldsymbol{\beta}}}$$

or by $e(t; \hat{\boldsymbol{\beta}}_r)$. The quantity $e(t; \boldsymbol{\beta})$ can be thought of as a weighted average of $\mathbf{z}$ values in the risk set at time t. Oakes (1977) suggests plotting $e(t; \hat{\boldsymbol{\beta}}_r)$ to obtain an empirical check on the efficiency of the rank analysis.

An explanation for the loss in efficiency of $\hat{\boldsymbol{\beta}}_r$ when the average $\mathbf{z}$ value over the risk set varies over the time axis is that such variations introduce asymptotic correlations between the estimator of $\boldsymbol{\beta}$ and that of λ. The parametric analysis can exploit such correlations but the rank procedure cannot. As an extreme situation, consider a two-sample problem in which one sample is uncensored and the second is totally censored at time t_0. The contribution of the rank based likelihood (4.6) at any time after t_0 is independent of $\boldsymbol{\beta}$ since all items in the risk set have the same covariate value. Parametric analyses, on the other hand, can utilize failures past t_0 to estimate the parameter in $\lambda_0(\cdot)$ more precisely and hence they yield more precise estimates of $\boldsymbol{\beta}$. One might expect that if the parametric model were allowed to become more flexible, the information on failures beyond t_0 would be of lesser value in $\boldsymbol{\beta}$ estimation and the loss in efficiency in using (4.6) would be less severe. This line of thought is continued below where the efficiency of the rank analysis is evaluated relative to more flexible parametric models. Before this, however, we consider in some detail a particular example of the type discussed above.

Efron (1977) examined the limiting ratio $\mathcal{I}_r\mathcal{I}_p^{-1}$ for the two sample problem in the absence of censoring where, without loss of generality, we may take $h_0(t) = 1$ so that the hazard for sample z is $\lambda \exp(z\beta)$, $z = 0, 1$. Let $n_i(t)$ represent the number of sample i items at risk at time t, $i = 0, 1$. Then,

$$N(t) = \sum_{l\in R(t)} \exp(z_l\beta) = n_0(t) + n_1(t)\exp(\beta)$$

from which $\eta(t)$, $\eta_1(t)$, and $\eta_2(t)$ can be calculated using

$$E[n_0(t)] = nq \exp(-\lambda t)$$

and

$$E[n_1(t)] = np \exp(-\lambda t e^{\beta})$$

where nq and np are the sample sizes in samples 0 and 1, respectively. After some calculation, we find from (4.42) that

$$\lim_{n\to\infty} \frac{\mathcal{I}_r}{n} = pq\alpha \int_0^1 (q\alpha + pu^{1-\alpha})^{-1}\, du$$

where $\alpha = \exp(-\beta)$. Straightforward calculation gives the information from the parametric analysis as $\mathcal{I}_p = npq$ so that the asymptotic relative efficiency is

$$\alpha \int_0^1 (q\alpha + pu^{1-\alpha})^{-1}\, du. \tag{4.44}$$

Figure 4.15 gives plots of this efficiency for several values of p. The efficiency of the rank analysis exceeds 75% in most situations of interest, for example, when the failure rate ratio e^{β} for the two samples is less than 3. A Taylor expansion for the logarithm of (4.44) about $\beta = 0$ gives an approximate expression for this efficiency as $\exp(-\beta^2 pq)$ which agrees reasonably closely with (4.44) for p not too near 0 or 1. Kalbfleisch (1974) showed that with a single regressor variable the efficiency is ap-

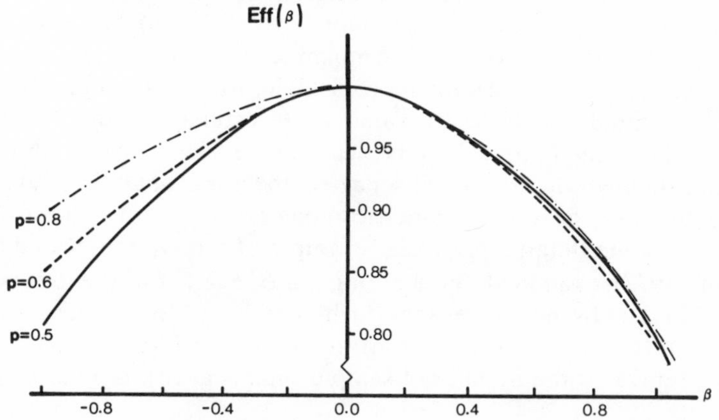

Figure 4.15 Relative efficiency of the proportional hazard estimation versus the exponential estimation of β [c.f. expression (4.44)].

proximately $\exp(-\beta^2\mu_2)$ where μ_2 is the second central moment of the z values.

Efron (1977) also gives some numerical results indicating that efficiencies in the two sample problem with censoring are in reasonable agreement with (4.44) at least for certain censoring patterns considered. Oakes (1977) also obtains the expression (4.44) with $p = q = \frac{1}{2}$ as a special case of an expression for the asymptotic efficiency of the rank analysis when censoring is exponential. Some of these results, and generalizations of them, are considered in Problems 39, 40, and 41 of Appendix 2.

Suppose now that, rather than specifying the $\lambda_0(\cdot)$ function up to a scale factor, we take

$$\lambda_0(t) = \exp[z_{s+1}(t)\beta_{s+1} + \cdots + z_{s+m}(t)\beta_{s+m}]h_0(t) \tag{4.45}$$

where $z_{s+1}(t), \ldots, z_{s+m}(t)$ are specified functions of t (usually with $z_{s+1}(t) = 1$), and $h_0(t)$ is assumed completely known. The overall likelihood can again be written as (4.37) and the notation has been chosen so that the (u, v) element in the observed information matrix $h_{uv} = -\partial^2 \log L/\partial\beta_u\,\partial\beta_v$, $u, v = 1, \ldots, s+m$, is again (4.39), where $\boldsymbol{\beta} = (\beta_1, \ldots, \beta_s)'$ and $\mathbf{z}_l = (z_{1l}, \ldots, z_{sl})$ as before, while $z_{ul} = z_u(t)$ for $u = s+1, \ldots, s+m$. Following the same steps as before, the information matrix can be written as a partitioned matrix

$$\mathscr{I}_1 = \begin{pmatrix} \int \boldsymbol{\eta}_2\lambda_0\,dt & \int \boldsymbol{\eta}_1\mathbf{w}\lambda_0\,dt \\ \int \mathbf{w}'\boldsymbol{\eta}_1'\lambda_0\,dt & \int \eta\mathbf{w}'\mathbf{w}\lambda_0\,dt \end{pmatrix}$$

where $\boldsymbol{\eta}_2$, $\boldsymbol{\eta}_1$, and η are defined as before and

$$\mathbf{w} = \mathbf{w}(t) = (z_{s+1}(t), \ldots, z_{s+m}(t)).$$

The marginal information on $\boldsymbol{\beta}$ when $\beta_{s+1}, \ldots, \beta_{s+m}$ have been eliminated can be written as (4.40) where $\mathbf{c}$ is replaced with the $s \times 1$ vector

$$\mathbf{c} = \int \boldsymbol{\eta}_1\mathbf{w}\lambda_0\,dt\left(\int \eta\mathbf{w}'\mathbf{w}\lambda_0\,dt\right)^{-1}\mathbf{w}'. \tag{4.46}$$

The information in the partial or marginal likelihood (4.38) is again given by (4.42) and as before, for purposes of evaluating asymptotic relative efficiency, the second term is typically asymptotically trivial and can be ignored. The asymptotic relative efficiency of the rank analysis is then given by the limit of $\mathscr{I}_r\mathscr{I}_p^{-1}$ which depends on the size of the second term in $\mathscr{I}_p$. This term is

$$\int \eta(\eta^{-1}\boldsymbol{\eta}_1 - \mathbf{c})(\eta^{-1}\boldsymbol{\eta}_1 - \mathbf{c})'\lambda_0\,dt$$

with $\mathbf{c}$ given by (4.46). This is recognizable as the deviation from regression sum of squares for a weighted regression of $\boldsymbol{\eta}_1(t)/\eta(t)$ on $\mathbf{w}(t) = (z_{s+1}(t), \ldots, z_{s+m}(t))$ with weights given by $\eta(t)\lambda_0(t)$. This shows that $\hat{\boldsymbol{\beta}}_r$ will be fully efficient with respect to $\hat{\boldsymbol{\beta}}_p$ from (4.45) whenever the elements of the limit of $\boldsymbol{\eta}_1/\eta$ are in the space spanned by the functions $z_{s+1}(t), \ldots, z_{s+m}(t)$. The rank analysis will have close to full efficiency provided the parametric modeling (4.45) is rich enough that a linear combination of $z_{s+1}(t), \ldots, z_{s+m}(t)$ closely approximates $\boldsymbol{\eta}_1(t)/\eta(t)$. Such closeness would occur, for example, if a parametric model were asserted in which $\lambda_0(\cdot)$ was taken to be arbitrary but constant within relatively narrow time intervals (see Holford, 1976, for some results related to this model). Such a model would allow the mean function to be well approximated by a linear function of the $z_j(t)$, $j = s+1, \ldots, m$, since such a function can be an arbitrary step function within the specified time intervals. This line of reasoning makes it clear that one cannot expect to find a $\boldsymbol{\beta}$ estimator that is more efficient than $\hat{\boldsymbol{\beta}}_r$ provided the $\lambda_0(\cdot)$ function is unrestricted.

Efron gives some numerical results for the two sample problem indicating that $\hat{\beta}_r$ is nearly fully efficient relative to $\hat{\beta}_p$ from a fairly simple parametric model in which $z_{s+1}(t) \equiv 1$, $z_{s+2}(t) = e^{-t}$.

All of the above discussion is concerned with *asymptotic* relative efficiency. Evaluation of small sample efficiencies would require the calculation, or simulation, of the actual sampling distribution of $\hat{\boldsymbol{\beta}}_r$ and $\hat{\boldsymbol{\beta}}_p$ under prescribed conditions. Some indication of finite sample relative efficiency can be obtained by comparing (expected) information matrices at fixed sample sizes. Such information matrices give the expected curvature of the likelihood function at its maximum and hence an indication of the precision with which $\boldsymbol{\beta}$ is estimated. We end this section with a few comparisons of this type based again on the parametric model in which $\lambda_0(t) = \lambda h_0(t)$, with $h_0(t)$ known. For variety, following Kalbfleisch (1974), we compare the information matrix from the rank analysis with that based on a marginal likelihood analysis of this parametric model. Information calculations for the marginal and full likelihood analyses will be virtually identical.

First define $t' = \int_0^t h_0(x)\,dx$ so that the hazard function for t' is $\lambda e^{\mathbf{z}\boldsymbol{\beta}}$. This model for t' is invariant under the group of scale transformations and, in the absence of censoring, the variates $a_i = t'_i/t'_1$, $i = 2, \ldots, n$ form the maximum invariant in the sample space; their density function generates the marginal likelihood for $\boldsymbol{\beta}$. For simplicity suppose there is a scalar regressor variable with values $z_1, \ldots, z_n$, corresponding to the uncensored failure times $t_1, \ldots, t_n$. Without loss of generality we may take $z_1 + \cdots + z_n = 0$. The density function for $\mathbf{a} = (a_2, \ldots, a_n)$ is

$$f_a(\mathbf{a}; \boldsymbol{\beta}) = (n-1)!\Big(\sum_{i=1}^{n} a_i e^{\beta z_i}\Big)^{-n}, \qquad a_2, \ldots, a_n > 0$$

where $a_1 = 1$. The probability distribution for the rank vector $\mathbf{r} = [(1), \ldots, (n)]$ is, from (4.4),

$$f_r(\mathbf{r}; \beta) = \prod_{i=1}^{n} \Big(\sum_{j=i}^{n} e^{\beta z_{(j)}}\Big)^{-1}.$$

The information in the rank vector about β is

$$\mathcal{I}_r(\beta) = -E\Big(\frac{\partial^2}{\partial\beta^2} \log f_r(\mathbf{r}; \beta)\Big),$$

which at $\beta = 0$ reduces to

$$\mathcal{I}_r(0) = \sum_{i=1}^{n} E_p\Big(\frac{\sum_{j=i}^{n} \sum_{k=i}^{n} z_{(j)}(z_{(j)} - z_{(k)})}{(n-i+1)^2}\Big)$$

$$= \sum_{i=1}^{n} E_p(m_{2,i} - m_{1,i}^2)$$

where

$$m_{k,i} = \frac{\sum_{j=i}^{n} z_{(j)}^k}{n-i+1} \qquad (k = 1, 2, \ldots).$$

Here E_p refers to the expectation over the permutation distribution on $\{(1), (2), \ldots, (n)\}$. It is easily verified that $E_p(m_{2,i}) = \mu_2$ and that

$$E_p(m_{1,i}^2) = \frac{\mu_2}{n(n-1)} \sum_{i=1}^{n} \frac{i-1}{n-i+1}$$

so that

$$\mathcal{I}_r(0) = \frac{n\mu_2}{n-1} \sum_{i=1}^{n} \frac{n-i}{n-i+1}$$

where μ_k is the kth central moment of $z_1, \ldots, z_n$.

On the other hand, straightforward calculation verifies that the information on β contained in the variate $\mathbf{a}$ is

$$\mathcal{I}_a(\beta) = -E\Big(\frac{\partial^2}{\partial\beta^2} \log f_a(\mathbf{a}; \beta)\Big)$$

$$= \frac{n^2\mu_2}{n+1}.$$

The relative efficiency at $\beta = 0$ for a sample of size n of the rank statistic compared to the statistic $\mathbf{a}$ is

$$R_n(0) = \frac{\mathscr{I}_r(0)}{\mathscr{I}_a(0)}$$
$$= \frac{n+1}{n(n-1)} \sum_{i=1}^{n} \frac{n-i}{n-i+1}. \qquad (4.47)$$

As noted above, the asymptotic relative efficiency at $\beta = 0$ is

$$R(0) = \lim_{n\to\infty} R_n(0) = 1.$$

From (4.47) the relative efficiency for finite n can be evaluated and this has a simple interpretation in terms of variance. For example, the case $n = 2$ gives $R_n(0) = \frac{3}{4}$ $(\mu_2 \neq 0)$. This can be interpreted as the ratio of the asymptotic variances of the rank and parametric analyses in a twin study when the ith twin pair has its own failure rate λ_i and β is a regression parameter common to all twin pairs. Table 4.4 gives such relative efficiencies for several values of n and the approach to full efficiency is readily seen to be rapid.

Table 4.4 RELATIVE EFFICIENCY OF RANK ANALYSIS VERSUS EXPONENTIAL AT $\beta = 0$

n	2	3	5	7	10	15	20	40	60	100	∞
$R_n(0)$	.75	.78	.82	.84	.89	.91	.94	.95	.97	.99	1

Consider now the case of two regression variables β_1, β_2 where β_1 is of interest and suppose the variable z_2 takes on only a finite number of distinct values. It is easily seen that the asymptotic relative efficiency of the rank analysis for the estimation of β_1 is 1 at $\beta_1 = 0$ since the problem could be handled with full efficiency by considering the hazard for the jth possible value of z_2 as $\lambda_{0j}(t)e^{\beta_1 z_1}$ where the information regarding proportionality for the second variable is suppressed. Clearly, the small sample efficiencies will be poorer; for example, if β_2 is very large and z_2 takes only two values each with equal frequency, the small sample relative efficiencies for the rank analysis will increase at about half the rate stated in Table 4.4. The relative efficiency for $n = 20$ in this case would be about .89 as compared to .94 with no auxiliary variable. This reduction in small sample efficiency could conceivably be severe if z_2 took many values and β_2 were reasonably large. This same line of reasoning suggests that no additional loss of asymptotic efficiency is incurred through unnecessary stratification, but that small sample efficiencies based on q strata may approach asymptotic results at a rate as low as $1/q$ of that for the unstratified analysis. Further examination of the small sample efficiencies is needed.

4.8 ASYMPTOTIC DISTRIBUTION THEORY

4.8.1 The Regression Estimator

The standard likelihood methods outlined in Chapter 3 were based on independent "scores" to which the central limit theorem would apply. The nonstandard likelihood (4.6), however, requires special techniques to justify the earlier use of asymptotic distributional results. Such special techniques are outlined in this section. It should be noted, however, that this is a topic requiring further work and that explicit weak conditions that assure the desired asymptotic results have not yet been specified.

As usual the essential step in developing the asymptotic distribution for $\hat{\boldsymbol{\beta}}$ involves a Taylor expansion of the score function $\mathbf{U}(\boldsymbol{\beta}) = \partial \log L(\boldsymbol{\beta})/\partial\boldsymbol{\beta}$ about $\hat{\boldsymbol{\beta}}$. The expansion formula can then be rewritten to give

$$n^{1/2}(\hat{\boldsymbol{\beta}} - \boldsymbol{\beta}_0) = n^{1/2}\left(\frac{-\partial^2 \log L(\boldsymbol{\beta}_*)}{\partial\boldsymbol{\beta}_*\,\partial\boldsymbol{\beta}_*}\right)^{-1}\mathbf{U}(\boldsymbol{\beta}_0), \tag{4.48}$$

where $\boldsymbol{\beta}_0$ is the "true" value of $\boldsymbol{\beta}$, and $\boldsymbol{\beta}_*$ is "between" $\hat{\boldsymbol{\beta}}$ and $\boldsymbol{\beta}_0$. Development of the asymptotic distribution of $\hat{\boldsymbol{\beta}}$ then involves showing $n^{-1/2}\mathbf{U}(\boldsymbol{\beta}_0)$ to be asymptotically normal with mean $\mathbf{0}$ and variance $n^{-1}\mathcal{I}(\boldsymbol{\beta}_0) = n^{-1}E[-\partial^2 \log L(\boldsymbol{\beta}_0)/\partial\boldsymbol{\beta}_0\,\partial\boldsymbol{\beta}_0]$, and showing $n^{-1}(-\partial^2 \log L(\boldsymbol{\beta}_*)/\partial\boldsymbol{\beta}_*\,\partial\boldsymbol{\beta}_*) = n^{-1}I(\boldsymbol{\beta}_*)$ to be a consistent estimator of $n^{-1}\mathcal{I}(\boldsymbol{\beta}_0)$. It will then follow that $n^{1/2}(\boldsymbol{\beta} - \boldsymbol{\beta}_0)$ is asymptotically normal with mean $\mathbf{0}$ and variance matrix $nI^{-1}(\boldsymbol{\beta}_0)$.

The desired mean and variance of $\mathbf{U}(\boldsymbol{\beta}_0)$ are readily established, as by Cox (1975): Denote by H_i the complete history of failures and censorings prior to $t_{(i)}$ along with the failure at $t_{(i)}$. The probability (4.12) that individual (i) fails at $t_{(i)}$ given H_i is

$$P[(i)|H_i] = \frac{e^{\mathbf{z}_{(i)}\boldsymbol{\beta}_0}}{\sum_{l\in R(t_{(i)})} e^{\mathbf{z}_l\boldsymbol{\beta}_0}}. \tag{4.49}$$

Since (4.49) is a probability distribution

$$\mathbf{U}_i = \mathbf{U}_i(\boldsymbol{\beta}_0) = \frac{\partial \log P[(i)|H_i]}{\partial\beta_0}$$

has mean $\mathbf{0}$ and unconditional variance matrix

$$\mathcal{I}_i(\boldsymbol{\beta}_0) = E\left(\frac{-\partial^2 \log P[(i)|H_i]}{\partial\boldsymbol{\beta}_0\,\partial\boldsymbol{\beta}_0}\right).$$

Since the conditional mean $E(\mathbf{U}_i|H_i)$ is a zero vector it follows that the unconditional mean

$$E(\mathbf{U}_i) = E[E(\mathbf{U}_i|H_i)]$$

is also the zero vector. The mean vector of $\mathbf{U}(\boldsymbol{\beta}_0) = \mathbf{U}_1 + \cdots + \mathbf{U}_k$ is then zero as required. The variance matrix for $\mathbf{U}(\boldsymbol{\beta}_0)$ could be written

$$\mathcal{I}(\boldsymbol{\beta}_0) = \sum_{i=1}^{k} \mathcal{I}_i(\boldsymbol{\beta}_0)$$

provided the $\mathbf{U}_i$, $i = 1, \ldots, k$, are uncorrelated. But for $j < i$ the condition H_i implies that $\mathbf{U}_j$ is fixed, giving

$$E(\mathbf{U}_j\mathbf{U}_i') = E[\mathbf{U}_j E(\mathbf{U}_i'|H_i)] = 0$$

as required.

The intuitive requirements for asymptotic normality of $U(\boldsymbol{\beta}_0)$ and for consistency of $\hat{\boldsymbol{\beta}}$ are that the information quantities $\mathcal{I}_i(\boldsymbol{\beta}_0)$, $i = 1, \ldots, k$, should be not too disparate and that the scores $\mathbf{U}_i$, $i = 1, \ldots, k$, should possess "some degree of independence" (Cox, 1975) in addition to being uncorrelated. Although such requirements are likely to impose only mild conditions on the censoring mechanism and the covariate values, rigorous proofs of consistency of $\hat{\boldsymbol{\beta}}$ and asymptotic normality of $\mathbf{U}(\boldsymbol{\beta}_0)$ have been worked out only under somewhat strict conditions. The results of Tsiatis (1978) on this topic are discussed below. Because these proofs tend to be lengthy and notationally complex only an outline of the main ideas is given here.

Tsiatis assumes that values of a (scalar) regressor variable are generated from a density $f(z)$ such that $|z|$ is bounded by a constant $K < \infty$. Censoring times are assumed to be bounded and to arise from a density that, conditional on z, is independent of failure time.

The following argument shows consistency of $\hat{\boldsymbol{\beta}}$, under the above conditions. Since $\log L(\beta)$ is continuous and differentiable as a function of β, it takes a maximum on an interval $|\beta_* - \beta_0| \le \delta$, for any $\delta > 0$. If one can show that the maximum does not occur on the boundary with limiting probability one, then one has a local maximum at which $U(\beta) = 0$. A sequence of δ values that converge to zero then gives the desired consistency.

To show the maximum does not occur on the boundary consider the function

$$H(\beta) = n^{-1}E\left(\sum_{1}^{k} \log \sum_{l \in R(t_{(i)})} e^{z_l\beta}\right),$$

where expectations are taken with respect to the distribution at β_0. The first and second derivatives of $H(\beta)$ are, respectively,

$$H'(\beta) = n^{-1}E\left[\sum_1^k \left(\frac{\sum_{l\in R(t_{(i)})} z_l e^{z_l\beta}}{\sum_{l\in R(t_{(i)})} e^{z_l\beta}}\right)\right]$$

and $H''(\beta)$ obtained by differentiating inside the expectation once more. We note that

$$H'(\beta_0) = n^{-1}E\left(\sum_1^k z_{(i)}\right)$$

since $E[U(\beta_0)] = 0$. Thus for $\delta > 0$ and $|\beta_* - \beta_0| = \delta$ it follows that

$$(\beta_* - \beta_0)H'(\beta_0) < H(\beta_*) - H(\beta_0) \tag{4.50}$$

since $H''(\beta) > 0$. In addition, the strong law of large numbers shows the almost sure convergence

$$\left|\frac{\sum_1^k z_{(i)}}{n} - H'(\beta_0)\right| \to 0.$$

With somewhat more effort, Tsiatis (1978) has shown that with probability one,

$$\left|n^{-1}\sum_1^k \log \sum_{l\in R(t_{(i)})} e^{z_l\beta^*} - n^{-1}\sum_1^k \log \sum_{l\in R(t_{(i)})} e^{z_l\beta_0} - H(\beta_*) + H(\beta_0)\right| \to 0.$$

Substitution into (4.50) gives $\log L(\beta_*) < \log L(\beta_0)$ with limiting probability one, as required.

Returning to (4.48) the consistent estimation of $n^{-1}\mathcal{I}(\beta_0)$ by $-n^{-1}\partial^2 \log L(\beta_*)/\partial\beta_*\,\partial\beta_*$ for β^* between β_0 and $\hat\beta$ follows from the consistency of $\hat\beta$ and the boundedness of the third derivative of $\log L(\beta)$, under $|z| < K$.

The asymptotic normality of

$$n^{-1/2}U(\beta_0) = n^{-1/2}\left[\sum_1^k z_{(i)} - \sum_1^k \left(\frac{\sum_{l\in R(t_{(i)})} z_l e^{z_l\beta_0}}{\sum_{l\in R(t_{(i)})} e^{z_l\beta_0}}\right)\right] \tag{4.51}$$

remains to be shown. The right side of (4.51) can be rewritten

$$n^{-1/2}\left[\sum_1^n z_i I_{[\delta_i=1]} - \sum_{i=1}^n \left(\frac{I_{[\delta_i=1]}\sum_{l=1}^n I_{[t_l\ge t_i]} z_l e^{z_l\beta_0}}{\sum_{l=1}^n I_{[t_l\ge t_i]} e^{z_l\beta_0}}\right)\right], \tag{4.52}$$

where $I_{[\,]}$ denotes an indicator function with value 1 if the condition in brackets is met and value zero otherwise, and (t_i, δ_i) denotes the survival time (minimum of censoring and failure time) and censoring indicator for the ith study subject. The desired asymptotic normality of (4.52) follows from expanding (4.52) so that it is in the form of a linear combination of negligible remainder terms plus well-behaved integrals on the asymptotically Gaussian empirical processes:

$$n^{-1/2}\sum_{i=1}^{n} I_{[\delta_i=1,\, t_i\geq t]}, \qquad n^{-1/2}\sum_{i=1}^{n} I_{[\delta_i=1,\, t_i\geq t]}z_i,$$
$$n^{-1/2}\sum_{i=1}^{n} I_{[t_i\geq t]}e^{z_i\beta_0} \qquad \text{and} \qquad n^{-1/2}\sum_{i=1}^{n} I_{[t_i\geq t]}z_i e^{z_i\beta_0}. \tag{4.53}$$

The asymptotic normality of the finite dimensional distributions from (4.53) follows directly from the central limit theorem (in view of $|z|<K$), and tightness (Billingsley, 1968, p. 106) follows fairly easily under the condition $|z|<K$.

The results outlined above show that $\hat{\beta}$ has the asymptotic normal distribution with mean β_0 and variance consistently estimated by $I(\hat{\beta})$ at least under the conditions specified by Tsiatis. The proof for a vector regression parameter $\boldsymbol{\beta}$ would be similar. Liu and Crowley (1978) give similar results under slightly different conditions that require z to be discrete, with finite range. Further work to relax the conditions for the asymptotic likelihood results to hold, particularly in respect to the regression variable, would be useful.

4.8.2 The Survivor Function

The weak convergence of the empirical processes (4.53) to Gaussian processes can be used to show the convergence to a Gaussian process of the estimator

$$\hat{F}_0(t) = \exp[-\hat{\Lambda}_0(t)],$$

where

$$\hat{\Lambda}_0(t) = \sum_{i|t_{(i)}<t}\Bigg(\sum_{l\in R(t_{(i)})} e^{z_l\hat{\beta}}\Bigg)^{-1} \tag{4.54}$$

is the "cumulative hazard" estimator arising directly from the approach of Breslow (Section 4.2.4), or from the nonparametric maximum likelihood approach (4.24) following first order expansions of $\log\hat{\alpha}_i$, $i = 1, \ldots, k$.

A Taylor expansion of (4.54), again with scalar regression variable, gives

$$n^{1/2}\hat{\Lambda}_0(t) = n^{1/2}\sum_{t_{(i)}<t}\Bigg(\sum_{l\in R(t_{(i)})} e^{z_l\beta_0}\Bigg)^{-1} - n^{1/2}(\hat{\beta}-\beta_0)\sum_{t_{(i)}<t}\Bigg(\frac{\sum_{l\in R(t_{(i)})} z_l e^{z_l\beta_0}}{(\sum_{l\in R(t_{(i)})} e^{z_l\beta_0})^2}\Bigg)$$
$$+\, n^{1/2}(\hat{\beta}-\beta_0)^2 R(t, \beta_*). \tag{4.55}$$

Under the conditions of Tsiatis mentioned above, the first term in (4.55) and the coefficient of $n^{1/2}(\hat{\beta}-\beta_0)$ can be written as well-behaved integrals on the empirical processes (4.53). This along with the asymptotic normality of $\hat{\beta}$ and the convergence to zero of $R(t, \beta_*)$ shows $n^{1/2}\hat{\Lambda}_0(t)$ to be an asymptotically Gaussian process with mean

$$-\log F_0(t) = \int_0^t \lambda_0(u)\, du.$$

The asymptotic variance of $n^{1/2}\hat{\Lambda}_0(t)$, at specified t, is consistently estimated by

$$n \sum_{t_{(i)}<t} \left(\sum_{l\in R(t_{(i)})} e^{z_l\hat{\beta}} \right)^{-2} + \frac{n}{I(\hat{\beta})} \left[\sum_{t_i<t} \left(\frac{\sum_{l\in R(t_{(i)})} z_l e^{z_l\hat{\beta}}}{(\sum_{l\in R(t_{(i)})} e^{z_l\hat{\beta}})^2} \right) \right]^2.$$

It follows readily that $e^{\tilde{z}\hat{\beta}}\hat{\Lambda}_0(t)$, $\hat{F}_0(t)$, and $[\hat{F}_0(t)]^{\exp(\tilde{z}\hat{\beta})}$ will be asymptotically Gaussian for specified $\tilde{z}$. Similar results will hold for vector regression variables.

BIBLIOGRAPHIC NOTES

The proportional hazards model was first proposed by Cox (1972). In his formulation, he allowed the covariates **z** in the model to be functions of time and gave the partial likelihood (Cox, 1975) analysis of Section 4.2.3. This argument and the more general framework of time dependent covariates are discussed in more detail in Chapter 5. The marginal likelihood derivation of Section 4.2.2 was given by Kalbfleisch and Prentice (1973) which generalized, to censored data, a result due to Savage (1957). Breslow (1974) proposed the maximum likelihood approach of Section 4.2.4 and Thompson and Godambe (1974) also considered maximum likelihood methods in the two sample problem. Meier (1978, personal communication) has carried the maximum likelihood approach based on (4.21) through to obtain a maximized likelihood of $\boldsymbol{\beta}$. Review papers have been given by Breslow (1975), Ware and Byar (1979), and Holford (1976). Kay (1977) gives a review and an extensive analysis of a set of data collected to investigate the use of prednisone in the treatment of cirrhosis.

The discrete model analogue (4.30) to the proportional hazards model was proposed by Kalbfleisch and Prentice (1973), who derived the maximum likelihood estimate (4.24) of the survivor function from it. Cox (1972) made use of the discrete logistic model (4.12) both in the analysis of tied failure times and to obtain an estimate of the survivor function. This estimate differs slightly from that given in Section 4.3. Breslow (1974) estimates the survivor function by maximum likelihood techniques within his step function model for the hazard function. The problems of estimation based on data from the discrete model (4.30) have been considered by Prentice and Gloeckler (1978). The discrete model (4.12) has been considered by Thompson (1977). The log-rank test has been extensively discussed in the literature and references are given in the bibliographic notes for Chapter 6.

Efficiency properties of the estimation of $\boldsymbol{\beta}$ from the proportional hazards model have been considered by Kalbfleisch (1974), Kalbfleisch and McIntosh (1977), Efron (1977), and Oakes (1977). The latter two papers obtained general efficiency expressions and can usefully be read in parallel with the material in Section 4.7, which draws heavily on this work. Kay (1979) considers further efficiency properties. Little work has been done on small sample efficiency properties. Efficiency properties of the log-rank statistic have been considered by Crowley and Thomas (1975).

As mentioned above, the asymptotic arguments of Section 4.8 are closely related to those of Tsiatis (1978), which contain a much more detailed treatment both for the estimation of $\boldsymbol{\beta}$ and of the survivor function. Cox (1975) has given a heuristic argument which applies to the partial likelihood derivation. These arguments are discussed in Chapter 5.

CHAPTER 5

Likelihood Construction and Further Results on the Proportional Hazards Model

5.1 INTRODUCTION

The construction of the likelihood for censored failure time data given in Chapter 3 assumes independence between the censoring or failure times of different individuals and so is valid for the special, although important, case of a random censorship model. In addition, the covariates that have been considered to this point are measured and fixed in advance for each individual under study. In this chapter, the previous results are generalized to accommodate more general censoring schemes and covariates that vary with time. The construction of the likelihood in the parametric case is considered in Section 5.2 for independent censoring schemes; time dependent covariates are introduced in Section 5.3 and the likelihood construction is generalized to accommodate these. Partial likelihood, which was considered briefly in Chapter 4, is developed more thoroughly in Section 5.4 and its application to the proportional hazards model is discussed. The remainder of the chapter contains a numerical example and some comments on efficiency.

5.2 INDEPENDENT CENSORING AND LIKELIHOOD CONSTRUCTION

As in Section 3.2, consider n individuals to have been placed on test at time 0 and suppose that the risk of failure at time t is determined by the hazard $\lambda(t; \mathbf{z}, \boldsymbol{\theta})$ where $\mathbf{z}$ is, as usual, a vector of fixed covariates measured in advance and $\boldsymbol{\theta}$ is a vector of unknown parameters. The data for the ith individual are t_i, δ_i, $\mathbf{z}_i$ where t_i is the failure time ($\delta_i = 1$) or censored time ($\delta_i = 0$). In what follows, we suppose t_i to be continuous.

In Section 3.2, a derivation of the likelihood was given which assumed a random censorship model. It was noted there, however, that this model is not sufficiently general to encompass many censoring schemes (e.g., type II censoring) which are frequently used in some applications. In this section it is shown that the likelihood (3.2) given by

$$L = \prod_{i=1}^{n} [f(t_i; \mathbf{z}_i, \boldsymbol{\theta})^{\delta_i} F(t_i; \mathbf{z}_i, \boldsymbol{\theta})^{1-\delta_i}] \tag{5.1}$$

$$= \prod_{i=1}^{n} \lambda(t_i; \mathbf{z}_i, \boldsymbol{\theta})^{\delta_i} \exp\left[-\int_0^\infty \sum_{l \in R(t)} \lambda(u; \mathbf{z}_l, \boldsymbol{\theta})\, du\right] \tag{5.2}$$

in the continuous case is the appropriate likelihood under a very broad class of censoring mechanisms. As usual, $R(t)$ is the risk set at $t-0$.

For random censorship, the likelihood (5.1) or (5.2) can be derived by considering the independent contributions of $(t_i, \delta_i, \mathbf{z}_i)$, $i = 1, \ldots, n$. More generally, however, the likelihood is formed as the product of conditional contributions, of the whole study group, over successive infinitesimal time intervals. Thus we let $H(t)$ represent the complete history of the study up to time t so that $H(t)$ records all failure and censoring information as well as complete information on all covariates. Since $H(t)$ is a Markov process the likelihood can be constructed as a product of the conditional terms

$$\begin{aligned} P[H(t+dt)|H(t)] &= P[D_t(dt), C_t(dt)|H(t)] \\ &= P[D_t(dt)|H(t)]P[C_t(dt)|H(t), D_t(dt)] \end{aligned} \tag{5.3}$$

where $D_t(dt)$ and $C_t(dt)$ are the set of labels associated with individuals that have failed or are censored in $[t, t+dt)$, respectively. It then follows that

$$P[D_t(dt)|H(t)] = \prod_{l \in D_t(dt)} \lambda(t; \mathbf{z}_l, \boldsymbol{\theta})\, dt \prod_{l \in R(t)-D_t(dt)} [1 - \lambda(t; \mathbf{z}_l, \boldsymbol{\theta})dt] \tag{5.4}$$

where the following has been assumed:

1. Given $H(t)$, the failure mechanisms act independently over $[t, t+dt)$.
2. For each individual in $R(t)$, and conditional on covariates $\mathbf{z}$,

$$P\{\text{failure in } [t, t+dt)|H(t)\} = P\{\text{failure in } [t, t+dt)|\text{survival to } t\}. \tag{5.5}$$

Censorings are presumed to follow failures in the interval $[t, t+dt)$. Such a convention is necessary to deal with schemes like type II censoring where items are withdrawn when a certain failure occurs. In effect ties between censorings and failures are broken by placing failures first.

Assumption 2 above is a kind of conditional independence between censoring and failure mechanisms. Censoring mechanisms that satisfy this are called *independent* and include, for example, type I and II censoring along with progressive type II censoring or indeed any censoring rule that, when applied at time t, depends only on $H(t)$ and on random mechanisms external to the failure process. Essentially we require that, conditional on z, the items withdrawn from risk at time t should be "representative" of the items at risk. In particular, items cannot be censored because they appear to be at unusually high or low risk of failure.

The total likelihood can now be written as a product integral

$$L = P[H(0)] \mathop{\mathscr{P}}_{0}^{\infty} P[H(t+dt)|H(t)]$$
$$= P[H(0)] \exp\left\{\lim \sum_{i=1}^{m} \log P[H(\tau_i + \Delta\tau_i)|H(\tau_i)]\right\}$$

where $0 = \tau_0 < \cdots < \tau_m < \infty$, $\Delta\tau_i = \tau_i - \tau_{i-1}$, $\tau_m \to \infty$ as $m \to \infty$, and the limit is taken as $m \to \infty$ and $\Delta\tau_i \to 0$. Using (5.3) and assuming $P[H(0)] = 1$, we have

$$L = \mathop{\mathscr{P}}_{0}^{\infty} P[D_t(dt)|H(t)] \mathop{\mathscr{P}}_{0}^{\infty} P[C_t(dt)|H(t), D_t(dt)].$$

The first factor on the right side arises from the failure information and, apart from differential elements, reduces to

$$\prod_{i=1}^{n} \lambda(t_i; \mathbf{z}_i, \boldsymbol{\theta})^{\delta_i} \mathop{\mathscr{P}}_{0}^{\infty} \sum_{l \in R(t)-D_t(dt)} [1 - \lambda(t; \mathbf{z}_l, \boldsymbol{\theta})\, dt].$$

This reduces further to (5.2) and thence to (5.1).

The remaining factor in L is due to the censoring contributions and can be written

$$\mathop{\mathscr{P}}_{0}^{\infty} P[C_t(dt)|H(t), D_t(dt)]. \tag{5.6}$$

If (5.6) depends on $\boldsymbol{\theta}$, the censoring mechanism is said to be *informative*, and otherwise *noninformative*. If the censoring is informative, the likelihood (5.1) is not complete since it ignores the censoring contributions. Even in this case, however, (5.1) has the interpretation of a partial likelihood (see Section 5.4) and so can still be used for inference though there will be an associated loss in efficiency. This distinction between informative and noninformative censoring schemes is similar to the distinction made in stopping rules as discussed, for example, in Cox and Hinkley (1974, p. 40).

It is instructive to consider some specific censoring schemes in the above construction. For example, suppose it is decided to censor at the kth failure or time T_0, whichever occurs first. The censoring scheme is independent and, even though the marginal censoring probabilities depend upon θ, it is noninformative. This can be seen by considering the contributions to (5.6). Since the censoring scheme, given $H(t)$, is deterministic, each of these contributions is unity. Realistic examples of informative, but independent, censoring schemes are hard to construct. An artificial example is one in which the censoring time for each individual is determined as the failure time of a similar individual who is not included in the test. Censoring schemes that are not independent arise if items are selectively censored when they appear to be at high or low risk of failure. For example, the censoring might be allowed to depend on a time dependent covariate $x(t)$ which measures the health status of the individual at time t. If such censoring is in effect, the fact that an item is not censored will alter the conditional failure probabilities in (5.4) and violate assumption 2 in (5.5).

Not a great deal is known about asymptotic properties of the maximum likelihood estimator $\hat{\boldsymbol{\theta}}$ in the presence of general independent censoring mechanisms. The arguments for the partial likelihood in Section 5.4.2 suggest, however, that the usual asymptotic results hold under fairly mild conditions on the hazard and covariates $\mathbf{z}$.

An alternative derivation of the likelihood (5.2) which does not introduce the censoring explicitly arises by considering the total hazard of failure at time t,

$$N(t, \boldsymbol{\theta}) = \sum_{l \in R(t)} \lambda(t; \mathbf{z}_l, \boldsymbol{\theta})$$

as the failure rate at time t in a nonhomogeneous Poisson process. Standard arguments then show that the probability of no failures in (t_{i-1}, t_i) is

$$\exp\left[-\int_{t_{i-1}}^{t_i} N(t, \boldsymbol{\theta})\, dt\right].$$

The probability that the item with covariate $\mathbf{z}_i$ fails at t_i is $\lambda(t_i; \mathbf{z}_i, \boldsymbol{\theta})\, dt_i$. The likelihood is then built up as a product of these terms. This argument, given by Efron (1977), is essentially equivalent to the product integral derivation given above. The latter argument, however, justifies the conditioning on $R(t)$ at each time point t that is required in specifying the Poisson rate function.

5.3 TIME DEPENDENT COVARIATES AND FURTHER REMARKS ON LIKELIHOOD CONSTRUCTION

In this section, conditions on the covariates $\mathbf{z}$ are relaxed to allow them to vary over time. Thus $\mathbf{z}_i(t)$ denotes the covariate vector at time t for the

ith individual under study. It is convenient to introduce also $Z_i(t)$ to denote the covariate path up to time t, $\{\mathbf{z}_i(u); 0 < u < t\}$, and Z_i to denote the whole covariate process to the cessation of testing. The data for the ith individual are $(t_i, \delta_i, Z_i(t_i))$, $i = 1, \ldots, n$.

Time dependent covariates fall into two broad classifications. An *external* covariate is one that is not directly involved with the failure mechanism, and an *internal* covariate is a time measurement taken on the individual. The internal covariate has the property that it requires the survival of the individual for its existence and thus carries with its observed path information on the failure time variate.

5.3.1 External Covariates

One type of external covariate is the *fixed* covariate whose value is measured in advance and fixed for the duration of study. Up to this point attention has been directed entirely to fixed or time independent covariates. A second type of external covariate is *defined* in that its total path Z_i, although not constant, is determined in advance for each individual on study. One example would be a stress factor under control of the experimenter that is to be varied in a predetermined way. Another example would be the age of an individual in a trial of long duration. A third type of external covariate is termed *ancillary*. A covariate of this sort is the output of a stochastic process that is external to the individual under study and has the property that the marginal probability distribution of Z_i, $i = 1, \ldots, n$ does not involve the parameters of the failure time model. An example of such a covariate would be one that measures airborne pollution as a predictor for the frequency of asthma attacks. Ancillary covariates play the role of ancillary statistics for the failure time model and the conditionality principle would suggest conditioning on their whole observed path Z.

Each of fixed, defined, and ancillary covariates can thus be incorporated into the general framework of Section 5.2 by using such conditioning. The hazard function is defined as

$$\lambda(t; Z, \theta)dt = P\{T \in [t, t + dt) | Z, \theta, T \geq t\} \tag{5.7}$$

and the survivor function is

$$F(t; Z, \theta) = P[T \geq t | Z, \theta].$$

It can be seen that the usual relationships between survivor and hazard functions hold. It follows that (5.1) or (5.2) is the appropriate likelihood under independent censoring where $\mathbf{z}_i$, in these expressions and in their derivations, is replaced with Z_i.

Ancillary covariates are characterized by the condition

$$P[Z(t+dt)|H(t)] = P[Z(t+dt)|Z(t)] \tag{5.8}$$

which is a formalization of the idea that the path of the covariate process may influence, but is not influenced by, the failure experience of the trial.

As a simple example of a defined covariate, suppose that a voltage $\exp[z_i(t)]$, where $z_i(t) = a_i \log t$ for some $a_i > 0$, is to be applied at time t to the ith item in a test of insulation in electrical cable. Suppose further that the failure rate at time t for the ith item is the exponential regression model

$$\lambda(t; Z_i, \beta) = \exp[\beta_0 + \beta_1 z_i(t)].$$

It is then easily seen that

$$\lambda(t; Z_i, \beta) = \alpha t^{\gamma_i - 1}$$

of the Weibull form where $\alpha = \exp(\beta_0)$ and $\gamma_i = a_i\beta_i + 1$. To some extent, such covariates offer nothing new but merely give a regression interpretation to certain parameters in the model. The log likelihood function on data (t_i, δ_i, z_i) for independent censoring is, from (5.1),

$$\sum \delta_i[\beta_0 + \beta_1 z_i(t_i)] - \int_0^\infty \sum_{l \in R(u)} \exp[\beta_0 + \beta_1 z_l(u)]\, du \tag{5.9}$$

which may also be written as the product of Weibull densities and survivor functions. If the voltage applied at time t were randomly determined by some external process, essentially the same model would apply after conditioning and (5.9) would be the appropriate likelihood.

5.3.2 Internal Covariates

An internal covariate is the output of a stochastic process that is generated by the individual under study and so is observed only so long as the individual survives and is uncensored. In consequence, its observed value carries information about the survival time of the corresponding individual. A simple example arises in a clinical trial where some measure of a patient's general condition is made at regular intervals. Suppose at time t values of 0 and 4 are assigned to $z(t)$ for dead and no clinical evidence of disease, respectively, and 1, 2, 3 represent intermediate levels of decreasing disability. A patient typically moves from one state to another over time and the hazard at time t depends markedly on $Z(t)$. A second example would arise in an immunotherapy trial in cancer. In such a trial, it may be of interest to examine the effect of immunotherapy on the failure rate given a current measure of immune status such as white blood count. In this case, the covariate $z(t)$ may be taken to specify white blood count at time t and models could be constructed to evaluate treatment effects while adjusting for the current white blood count or to allow treatment effects to depend on current white blood count.

It is clear that such covariates must be handled differently than ancillary or defined covariates since **Z** determines the survival information for the corresponding individual. Accordingly, for internal covariates, the hazard function is defined by

$$\lambda(t; Z(t), \boldsymbol{\theta})dt = P\{T \in [t, t+dt) | Z(t), \boldsymbol{\theta}, T \geq t\} \tag{5.10}$$

which conditions on the covariate process up to time t, but not further. This hazard bears no relationship to a survivor function. Indeed, for a covariate $z(t)$ like general condition above,

$$P[T \geq t | Z(t), \theta] = 1$$

provided $z(t)$ does not indicate that failure has occurred.

The formal construction of the likelihood proceeds, in this case, in essentially the same way as in Section 5.2. The probability contribution of the interval $[t, t+dt)$ is written as

$$P[H(t+dt)|H(t)] = P[D_t(dt)|H(t)]P[\mathscr{X}(t+dt)|H(t), D_t(dt)] \times P[C_t(dt)|H(t), D_t(dt), \mathscr{X}(t+dt)]. \tag{5.11}$$

Here, $H(t)$ contains, in addition to the failure and censoring information, all information on the covariates up to time t while $\mathscr{X}(t)$ contains only the covariate information. The product integral of the first factor on the right side of (5.11) gives, as before, the likelihood (5.2) under assumptions similar to (5.5). In this case, (5.1) is not a legitimate expression in that F and f do not have survivor and density function interpretations. The other factors are

$$\mathop{\mathscr{P}}_{0}^{\infty} P[\mathscr{X}(t+dt)|H(t), D_t(dt)] \tag{5.12}$$

and

$$\mathop{\mathscr{P}}_{0}^{\infty} P[C_t(dt)|H(t), D_t(dt), \mathscr{X}(t+dt)] \tag{5.13}$$

which correspond to the contributions of the covariate processes and the censoring, respectively. If either (5.12) or (5.13) depends on the parameters θ in the model $\lambda(t; Z(t), \theta)$, the likelihood (5.2) is a partial likelihood (see Section 5.4). It can still be used for inference but may be seriously inefficient. It should be noted that the likelihood (5.2) gives information only on the instantaneous failure rate given $Z(t)$. Inferences about the marginal distribution of T would require an integration over the distribution of $\mathbf{z}(t)$, or a model for failure time in which $\mathbf{z}(t)$ is suppressed.

In a therapeutic trial, it should be kept in mind that an internal covariate process generally takes its values subsequent to the treatment assignment. As a consequence, such covariates may be "responsive" in that their values may be influenced by the treatment assignment. A conditional analysis gives information on instantaneous failure rates given the covariate values and the treatment assignment. If the effect of treatment is predominantly reflected in the covariate process, such an analysis will show little or no treatment differences. This can give useful information about the mechanism by which the treatment operates. Care must be exercised in interpretation, however, since treatment differences could be large in spite of the negative results of this analysis. Suppose, for example, that the covariate for general condition defined above is being used and that all individuals begin in the same state ($z(0) = 2$, say). If one treatment decelerates the passage through the levels of $z(t)$ but the failure rate within each state is the same for both treatments, an analysis conditional on $z(t)$ shows no treatment difference. It is still possible, however, that the one treatment is greatly superior to the other but this effect is predominantly through $z(t)$.

A censoring scheme that depends on the level of a time dependent covariate $z(t)$ (e.g., general condition) is, as noted before, not independent if $z(t)$ is not included in the model. One way to circumvent this is to include $z(t)$ in the model, but this may mask treatment differences of interest. An alternative method is to evaluate the effect of treatment directly on $z(t)$ instead of using the failure end point.

5.3.3 Other Types of Time Dependent Covariates

Although in what follows, time dependent covariates are taken to be either internal or external, other possibilities do exist. Consider again the voltage stress test of insulation in electrical cable but allow $z(t)$, the voltage applied at time t, to be adjusted according to the history $H(t)$ of the trial to that point. This covariate is not defined nor is it ancillary in that the distribution of $z(t)$ involves the parameters $\boldsymbol{\theta}$ in the model for the hazard. In this case, one could define the hazard

$$\lambda(t; Z(t+dt), \boldsymbol{\theta})\, dt = P\{T \in [t, t+dt) | Z(t+dt), \boldsymbol{\theta}, T \geq t\}$$

which is equivalent to (5.7) for the ancillary case. The likelihood derivation proceeds as in Section 5.2 but the factorization (5.3) of $P[H(t+dt)|H(t)]$ is replaced with

$$P[\mathscr{X}(t+dt)|H(t)]P[D_t(dt)|H(t), \mathscr{X}(t+dt)]P[C_t(dt)|H(t), \mathscr{X}(t+dt), D_t(dt)] \tag{5.14}$$

where $H(t)$ contains, in addition to the failure and censoring information, all information on the covariates up to time t while $\mathscr{X}(t)$ contains the information on the covariates up to time t. In the voltage example the product integral of the first term of (5.14) is free of θ. In effect, $Z(t)$ is, at each time and conditional on $H(t)$, conditionally ancillary for θ. The usual likelihood (5.2) arises as the product integral of the second term in (5.14) if the censoring is independent. For the voltage conditionally determined as above, the log likelihood (5.9) is clearly still valid.

5.4 PARTIAL LIKELIHOOD

5.4.1 Partial Likelihood Construction

As was remarked in Chapter 4, the partial likelihood analysis of the proportional hazards model applies in more general situations than are considered there. In particular, if the covariates are time dependent, the partial likelihood is still available and this immediate application of the method is discussed in Section 5.5. The general method of partial likelihood was proposed by Cox (1975) and although a brief discussion of the method is given here, the reader is referred to that paper for more detail. The partial likelihood, like the marginal likelihood, has as its purpose the development of techniques to make useful inference in the presence of many nuisance parameters.

Suppose that the data consist of a vector of observations $\mathbf{y}$ from the density $f(\mathbf{y}; \boldsymbol{\theta}, \boldsymbol{\beta})$, $\boldsymbol{\beta}$ is the vector of parameters of interest, and $\boldsymbol{\theta}$ is a nuisance parameter and typically of very high or infinite dimension. In some applications, $\boldsymbol{\theta}$ is in fact a nuisance function as, for example, the hazard function $\lambda_0(\cdot)$ in the proportional hazards model. Suppose now that the data $\mathbf{Y}$ are transformed into a set of variables $A_1, B_1, \ldots, A_m, B_m$ in a one to one manner and let $A^{(j)} = (A_1, \ldots, A_j)$ and $B^{(j)} = (B_1, \ldots, B_j)$. Suppose that the joint density of $A^{(m)}, B^{(m)}$ can be written

$$\prod_{j=1}^{m} f(b_j|b^{(j-1)}, a^{(j-1)}; \boldsymbol{\theta}, \boldsymbol{\beta}) \prod_{j=1}^{m} f(a_j|b^{(j)}, a^{(j-1)}; \boldsymbol{\beta}). \tag{5.15}$$

The second term is called that *partial likelihood* of $\boldsymbol{\beta}$ based on $\mathbf{A}$ in the sequence (A_j, B_j). The number of terms m could be random or fixed.

In certain applications one may argue that any information on $\boldsymbol{\beta}$ in the first term is inextricably tied up with information on the nuisance parameters $\boldsymbol{\theta}$, and as a simplification we might take for inference the simpler second term which involves only $\boldsymbol{\beta}$. There is in general no mathematical argument leading to a conclusion of marginal or conditional sufficiency of the sequence $A_j|B^{(j)}, A^{(j-1)}$, $j = 1, \ldots, m$, as group invariance supplies in the marginal case; in many situations, however, heuristic arguments can be put forward which suggest that little is lost through ignoring the first term.

It should be noted that the partial likelihood is not a likelihood in the ordinary sense. In fact,

$$L(\boldsymbol{\beta}) = \prod_{j=1}^{m} f(a_j|b^{(j)}, a^{(j-1)}; \boldsymbol{\beta}) \tag{5.16}$$

cannot in general, be given a direct probability interpretation as either a conditional or a marginal probability statement.

The general formulation of partial likelihood becomes clearer on consideration of a simple example. This example is redundant in the sense that a more general version of it has already been given in Section 4.2.4. This discussion is given simply to clarify further the concept of partial likelihood.

Suppose that the survival times for two samples of size n_1 and n_2 are observed to fall in one of three disjoint intervals $[0, c_1)$, $[c_1, c_2)$, $[c_2, \infty)$ with frequencies given in Table 5.1. Suppose that conditional upon entry of interval i, the odds of failure in that interval are γ_i for sample 0 and $\gamma_i e^{\beta}$ for sample 1. This is a simple example of the log-linear logistic model for a binary response. A more general form of this model is given by (4.13), which leads to the partial likelihood (4.14) and thence to the Mantel–Haensel or log-rank test of Sections 1.4 and 4.2.5. The full data set is equivalent to the set A_1, B_1, A_2, B_2, A_3, B_3 where A_i is the number of failures in sample 1 in the ith interval and B_i gives the total number of failures in interval i. The conditioning event $B^{(j)} = b^{(j)}$, $A^{(j-1)} = a^{(j-1)}$ is equivalent to conditioning on the row and column totals for the jth table. Thus the jth term in the partial likelihood (from 5.15) is obtained as the familiar conditional density of A_j given the row and column totals. Hence

$$P(A_1 = a_1|b_1) = \frac{\binom{n}{a_1}\binom{m}{b_1 - a_1} e^{a_1\beta}}{\sum \binom{n}{l}\binom{m}{b_1 - l} e^{l\beta}}, \tag{5.17}$$

$$P(A_2 = a_2|n - a_1, m - r_1, b_2) = \frac{\binom{n - a_1}{a_2}\binom{m - r_1}{b_2 - a_2} e^{a_2\beta}}{\sum \binom{n - a_1}{l}\binom{m - r_1}{b_2 - l} e^{l\beta}} \tag{5.18}$$

$$P(A_3 = n - a_1 - a_2|n - a_1 - a_2, m - r_1 - r_2, b_3) = 1.$$

Table 5.1 GROUPED SURVIVAL DATA FOR TWO SAMPLES

	$[0, c_1)$			$[c_1, c_2)$			$[c, \infty)$		
Sample	Fail	Surv.	Total	Fail	Surv.	Total	Fail	Surv.	Total
1	a_1	$n - a_1$	n	a_2	$n - a_1 - a_2$	$n - a_1$	$a_3 = n - a_1 - a_2$	0	$n - a_1 - a_2$
0	r_1	$m - r_1$	m	r_2	$m - r_1 - r_2$	$m - r_1$	$r_3 = m - r_1 - r_2$	0	$m - r_1 - r_2$
	b_1			b_2			b_3		

The partial likelihood of β is proportional to the product of (5.17) with (5.18). Note that this product cannot be interpreted as a conditional probability statement.

This example is easily extended to more general situations involving grouped survival data. Censoring at the ends of intervals is easily handled by allowing B_i to specify both the total number of failures in interval i and the numbers of individuals in each sample with survival times censored at the end of the $(i-1)$th interval. The general result for this case is given by the partial likelihood (4.14). If losses were to occur during the intervals, an exact partial likelihood would not be available. Various approximations could be introduced to handle this situation (Cox, 1975).

In this example, the part of the data ignored by the partial likelihood is the information about β contained in, for example, the total number of failures in interval j given the number entering the interval on each treatment group. That is the information in the term

$$f(b_j|b^{(j-1)}, a^{(j-1)}; \gamma_j, \beta). \tag{5.19}$$

It seems intuitively clear that the total number of failures can contribute little information about any differences between the two failure probabilities. This parallels the argument for the conditional test for independence in a single 2×2 contingency table and the related conditional estimation of the log odds ratio β. Formalization of the argument that no information on β is lost through conditioning on the column total is difficult. An interesting contribution in this direction is due to Barndorff-Nielson (1973). An interesting example with paired data which indicates that, at least in some instances, the column totals do contain information is due to Sprott (1975a).

A second example of the use of partial likelihood arises in the likelihood construction of Section 5.2. If the censoring mechanism is informative, then the likelihood (5.1) still has the interpretation of a partial likelihood based on the sequence $\{D_t(dt)\}$ given $H(t)$.

5.4.2 Asymptotic Results for Partial Likelihood

It is remarkable that the general method of construction of the partial likelihood leads to the usual asymptotic properties of likelihood based inferences under regularity conditions that are evidently quite mild.

Following Cox (1975) we consider the efficient scores

$$\mathbf{U}_j = \frac{\partial \log f(A_j|H_j; \boldsymbol{\beta})}{\partial \boldsymbol{\beta}} \qquad j = 1, 2, \ldots, m$$

where $H_j = (B^{(j)}, A^{(j-1)})$ is used to specify the conditioning variables for the jth term in (5.16). The total score for the partial likelihood is

$$\mathbf{U} = \sum_{j=1}^{m} \mathbf{U}_j = \frac{\partial \log L}{\partial \boldsymbol{\beta}}$$

and the usual asymptotic results would follow, under additional mild conditions, if it could be established that a central limit theorem applies to $\mathbf{U}$. Since $f(a_j|h_j;\boldsymbol{\beta})$ is a density, then under the usual regularity conditions we have

$$E(U_j|H_j = h_j) = 0$$

and hence,

$$E(\mathbf{U}_j) = \mathbf{0}.$$

Further, if $j < k$, the condition $H_k = h_k$ implies that U_j is fixed. Hence

$$E(\mathbf{U}_k|\mathbf{U}_j = \mathbf{u}_j) = \mathbf{0}, \qquad j < k$$

so that unconditionally, the score contributions are uncorrelated,

$$E(\mathbf{U}_j\mathbf{U}_k') = 0, \qquad j \neq k.$$

Finally,

$$\text{var}(\mathbf{U}_j) = E(\mathbf{U}_j\mathbf{U}_j') = \mathcal{I}_j$$

say, and for the total score $\mathbf{U} = \Sigma\, \mathbf{U}_j$ we have

$$E(\mathbf{U}) = \mathbf{0}, \qquad \text{var}\left(\sum \mathbf{U}_j\right) = \sum \mathcal{I}_j.$$

Thus the total score $\mathbf{U}$ is the sum of uncorrelated variables with expectation $\mathbf{0}$. The intuitive conditions for a central limit theorem to apply as $m \to \infty$ are that the score contributions $\mathbf{U}_j$ should exhibit a certain degree of independence, the $\mathcal{I}_j$ should not be too disparate, and $\Sigma\, \mathcal{I}_j$ should approach infinity at a suitable rate. Under these conditions, the observed information matrix from the partial likelihood, $I(\boldsymbol{\beta}) = (-\partial^2 \log L/\partial\boldsymbol{\beta}\, \partial\boldsymbol{\beta})$, can replace $\Sigma\, \mathcal{I}_j$ in the asymptotic results since $I(\boldsymbol{\beta}) \times (\Sigma\, \mathcal{I}_j)^{-1}$ converges to an identity matrix as $m \to \infty$ (assuming $\Sigma\, \mathcal{I}_j$ is of full rank). If, in addition, $\hat{\boldsymbol{\beta}}$ is consistent for $\boldsymbol{\beta}$, it follows that $I(\hat{\boldsymbol{\beta}})$ can replace $\Sigma\, \mathcal{I}_j$. This leads to the usual asymptotic result that $\mathbf{U}$ is asymptotically normal with mean $\mathbf{0}$ and covariance matrix estimated by $I(\hat{\boldsymbol{\beta}})$. Additional mild conditions on the third partial derivatives of the log partial likelihood now yield, from a Taylor series expansion, that $\hat{\boldsymbol{\beta}}$ is asymptotically normal with mean $\boldsymbol{\beta}$ and covariance estimated by $I(\hat{\boldsymbol{\beta}})^{-1}$, as is true for the maximum likelihood estimator in the usual case. The usual asymptotic results for likelihood ratio tests based on the partial likelihoods are a direct consequence. The above argument is only an outline and more work is needed to determine weak conditions on the score which are sufficient for the asymptotic results. Little work has been done on conditions that ensure the consistency of the maximum likelihood estimator from the partial likelihood.

The above argument provides some justification for the use of asymptotic properties of partial likelihood when the likelihood is composed of many terms m of which each contributes a small amount of information on $\boldsymbol{\beta}$. This is the situation with its application to the continuous proportional hazards model discussed in the next section and in Section 4.2.3. Sometimes, however, the value m is fixed and as the sample size increases each score statistic is based on a large amount of data. This is the case, for example, with the grouped failure time example of the last section. In that case, a central limit theorem often applies to each score vector $\mathbf{U}_j$ individually and hence to their sum. The same asymptotic results as given above hold in this case also.

5.5 APPLICATIONS TO THE PROPORTIONAL HAZARDS MODEL

5.5.1 General

In Section 4.1, the regression model with hazard function

$$\lambda(t; \mathbf{z}) = \lambda_0(t)e^{\mathbf{z}\boldsymbol{\beta}}$$

was analyzed from several different viewpoints for fixed covariates. In the original formulation, Cox (1972) proposed that the covariates be allowed to be time dependent. Accordingly, we consider a vector $\mathbf{z}(t) = (\mathbf{y}(t), \mathbf{x}(t))$ of time dependent covariates where $\mathbf{y}(t)$ is the vector of all fixed, defined or ancillary covariates and $\mathbf{x}(t)$ is the vector of all internal covariates. Let $Y = \{\mathbf{y}(u), u > 0\}$ denote the complete covariate function over the whole study period for the external covariates and $X(t) = \{\mathbf{x}(u), 0 < u < t\}$ denote the covariate process up to time t of the internal covariates. Let $Z(t) = (Y, X(t))$. In the special, and usual, case where no internal covariates are present, $\mathbf{z}(t) = \mathbf{y}(t)$ and $Z(t)$ specifies the full covariate path. The hazard function is defined by

$$\lambda(t; Z(t))\, dt = P\{T \in [t, t + dt) | Z(t), T \geq t\},$$

and as a special case of interest, we suppose

$$\lambda(t; Z(t)) = \lambda_0(t)e^{\mathbf{z}(t)\boldsymbol{\beta}} \tag{5.20}$$

in which the hazard at t depends only on the current value, $\mathbf{z}(t)$.

As in Section 4.1 we consider the sample to consist of k failure times $t_{(1)} < t_{(2)} < \cdots < t_{(k)}$ and ignore for the moment the case of ties. The remaining $n-k$ observations are right censored. Let $\mathbf{z}_{(j)}(t)$ be the value of $\mathbf{z}(t)$ for the item failing at $t_{(j)}$ and let $\mathbf{z}_l(t)$ be the value for the lth item at the same elapsed time t. In the notation of the last section, we let B_j specify the censoring and covariate information in $[t_{(j-1)}, t_{(j)})$ plus the information that an individual fails at $t_{(j)}$, A_j specifies the particular individual that fails. In this case, the jth term in the partial likelihood (5.16) is

$$L_j(\boldsymbol{\beta}) = f(a_j|b^{(j)}, a^{(j-1)})$$
$$= \frac{\lambda(t_{(j)}; Z_{(j)}(t_{(j)}))}{\sum_{l \in R(t_{(j)})} \lambda(t_{(j)}; Z_l(t_{(j)}))}.$$

For the proportional hazards model (5.20), this gives the partial likelihood

$$L(\boldsymbol{\beta}) = \prod_{j=1}^{k} \left(\frac{\exp[\mathbf{z}_{(j)}(t_{(j)})\boldsymbol{\beta}]}{\sum_{l \in R(t_{(j)})} \exp[\mathbf{z}_l(t_{(j)})\boldsymbol{\beta}]} \right) \tag{5.21}$$

which is of the same form as before. The inclusion of ties in the data set can be accounted for by allowing B_j to include the multiplicity of failures at $t_{(j)}$; A_j specifies the particular items that fail. The result of this calculation is the partial likelihood with general term given by (4.14) except that now the covariables are time dependent. Again, a suitable approximation to the log partial likelihood, if the ties are not too frequent, is

$$\sum_{j=1}^{k} \left\{ \mathbf{s}_j(t_{(j)})\boldsymbol{\beta} - d_j \log \sum_{l \in R(t_{(j)})} \exp[\mathbf{z}_l(t_{(j)})\boldsymbol{\beta}] \right\} \tag{5.22}$$

where $\mathbf{s}_j(t_{(j)})$ is the sum of covariates $\mathbf{z}_l(t_{(j)})$ over the d_j individuals failing at $t_{(j)}$. Formulas for first and second derivatives of (5.22) are immediate generalizations of the results given in Section 4.2 (see expressions (4.9) and (4.10)).

The partial likelihood (5.21) ignores the contributions to the likelihood (or partial likelihood) (5.2) that are made by the process $K(t)$ that records the number of failures up to time t. Thus the term in (5.11) that generates the likelihood (5.2) is being factored as

$$P[D_t(dt)|H(t), \mathscr{X}(t)] = P[D_t(dt)|H(t), \mathscr{X}(t), K(t+dt)]$$
$$\times P[K(t+dt)|H(t), \mathscr{X}(t)].$$

The product integral of the first term on the right is the partial likelihood (5.21) assuming no ties. The product integral of the second term gives the conditional contributions of the process $K(t)$ to the likelihood (5.2). This second term is informative if a particular parametric form for $\lambda_0(\cdot)$ is specified. Without any restriction on $\lambda_0(\cdot)$, however, the observed process

$K(t)$ can be accounted for by a hazard near 0 in the intervals containing no failures and large at the observed failure times. In this case, it would appear that little information about $\boldsymbol{\beta}$ is lost through ignoring the second term.

The estimation of the survivor function

$$F_0(t) = \exp\left[-\int_0^t \lambda_0(u)\,du\right]$$

can also be generalized to the case where the covariates $\mathbf{z}$ are time dependent. However, it would usually be meaningful to do this only if the covariates were fixed, defined, or ancillary. With internal covariates, the function $F_0(t)$ has no simple interpretation. The estimation can be accomplished in the same way as was done in Section 4.3 for fixed covariates. The details of the derivations are left to the reader, but the results parallel those given in (4.24) and (4.25) with the covariates in the ith term of each expression evaluated at $t_{(i)}$.

Some uses of internal covariates are discussed in Chapter 7 in relation to the competing risks problem and the analysis of multivariate failure time data. The remainder of this section discusses some particular uses of defined or ancillary covariates.

5.5.2 Defined Covariates in the Proportional Hazards Model

As noted earlier, defined covariates arise in a straightforward way when a stress variable is varied in a predetermined way over time. If the stress applied at time t, $z(t)$, acts multiplicatively on the hazard then this fits naturally into the proportional hazards framework.

An important use of defined covariates is in checking the proportional hazards assumption within a model involving only fixed covariates. Consider, for example, a two sample problem and let $z_1 = 0, 1$ denote the sample. Suppose the conditional hazard to be

$$\lambda_0(t) \exp[z_1\beta_1 + z_2(t)\beta_2] \tag{5.23}$$

where $z_2(t) = z_1 \log t$ is a defined time dependent covariate. A test of $\beta_2 = 0$ provides a check of the proportional hazards model for the levels of z_1 versus one in which the hazard ratio of sample 1 to sample zero is increasing ($\beta_2 > 0$) or decreasing ($\beta_2 < 0$) with time. Various other types of alternatives could be checked by changing the definition of $z_2(t)$.

The model (5.23) was fitted to the carcinogenesis data in order to check the proportionality assumptions of Chapter 4. On examining the second derivatives of the log partial likelihood, it becomes clear that the estimates of β and γ are highly correlated owing to lack of centering of the

$\log t_{(i)}$. Accordingly, the model is replaced by

$$\lambda_0(t)\exp[z_1\beta_1' + z_1(\log t - c)\beta_2]$$

where c is taken to be the average of the $\log t_{(i)}$'s so that $\beta_1' = \beta_1 + \beta_2 c$. If such an adjustment is not made, then the high correlation between $\hat{\beta}_1$ and $\hat{\beta}_2$ makes maximization of the log partial likelihood difficult.

Convergence is reached in four iterations from initial values $\beta_1' = \beta_2 = 0$ and one obtains the estimates $\hat{\beta}_1' = -.599$, $\hat{\beta}_2 = -.230$, with the estimated covariance matrix

$$I^{-1} = \begin{pmatrix} .1211 & .0487 \\ .0487 & 3.3303 \end{pmatrix}.$$

A test of $\beta_2 = 0$ based on the asymptotic distribution of $\hat{\beta}_2$ gives a standard normal statistic $-.230/\sqrt{3.3303} = -.1258$ which is not significant. There is no evidence to suggest inadequacy of the proportional hazards assumption (at least in the direction of an increasing or decreasing hazard ratio over time).

The immediate extension of this procedure gives a check on the proportional hazards assumption in the model for fixed covariates, $\lambda_0(t)\exp(\mathbf{z}\boldsymbol{\beta})$. The procedure involves specification of a more general model

$$\lambda_0(t)\exp[\mathbf{z}\boldsymbol{\beta} + \mathbf{x}(t)\boldsymbol{\gamma}]$$

where $x_i(t) = z_i g_i(t)$, with $g_i(t)$ some specified function of t, and γ_j is taken to be zero except for components of $\mathbf{z}$ for which the proportionality assumption is being questioned. Tests of the hypothesis $\boldsymbol{\gamma} = \mathbf{0}$, or that particular components of $\boldsymbol{\gamma}$ are zero, provide the checks on proportionality. Further investigation of this procedure is needed; in particular, little is known about the efficiency of such tests although one particular instance is discussed in Section 5.6.

5.5.3 Ancillary Covariates and the Stanford Heart Transplant Data

An interesting data set that can be modeled either as a bivariate survival distribution or by using time dependent covariables is the heart transplant data taken from Crowley and Hu (1977) and reproduced in Appendix 1. A brief description of these data is given in Section 1.2.

From the time of admission to study until the time of death a patient was eligible for a heart transplant. The time to transplant (the waiting time) is denoted by W and the time to death by T. One possibility would be to consider T and W as variables representing the times to two different types of failure. Of interest is a comparison of survival experience of transplanted and nontransplanted patients. More specifically, let

$$\lambda_1(t)dt = P\{T \in [t, t+dt) | T \geq t, W \geq t\}$$

and, for $t > w$,

$$\lambda_1(t|w)dt = P\{T \in [t, t+dt) | T \geq t, W = w\}.$$

Then interest centers on the comparison of these two hazard functions. In order to make such comparisons, it is essential that there be no selection in the assignment of hearts to individuals; that is, the assignment must be randomly made to eligible individuals. The randomization may be restricted by taking account of covariates included in the model or of covariates not correlated with subsequent survival (in the absence of transplantation). The donor–recipient tissue matching variables would presumably be variables used in the assignment, and it must be supposed that they fall into the latter category. Since assignments were not made at random, the possibility of selection exists and results should be viewed with some caution.

One possible approach to the analysis of such data is to specify particular parametric forms for the hazards, $\lambda_1(t)$ and $\lambda_1(t|w)$. Alternatively, one might adopt nonparametric models such as

$$\lambda_1(t) = \lambda^*(t); \qquad \lambda_1(t|w) = \lambda^{**}(t), \qquad t > w \tag{5.24}$$

where the hazard is affected by transplantation, but no further role is played by the waiting time w. Either approach could be modified to incorporate covariates. It should be noted that in the nonparametric case, an assumption like (5.24) is necessary. Otherwise, $\lambda_1(t|w)$ will require separate estimation at each value of w and this could be handled, only with sufficient data, by grouping on the w variable.

Let $\mathbf{z}$ be a vector of s fixed, defined, or ancillary covariates and consider the model [a generalization of (5.24)] with hazards

$$\begin{aligned} \lambda_1(t|\mathbf{z}) &= \lambda^*(t)e^{\mathbf{z}\boldsymbol{\beta}_1} \\ \lambda_1(t|w, \mathbf{z}) &= \lambda^{**}(t)e^{\mathbf{z}\boldsymbol{\beta}_2}, \qquad t > w \end{aligned} \tag{5.25}$$

where $\boldsymbol{\beta}_1$ and $\boldsymbol{\beta}_2$ are vectors of regression parameters measuring, respectively, the effect of $\mathbf{z}$ on the marginal and conditional hazards. Note that different covariates can be included in the two parts of (5.25). For example, if age is to appear in the first hazard but not the second, we merely take the coefficient in $\boldsymbol{\beta}_2$ corresponding to age to be zero. The parameters $\boldsymbol{\beta}_1$ and $\boldsymbol{\beta}_2$ can be estimated by partial likelihood. Let $t_{(1)} < \cdots < t_{(k)}$ be the observed failure times of the k individuals dying prior to transplantation. Denote by $R_1(t_{(i)})$ the set of labels associated with individuals not transplanted and at risk at $t_{(i)}-0$. The failure times of individuals who have received a transplant are $t'_{(1)}<\cdots<t'_{(m)}$ and $R_2(t'_{(j)})$ denotes the set of labels associated with individuals who have been transplanted and are at risk at $t'_{(j)}-0$. We let $\mathbf{z}_{1(i)}$ denote the covariate vector for the individual failing at $t_{(i)}$ and $\mathbf{z}_{2(j)}$ denote the covariate vector for the individual failing at $t'_{(j)}$. Considering the situation at $t_{(i)}$ and conditioning on $R_1(t_{(i)})$ and that one nontransplanted patient fails at $t_{(i)}$, we find the ith term in the partial likelihood for $\boldsymbol{\beta}_1$ to be

$$\frac{\exp(\mathbf{z}_{1(i)}\boldsymbol{\beta}_1)}{\sum_{l\in R_1(t_{(i)})}\exp(\mathbf{z}_l\boldsymbol{\beta}_1)}$$

with a similar term contributing to the partial likelihood of $\boldsymbol{\beta}_2$ at each $t'_{(j)}$. The log partial likelihood (if there are no ties) is then

$$L(\boldsymbol{\beta}_1,\boldsymbol{\beta}_2)=\sum_{i=1}^{k}\left\{\mathbf{z}_{1(i)}\boldsymbol{\beta}_1-\log\left[\sum_{l\in R_1(t_{(i)})}\exp(\mathbf{z}_l\boldsymbol{\beta}_1)\right]\right\}$$
$$+\sum_{j=1}^{m}\left\{\mathbf{z}_{2(j)}\boldsymbol{\beta}_2-\log\left[\sum_{l\in R_2(t'_{(j)})}\exp(\mathbf{z}_l\boldsymbol{\beta}_2)\right]\right\}.$$

Crowley and Hu (1977) considered a model similar to (5.18) but with proportional hazards. The hazard for a nontransplanted patient is taken as $\lambda^*(t)\exp(\mathbf{z}\boldsymbol{\beta}_1)$ whereas for a transplanted patient it is

$$\lambda^*(t)\exp(\mathbf{z}\boldsymbol{\beta}_1+\beta_0+\mathbf{z}\boldsymbol{\beta}_2).$$

The hazard can then be written

$$\lambda^*(t)\exp[\mathbf{z}\boldsymbol{\beta}_1+H(t-w)(\beta_0+\mathbf{z}\boldsymbol{\beta}_2)] \tag{5.26}$$

which is of the time dependent variety with $2s+1$ regression variables given by $y_0, y_1, \ldots, y_{2s}$ where

$$y_i = z_i, \qquad i=1,\ldots,s.$$
$$y_0(t)=H(t-w)$$
$$y_{i+s}(t)=H(t-w)z_i, \qquad i=1,\ldots,s.$$

and $H(t-w)$ is the Heaviside function. The methods of Section 5.2 are now available and the parameters $\boldsymbol{\beta}_1$, β_0, and $\boldsymbol{\beta}_2$ can be estimated through their joint partial likelihood. In this model, $\beta_0+\mathbf{z}\boldsymbol{\beta}_2$ measures the effect of transplantation at the covariate value $\mathbf{z}$ and $\boldsymbol{\beta}_2$ can be viewed as measuring the interactions between transplantation and the components of $\mathbf{z}$. No absolute interpretation can be given β_0 since its meaning is affected by location changes in the covariates $\mathbf{z}$. The assumption is being made in the above discussion that a main effect for z_i has been estimated in β_{1i}. If a covariate is included in the time dependent part and not in the constant part, the interpretation of its coefficient is unclear; its value is influenced both by a dependence of survival overall and by a dependence of post-transplant survival on that variable. Thus such variables as waiting time or matching variables are difficult to interpret except as post-transplant prognostic factors; there is no comparable control group. It might, however, be reasonable to assume that the matching variables are not in themselves correlated with subsequent survival experiences and would have an effect only when combined with transplantation.

The model (5.26) was fitted to the transplantation data. In order to remove ambiguity in the data the following conventions were adopted: ties between waiting times and failure times were broken conservatively by placing the failure first except in the one case where the patient died on the day of transplant. As usual, ties between censored times and failure times were broken by placing failure times first. Patients that were deselected (taken off study before transplantation because of improved prognosis) were treated as censored at the time of removal. Patients lost to follow-up were treated in the same way.

Table 5.2 gives estimates of the regression coefficients and estimated standard errors for a number of models that were fitted to these data. As well as those reported here, a number of models using measures of mismatch were fitted, but none of these measures was found to correlate with subsequent survival.

The most remarkable feature of these data is the dependence of survival time on the time of acceptance to the study. The main effect of year of acceptance is significant at less than the 5% level in all the models in which it is included. Unfortunately, the year of acceptance interaction with transplantation is also approaching significance but in the opposite direction. Together, these suggest that the overall quality of patient being admitted to the study is improving with time (possibly due to relaxation of admission requirements or to improving patient management) but the survival time of the transplanted patients is not improving at the same rate. In fact, the sum of the two coefficients for year of acceptance would suggest a nearly constant survival pattern for the transplanted patients. (This was the finding of Crowley and Hu, who fitted a model that included no main effect term but only an interaction term.) It may be possible to explain such a dependence on time of admission if there were some measure of general condition of the patients placed on study. An examination of such a variable could determine whether the selection for transplant moved in the direction of the poorer risk patients as time progressed while the overall general condition of patients improved.

If it is assumed that the interaction of year of acceptance and transplantation is zero, models can be fitted which involve only the main effect term for z_2. From model 5 one finds that a test that the age interaction

Table 5.2 REGRESSION COEFFICIENTS AND (ESTIMATED) STANDARD ERRORS FOR MODELS OF THE FORM (6.21) FITTED TO THE HEART TRANSPLANT DATA

	Main Effects ($\boldsymbol{\beta}_1$)			Transplant Status	Interactions ($\boldsymbol{\beta}_2$)		
Model	z_1	z_2	z_3	β_0	z_1	z_2	z_3
1	.0138 (.018)		−.546 (.611)	.118	.035 (.027)		−.291 (.758)
2		−.265 (.105)		−.282		.135 (.140)	
3	.0155 (.017)	−.274 (.105)		−.588	.033 (.028)	.201 (.142)	
4		−.254 (.107)	−.236 (.628)	−.292		.164 (.141)	−.550 (.775)
5	.0150 (.018)	−.135 (.071)	−.420 (.615)	.077	.027 (.027)		−.298 (.758)
6	.0152 (.018)	−.136 (.071)	−.621 (.368)	.048	.027 (.027)		

Estimated covariance matrix for model 6:

$$I^{-1} = 10^{-4}\begin{pmatrix} 3.06 & -.41 & -2.30 & -3.03 & -19.39 \\ -.41 & 50.27 & -50.99 & 2.85 & 14.12 \\ -2.30 & -50.99 & 1353.28 & 5.40 & -34.43 \\ -3.03 & 2.85 & 5.40 & 7.37 & 15.07 \\ -19.39 & 14.12 & -34.43 & 15.07 & 1037.97 \end{pmatrix}$$

z_1 = age of acceptance − 48; z_2 = year of acceptance − 1900; z_3 = surgery (1 = yes, 0 = no).

term is zero is nonsignificant ($p = .32$ on a two tailed test). If the interaction is set to zero and only a main effect is estimated, this coefficient is also not significantly different from zero. If, however, the hypothesis that both the main effect β_{11} and the interaction β_{21} are zero is considered, we find, from the asymptotic normal distribution of the estimate,

$$\begin{pmatrix} .0150 \\ .027 \end{pmatrix}' \begin{pmatrix} .018^2 & -.000306 \\ -.000306 & .027^2 \end{pmatrix}^{-1} \begin{pmatrix} .0150 \\ .027 \end{pmatrix} = 4.75$$

which when compared with the $\chi^2_{(2)}$ tables gives a significance level of .09. This effect is further suggested in that the hypothesis $\beta_{11} + \beta_{12} = 0$ is rejected at the 5% level. It would seem then that the effect of age is complicated and that a final model should include both β_{11} and β_{21}. It would be dangerous to fit a model with only the β_{21} term with no main effect estimated since $\hat{\beta}_{11}$ and $\hat{\beta}_{21}$ are negatively correlated and setting β_{11} to zero will lead to an overestimate of β_{21}.

Surgery is an easier variable. There is no evidence of an interaction between surgery and transplantation. It is further the case that there is no evidence against the hypothesis $\beta_{13} = \beta_{23} = 0$. In fitting a model with just the main effect (model 6) it is found that the estimate $\hat{\beta}_{13} = -.621$ with estimated standard deviation .368 and the data are significant with reference to the hypothesis $\beta_{13} = 0$ at the 10% level on a two tailed test. There is then mild evidence of a better prognosis overall for patients who have had previous heart surgery.

This analysis would suggest that transplantation is beneficial for younger patients. The critical age is $48 + c$, with c estimated by $\hat{c}$ where $.048 + \hat{c}.027 = 0$ or $\hat{c} = -1.8$. Thus it might be argued that for patients under 46 years of age transplantation is beneficial but for those over 46, there is no evidence of such benefit as far as survival is concerned. Note, however that Fieller's theorem could be used to place an approximate confidence limit on c. The 95% confidence interval for c is the set of solutions to

$$\frac{|.048 + .027c|}{\sqrt{.10397 + .003014c + .00737c^2}} \leq 1.96.$$

This gives the interval $(-\infty, \infty)$, which would suggest that c is estimated with considerable imprecision. This rather surprising result is due to the fact that the estimation of c is asymptotically equivalent to the estimation of the ratio of two normal means. The interval $(-\infty, \infty)$ arises since the data on the denominator mean are not significantly different from 0.

The above conclusions based on the model (5.26) are in close agreement with those reached when (5.25) is used. The difficulty with (5.25) is that there is no easily used measure of the difference between the transplanted and nontransplanted groups. Conclusions must be based on comparisons of the estimated survivor functions.

5.6 EFFICIENCY OF THE REGRESSION ESTIMATOR

Assuming conditions hold for the asymptotic likelihood results as outlined in Section 5.4, one can proceed to study the (asymptotic) efficiency of $\hat{\boldsymbol{\beta}}$. The notation has been selected so that the same expressions as in Section 4.6 apply to the likelihood, the likelihood derivatives, and the information matrices $\mathcal{I}_r$ and $\mathcal{I}_p$ in the presence of defined or ancillary time dependent covariates. In fact the efficiency results of both

Efron (1977) and Oakes (1977) purport to cover time dependent covariates in general, though both authors implicitly exclude the possibility of internal covariates in their specification of the likelihood function.

The efficiency discussion of Section 4.6 then carries over directly to defined or ancillary time dependent covariates. Specifically, if the true parametric model has hazard $\lambda_0(t)\exp[\mathbf{z}(t)\boldsymbol{\beta}]$ where $\lambda_0(t) = h_0(t)\exp[z_{s+1}(t)\beta_{s+1} + \cdots + z_{s+m}(t)\beta_{s+m}]$ with $h_0(t)$ and $z_i(t)$, $i = s+1, \ldots, s+m$ completely specified, a loss of asymptotic efficiency results if the elements of the limit of

$$\frac{\boldsymbol{\eta}_1(t)}{\eta(t)} = \frac{E\{\sum_{i\in R(t)} \mathbf{z}_i(t)\exp[\mathbf{z}_i(t)\boldsymbol{\beta}]\}}{E\{\sum_{i\in R(t)} \exp[\mathbf{z}_i(t)\boldsymbol{\beta}]\}}$$

are not well approximated by a linear function of $z_{s+1}(t), \ldots, z_{s+m}(t)$. If $m = 1$ and $z_{s+1}(t) = 1$ so that the parametric model involves only a scale parameter, a loss in efficiency corresponds to variation over time in $\boldsymbol{\eta}_1(t)/\eta(t)$. As with fixed covariates, such variations can arise because of regression coefficients that differ from zero or because of a dependence of the censoring on $\mathbf{z}$. In addition, in the current setting, variation in this expectation function may arise because of changes in the distribution of values of the time dependent covariates over time. For example, in the above heart transplant illustration, the expected z value for the transplant indicator would range from zero at the time of acceptance into the transplant program to a value that approaches one as all surviving patients in the program receive hearts.

Kalbfleisch and McIntosh (1977) numerically investigated the efficiency of the regression coefficient $\hat{\beta}_2$ where z_1 is a fixed covariate distinguishing two samples and $z_2 = z_1 \log t$ is a time dependent covariate that permits the hazard ratio to increase or decrease in a smooth manner over time. Their results, under Weibull sampling with no censoring, indicate that efficiency losses for samples of size 50 are typically not severe for testing $\beta_2 = 0$ (e.g., .94 at $\beta_1 = 0$) though the efficiency declines somewhat as $|\beta_1|$ increases (e.g., .69 at $e^{\beta_1} = 3$). Efficiency losses for $\hat{\beta}_2$ at $\beta_2 \neq 0$ may be greater or less than those at $\beta_2 = 0$ depending on whether β_1 and β_2 values combine to give a more, or less, dramatic variation in the expected value of $z_2(t)$ for failures, as a function of time.

With internal time dependent covariates the efficiency properties of the partial likelihood estimator can evidently be very poor in circumstances in which virtually all the failure time "information" is carried by the covariate process. A characterization of situations in which the partial likelihood has reasonable efficiency properties is necessary before the routine use of this partial likelihood can be advocated.

BIBLIOGRAPHIC NOTES

The likelihood construction for general independent censoring mechanisms is an extension of the work of Cox (1975). He generates the partial likelihood construction for censored data and defines and discusses independent censoring. Other general constructions have been given by Efron (1977) and Cornfield and Detre (1977) by considering the failure mechanism to be a nonhomogeneous Poisson process where at each time t the risk set $R(t)$ is taken as given. Kalbfleisch and MacKay (1978b) introduce the role of censoring explicitly as is done in Section 5.2.

Cox (1972) suggests the use of time dependent covariates in proportional hazards regression models. The partial likelihood analysis of these models is given by Cox (1972) and more formally by Cox (1975). The categorization of time dependent covariates is discussed by Prentice and Kalbfleisch (1979) and, with regard to the likelihood construction, by Kalbfleisch and MacKay (1978b). The particular application of time dependent covariates to the heart transplant data was considered by Crowley and Hu (1977), who amplified a remark by Breslow (1975).

The heart transplant data have been considered by several authors. A detailed account of the clinical trial from which the data arose is given by Clark et al. (1971). Gail (1975) gives a critical view of their findings. The analyses most closely related to those considered in Section 5.5 are those by Crowley (1974a), Mantel and Byar (1974), and especially Crowley and Hu (1977). Crowley (1974a) generalizes the log-rank test to accommodate such time varying covariates as transplant status. Other papers dealing with models and analyses of these data are by Brown et al. (1974) and Turnbull et al. (1974).

Williams and Lagakos (1977) consider a random censorship model with no covariates and define a constant sum relationship between the failure and censoring mechanisms. They show that under this condition the likelihood (5.1) is appropriate. Kalbfleisch and MacKay (1979) show the constant sum condition to be equivalent to the condition for independent censoring in Section 5.2.

CHAPTER 6

Inference Based on Ranks in the Accelerated Failure Time Model

6.1 INTRODUCTION

In the comparison of two or more samples with uncensored data, linear rank tests have often been proposed as alternatives to parametric tests. Although the rank tests themselves are derived with certain alternatives in mind for which optimum parametric procedures exist, they generally possess greater efficiency robustness than the corresponding parametric tests and are generally less sensitive to outliers. In addition, for testing the null hypothesis, these tests generally involve only a small loss in efficiency compared to the parametric procedure when such a procedure is appropriate.

During the past decade, censored data generalizations of some popular rank tests have been developed. One example of such a test has already been met in the form of the log-rank test or generalized Savage test of Sections 1.4 and 4.2.5. In addition, generalized forms of the Wilcoxon and Kruskal–Wallis tests (Gehan, 1965a; Breslow, 1970) have frequently been used for survival comparisons in clinical trials. Such tests can be conveniently generated from the accelerated failure time model of Section 2.3.3.

The accelerated failure time model specifies that the effect of the fixed covariate $\mathbf{z}$ is to act multiplicatively on the failure time T or additively on $Y = \log T$. A linear modeling gives

$$y = \alpha + \mathbf{z}\boldsymbol{\beta} + \sigma e \tag{6.1}$$

with error density $f(e)$. As usual, $\mathbf{z}$ is a row vector of dimension s and $\boldsymbol{\beta}$ is a column vector of regression coefficients. The linear rank statistics and their censored counterparts are derived in Sections 6.3 and 6.4 as score function tests based on data from the model (6.1). Methods of estimation of $\boldsymbol{\beta}$ based on these procedures are also considered for both the censored and uncensored cases, though numerical procedures are complicated and generally require further development.

The rank procedures developed here offer alternatives to the parametric procedures based on (6.1) and discussed in Sections 3.6 and 3.8. These alternatives should be less sensitive to misspecification of f or to extreme observations in the sample. Moreover, even though the procedures of Chapter 4 based on the proportional hazards model may well be regarded as the current methods of choice, these methods may be criticized on the basis of robustness and efficiency. Samuels (1977) shows that the possible influence of an outlying $\mathbf{z}$ on the proportional hazards estimator $\hat{\boldsymbol{\beta}}$ can be of any magnitude whatsoever. In addition, as is shown in Section 4.7, the efficiency of this estimator decreases as $\boldsymbol{\beta}$ departs from zero. This is a situation unfamiliar to ordinary linear regression; the rank procedures of this chapter retain their efficiency properties for all values of $\boldsymbol{\beta}$, though, as in Chapter 4, some efficiency is lost if censoring rates depend on $\mathbf{z}$.

This chapter begins with a brief discussion of linear rank statistics and their uses in regression problems. Some particular tests for both uncensored and censored data are introduced and exemplified. This is done in Section 6.2, which contains no formal development of the test procedures. The remainder of this chapter gives derivations of the procedures and considers their extensions to the more general inference problem of estimation. The reader who wishes only an acquaintance with the rank tests for censored data may wish to read only Section 6.2 and move then to Chapter 7. The derivations of the remainder of this chapter are not used in either Chapter 7 or Chapter 8.

6.2 SOME EXAMPLES: LINEAR RANK TESTS AND THEIR CENSORED DATA COUNTERPARTS

6.2.1 General

Let $y_1 = \log t_1, \ldots, y_n = \log t_n$ be an uncensored sample of log failure times with corresponding covariates $\mathbf{z}_1, \ldots, \mathbf{z}_n$ where $\mathbf{z}_i$ is a vector of time independent (fixed) covariates for the ith individual. Let $y_{(1)} < \cdots < y_{(n)}$ be the order statistic with corresponding covariates $\mathbf{z}_{(1)}, \ldots, \mathbf{z}_{(n)}$. A linear rank statistic is one of the form

$$\mathbf{v} = \sum_{1}^{n} \mathbf{z}'_{(i)} c_i \qquad (6.2)$$

where c_i is a score attached to the ith ordered sample value and we take $\Sigma\, c_i = 0$ for convenience. It is convenient to think of the data as arising from the accelerated failure time model (6.1) and the scores are chosen, by methods discussed in Section 6.3, to be efficient for certain specifications of the error density $f(e)$. We consider the null hypothesis $\boldsymbol{\beta} = \mathbf{0}$ for which the failure times are unrelated to the covariates and so are independent and identically distributed. Under this hypothesis, the mean and variance of $\mathbf{v}$ can be obtained by consideration of the permutation distribution of the rank labels $(1), \ldots, (n)$. Let E_p denote expectation over this permutation distribution. Then

$$E_p(\mathbf{v}) = \sum_{i=1}^{n} c_i E_p(\mathbf{z}'_{(i)})$$
$$= \sum_{i=1}^{n} c_i \bar{\mathbf{z}}' = \mathbf{0}$$

where $\bar{\mathbf{z}} = \Sigma\, \mathbf{z}_i / n$. The covariance matrix is

$$V = E_p(\mathbf{v}\mathbf{v}') = \sum \sum c_i c_j E_p(\mathbf{z}_{(i)}\mathbf{z}'_{(j)})$$
$$= \sum_{i=1}^{n} c_i^2 E_p(\mathbf{z}_{(i)}\mathbf{z}'_{(i)}) + \sum_{i \neq j} c_i c_j E_p(\mathbf{z}_{(i)}\mathbf{z}'_{(j)})$$

Now $\Sigma_{i \neq j}\, c_i c_j = -\Sigma\, c_i^2$ and $E_p(\mathbf{z}_{(i)}\mathbf{z}'_{(j)}) = [n(n-1)]^{-1}(n^2 \bar{\mathbf{z}}\bar{\mathbf{z}}' - \Sigma\, \mathbf{z}_i \mathbf{z}'_i)$ so that

$$V = (n-1)^{-1} \sum c_i^2 Z'Z \tag{6.3}$$

where Z is the $n \times s$ matrix of regression vectors with columns standardized to add to zero.

The use of (6.2) in a significance test involves calculation of probabilities over the permutation distribution. Thus if $\mathbf{v} = \mathbf{v}_0$ is observed, an exact computation would involve summing the probabilities of all outcomes $\mathbf{v}$ for which $\mathbf{v}'V^{-1}\mathbf{v} \geq \mathbf{v}'_0 V^{-1}\mathbf{v}_0$. If only one covariate is included, this procedure is equivalent to evaluating $P(|v| \geq |v_0|)$. Except in the simplest of problems, this computation is very laborious. Fortunately for most scoring procedures, an accurate approximation is obtained by comparing $\mathbf{v}'V^{-1}\mathbf{v}$ with the $\chi^2_{(s)}$ distribution assuming $Z'Z$ is nonsingular. This approximation is discussed further in later sections.

In the special case where $s+1$ samples are being compared, $\mathbf{z}$ is a vector of 0, 1 indicators for s of the samples so that $\mathbf{v}$ records the sum of the scores c_i for each of the s samples. In the simplest case, $s = 1$, (6.2) gives the sum of the scores for sample 1 and the variance, from (6.3), is $m_0 m_1 n^{-1}(n-1)^{-1} \Sigma\, c_i^2$ where m_0, m_1 are the sample sizes. The log-rank or Savage exponential scores test (4.18) is one example. It has previously been derived as a score function test in the proportional hazards model but also arises from the accelerated failure time model (6.2) when the error distribution is extreme value. In this case (6.1) corresponds to a model for failure time with proportional hazards. The test statistic is of the form (6.2) with scores

$$c_i = \frac{1}{n} + \frac{1}{n-1} + \cdots + \frac{1}{n-i+1} - 1, \qquad i = 1, \ldots, n. \tag{6.4}$$

in which $c_i + 1$ is the expected value of the ith order statistic in a sample of size n from the unit exponential. The covariance matrix is obtained by direct substitution in (6.3).

The Wilcoxon test provides another example of a linear rank test for which the scores are

$$c_i = 2i(n+1)^{-1} - 1, \qquad i = 1, \ldots, n. \tag{6.5}$$

This test, as will be seen, is the optimum rank test if the error distribution in (6.1) is logistic. The particular case of this test which corresponds to the comparison of $s+1$ samples is known as the Kruskal–Wallis test.

The generalization of these tests to the censored data situation is conceptually simple. Suppose that $y_{(1)} < y_{(2)} < \cdots < y_{(k)}$ are the observed log failure times and that $y_{i1}, \ldots, y_{im_i}$ are censored values in $[y_{(i)}, y_{(i+1)})$, $i = 0, \ldots, k$, where $y_{(0)} = -\infty$ and $y_{(k+1)} = \infty$. We consider test statistics of the form

$$\sum_{i=1}^{k} \left(c_i \mathbf{z}'_{(i)} + \sum_{j=1}^{m_j} C_i \mathbf{z}'_{ij} \right) \tag{6.6}$$

where c_i and C_i are the scores for the uncensored and censored data points, respectively. Note that all censored data points in the interval $[y_{(i)}, y_{(i+1)})$ receive the same score C_i. Variance formulas are in this case difficult and are discussed in detail in Section 6.4. In this section we consider only the formulas for the generalized exponential scores and Wilcoxon tests.

For the exponential scores test, the appropriate scores are obtained from the score function at $\boldsymbol{\beta} = \mathbf{0}$ in the proportional hazards model (see Section 4.2.5). We find that

$$c_i = \sum_{j=1}^{i} n_j^{-1} - 1, \qquad C_i = \sum_{j=1}^{i} n_j^{-1} \tag{6.7}$$

where n_j is the number of items at risk at $t_{(j)} - 0$. The test statistic (6.6) can then be written as

$$\mathbf{v} = \sum_{i=1}^{k} \left(n_i^{-1} \sum_{l \in R(t_{(i)})} \mathbf{z}'_l - \mathbf{z}'_{(i)} \right)$$

which is the negative of the score statistic from the proportional hazards model [see (4.19)]. The appropriate estimate of the variance can also be obtained as the variance estimate for the log-rank statistic as in (4.10) and can be written

$$V_0 = \sum_{i=1}^{k} n_i^{-1} \prod_{l \in R(t_{(i)})} (\mathbf{z}_l - \bar{\mathbf{z}}_i)(\mathbf{z}_l - \bar{\mathbf{z}}_i) \tag{6.8}$$

where $\bar{\mathbf{z}}_i = n_i^{-1} \sum_{l \in R(t_{(i)})} \mathbf{z}_l$ is the average of the covariates of items at risk at $t_{(i)} - 0$. Thus V_0 is the sum of covariance matrices of the covariates of items at risk at each failure time.

The Wilcoxon test can be generalized in a similar way. The scores are

$$c_i = 1 - 2\prod_{j=1}^{i}\left(\frac{n_j}{n_j+1}\right), \qquad C_i = 1 - \prod_{j=1}^{i}\left(\frac{n_j}{n_j+1}\right). \tag{6.9}$$

Variance estimates are again available and the result is given in (6.21) following its derivation.

6.2.2 Illustration

Before proceeding with a detailed development of these rank tests and their generalizations, we give an example of their uses with censored and uncensored data.

An extremely simple illustration can be based on the data of Table 6.1. Data are given there for the log failure times Y for two samples and we consider testing the null hypothesis of no sample difference ($\beta = 0$ in the accelerated failure time distribution). Following the data, the next two lines in the table give the Wilcoxon and exponential ordered scores for

Table 6.1 SCORES FOR THE COMPARISON OF TWO UNCENSORED SAMPLES

	Individual								
	1	2	3	4	5	6	7	8	9
Log failure time									
Sample 1	2.1	4.7	6.8		7.9	8.6			
Sample 2				7.5			8.9	9.2	9.3
Wilcoxon scores	−.8	−.6	−.4	−.2	0	.2	.4	.6	.8
Exponential ordered scores	−.889	−.746	−.621	−.454	−.254	−.004	.329	.829	1.829
Individuals 4, 5 Censored									
Wilcoxon scores	−.80	−.60	−.40	.30	.30	−.12	.16	.44	.72
Exponential ordered scores	−.889	−.746	−.621	.379	.379	−.596	−.266	.334	1.334

each of the nine individuals entered. The last two lines in the table give the scores when two of the items are censored and can, for the moment, be ignored.

Let $z = 0$ for observations in sample 1 and $z = 1$ for sample 2. The Wilcoxon test statistic has value

$$\begin{aligned} v &= -.2 + .4 + .6 + .8 \\ &= 1.6. \end{aligned}$$

Each of the 9! possibilities is equally likely under $\beta = 0$ so that the (two sided) significance level of the hypothesis is the number of sequences for which $|v| \geq 1.6$ divided by 9!. Because Wilcoxon scores are symmetric this probability can be written

$$P(|v| \geq 1.6) = 2\left(\frac{4!\,5!\,(4)}{9!}\right) = .063.$$

Sample sizes are too small to expect a normal approximation to be warranted. For comparative purposes, however, the Wilcoxon test variance is

$$\begin{aligned} V &= (8)^{-1} \sum_1^9 \left(\frac{2i}{10} - 1\right)^2 \sum_1^9 (z_i - \bar{z})^2 \\ &= \tfrac{2}{3}. \end{aligned}$$

Thus $1.6(\frac{2}{3})^{-1/2} = 1.96$ is a realization on an "approximate" standard normal statistic under $\beta = 0$. The associated significance level of .05 is in surprisingly good agreement with the exact test.

Similarly the exponential scores test statistic has value

$$\begin{aligned} v &= -.454 + .329 + .829 + 1.829 \\ &= 2.53. \end{aligned}$$

The associated exact significance level is $P(|v| \geq 2.53) = 4(4!)(5!)/(9!) =$ .032 and again in good agreement, the approximate standard normal statistic has value 1.93.

In order to illustrate the censored versions of the test, we suppose that items 4 and 5 are censored. The last two rows of Table 6.1 give the revised scores. The censored Wilcoxon statistic has the value

$$v = .30 + .16 + .44 + .72 = 1.62$$

and the estimated variance is, from (6.21), $V_0 = .810$.

Although sample sizes are very small, for purposes of illustration one might compare the observed value $v/\sqrt{V_0} = 1.80$ with the $N(0, 1)$ distribution to obtain an approximate significance level of 7%. The exponential ordered scores gives, in this case,

$$v = .379 - .266 + .334 + 1.334 = 1.781.$$

The estimated variance, from (6.8) is $V_0 = 1.030$. Again, $v/\sqrt{V_0} = 1.75$ is in close agreement with the Wilcoxon test.

Additional examples of these procedures are given in Sections 6.6 and in 8.5.

6.2.3 Stratification

A useful extension of the rank test procedures is provided when the study individuals are divided into strata on some auxiliary variable and comparisons are to be made on $\mathbf{z}$ which is the variable of primary interest. For example, patients in a clinical trial may be subdivided on age and then the treatments (given by $\mathbf{z}$) compared within each age group. For either the log-rank or the Wilcoxon test discussed above, we can let $\mathbf{v}_j$ be the test statistic, from (6.2) or (6.6), computed within the jth stratum, and V_{0j} is the corresponding covariance matrix. The total score is then $\Sigma\, \mathbf{v}_j$ with covariance matrix estimated by $\Sigma\, V_{0j}$, and the stratified test statistic is

$$\left(\Sigma\, \mathbf{v}_j\right)'\left(\Sigma\, V_{0j}\right)^{-1}\left(\Sigma\, \mathbf{v}_j\right)$$

which again can be compared with a $\chi^2_{(s)}$, where it has been assumed that $\Sigma\, V_{0j}$ is nonsingular.

6.3 RANK REGRESSION WITH UNCENSORED DATA

6.3.1 Test Statistics

Suppose that an uncensored sample of size n has been observed from the accelerated failure time model (6.1) with $f(e)$ of known form and consider the problem of testing the hypothesis $\boldsymbol{\beta} = \boldsymbol{\beta}_0$. The methods in the preceding section were for the case $\boldsymbol{\beta}_0 = \mathbf{0}$. We consider here the hypothesis $\boldsymbol{\beta} = \boldsymbol{\beta}_0$ instead of simply $\boldsymbol{\beta} = \mathbf{0}$ since the treatment of the former is no more difficult and it allows a direct extension to problems of estimation.

Let $w = y - \mathbf{z}\boldsymbol{\beta}_0$ represent the residuals about the hypothesized value. From (6.1) we have that

$$w = \alpha + \sigma \mathbf{z}\boldsymbol{\gamma} + \sigma e$$

where $\gamma = \sigma^{-1}(\boldsymbol{\beta} - \boldsymbol{\beta}_0)$ and the hypothesis $\boldsymbol{\beta} = \boldsymbol{\beta}_0$ is equivalent to the hypothesis $\boldsymbol{\gamma} = \mathbf{0}$. Let $w_{(1)} < \cdots < w_{(n)}$ represent ordered residuals with corresponding regression vectors $z_{(1)}, \ldots, z_{(n)}$, respectively. As in Chapter 4, the rank vector $\mathbf{r} = \mathbf{r}(\mathbf{w})$ is given by the corresponding labels $(1), \ldots, (n)$. That is,

$$\mathbf{r} = [(1), (2), \ldots, (n)]$$

and the rank vector probability can be computed as

$$P(\mathbf{r}) = \int \cdots \int_{\tau_{(1)} < \cdots < \tau_{(n)}} \prod_1^n f(\tau_{(i)} - \mathbf{z}_{(i)}\boldsymbol{\gamma})\, d\tau_i \tag{6.10}$$

where $\tau_{(i)} = (w_{(i)} - \alpha)/\sigma$. This is the same calculation as was used in the marginal likelihood derivation in the proportional hazards model and (6.10) might be viewed as a marginal likelihood of $\boldsymbol{\gamma}$ based on the ranks. Note that (6.10) is independent of α, σ which is in contrast to the fully parametric likelihood for $\boldsymbol{\gamma}$. A locally most powerful rank test of $\boldsymbol{\gamma} = \mathbf{0}$, or equivalently of $\boldsymbol{\beta} = \boldsymbol{\beta}_0$, can be based on the score statistic from (6.10); see, for example, Puri and Sen (1971, p. 108) and Cox and Hinkley (1974, p. 188). Straightforward calculation gives

$$\begin{aligned} \mathbf{v} &= \left.\frac{d \log P(r)}{d\boldsymbol{\gamma}}\right|_{\boldsymbol{\gamma}=\mathbf{0}} \\ &= \sum c_i \mathbf{z}'_{(i)}, \end{aligned}$$

a linear rank statistic, with

$$\begin{aligned} c_i &= n! \int \cdots \int_{\tau_{(1)} < \cdots < \tau_{(n)}} \left(\frac{-d \log f(\tau_{(i)})}{d\tau_{(i)}} \prod_{j=1}^n f(\tau_{(j)})\, d\tau_{(j)} \right) \\ &= n! \int \cdots \int_{u_1 < \cdots < u_n} \phi(u_i) \prod_1^n du_j \\ &= E[\phi(u_i)], \end{aligned} \tag{6.11}$$

where $u_j = 1 - F(\tau_{(j)})$ is the jth order statistic in a uniform (0, 1) sample of size n, $\phi(u)$ $(0 < u < 1)$ is given by

$$\phi(u) = \phi(u, f) = \frac{-f'(F^{-1}(1-u))}{f(F^{-1}(1-u))}$$

and as usual $F(\tau) = \int_\tau^\infty f(w)\, dw$. It is easy to see that the sum of scores is zero since

$$\sum_1^n c_i = \sum_1^n E\left(\frac{-d \log f(\tau_{(i)})}{d\tau_{(i)}}\right)$$
$$= \sum_1^n E\left(\frac{-d \log f(\tau_i)}{d\tau_i}\right) = 0.$$

The fact that u_i has expectation $i(n+1)^{-1}$ and variance $i(n-i+1)(n+1)^{-2}$, $i = 1, \ldots, n$, leads to an asymptotically equivalent system of scores

$$c_i = \phi[i(n+1)^{-1}]. \tag{6.12}$$

Some interesting special cases of (6.11) and (6.12) are as follows: a logistic density $f(\tau) = e^{\tau}(1+e^{\tau})^{-2}$ gives Wilcoxon scores (6.6) for both (6.11) and (6.12). A standard normal density gives for (6.11) and (6.12), respectively, $c_i = E(\tau_{(i)})$, the normal scores test of Fisher and Yates (1938), and $c_i = G^{-1}[i/(n+1)^{-1}]$, the van der Waerden (1953) test, where G represents the standard normal distribution function. Similarly an extreme value p.d.f., $f(\tau) = \exp(\tau - e^{\tau})$ yields the exponential scores (6.5) from (6.11). The double exponential density $f(\tau) = 2^{-1} e^{-|\tau|}$ gives, from (6.12), the sign (median) scores, $c_i = \text{sign}[2i - (n+1)]$, with $c_i = 0$ if $2i = (n+1)$.

As illustrated in Section 6.2.2, the significance level based on the statistic $\mathbf{v}$ can be assessed in terms of its permutation distribution in that any of the $n!$ possible realizations of the rank vector for $\mathbf{w}$ are equally probable under $\boldsymbol{\beta} = \boldsymbol{\beta}_0$. As outlined there, an approximate significance test may be carried out upon noting that $\mathbf{v}$ has mean zero and variance matrix V given by (6.3). Under mild restrictions on the regression vectors (Hájek and Sĭdák, 1967, p. 159), $\mathbf{v}$ is asymptotically normal so that, if $\boldsymbol{\beta} = \boldsymbol{\beta}_0$,

$$\mathbf{v}' V^{-1} \mathbf{v} \tag{6.13}$$

is asymptotically $\chi^2_{(s)}$, where it has been assumed that V is nonsingular.

6.3.2 Properties and More General Inference

The rank tests considered above are asymptotically fully efficient (Pitman efficiency 1; see, for example, Cox and Hinkley, 1975, p. 338) if the assumed score generating distribution, f, and "actual" sampling distribution, say, f_0, agree up to location and scaling. More generally, the asymptotic relative efficiency is the square of the limiting correlation of $\mathbf{v}$ with the locally optimum test based on f_0. This efficiency may be written (Hájek and Sĭdák, 1967, p. 268)

$$\frac{[\int_0^1 \phi(u, f)\phi(u, f_0)\, du]^2}{\int_0^1 \phi^2(u, f)\, du \int_0^1 \phi^2(u, f_0)\, du}, \tag{6.14}$$

where it has been assumed that the Fisher information terms in the denominator are both finite. Under $f \neq f_0$, (6.14) typically indicates substantial improvement over the corresponding parametric test. For example, the normal scores test, under mild conditions on f_0 (Puri and Sen, 1971, p. 118), has efficiency equal to or greater than that of the corresponding least squares test. In particular, under Cauchy sampling the least squares procedure has efficiency zero, whereas the normal scores test has efficiency .43. Rank tests themselves differ somewhat in efficiency properties. For example, the Wilcoxon test has asymptotic efficiency .61 under Cauchy sampling while the sign test has an even higher efficiency of .81. It is then important to consider the class of plausible sampling distributions in selecting a rank test. Birnbaum and Laska (1967) and Gastwirth (1970) give some consideration to the selection of robust rank tests when f_0 is restricted to an indexed family.

The approximate χ^2 statistic (6.13) can form the basis for more general inference on $\boldsymbol{\beta}$. For example, an approximate confidence region for $\boldsymbol{\beta}$ is given by those $\boldsymbol{\beta}_0$ values for which (6.13) does not exceed specified percentage points of the χ^2 distribution. In principle, exact significance levels could also be used to generate confidence regions though the computation would usually be prohibitive. An estimator $\tilde{\boldsymbol{\beta}}$ may be defined as the $\boldsymbol{\beta}_0$ value, or values, for which (6.13) is minimized. Jurečková (1969, 1971) has shown, under mild restrictions, that such a $\tilde{\boldsymbol{\beta}}$ is fully efficient if the assumed form for the score generating density obtains and that its efficiency, more generally, is the same as that of the corresponding test, as given in (6.14).

Unfortunately, numerical aspects relating to the calculation of $\tilde{\boldsymbol{\beta}}$ and its variance have received rather little attention. If there is only a single regressor variable, however, a straightforward search technique can be used. In that case, the statistic $v = v(\beta_0)$ is a monotone increasing function of β_0 provided the scores c_i, $i = 1, \ldots, n$ are monotone increasing. Some recent work on the numerical aspects of the calculation of $\tilde{\boldsymbol{\beta}}$ has been done by Hettmansperger and McKean (1977), Anderson (1978), and McKean and Hettmansperger (1978).

6.4 RANK REGRESSION WITH CENSORED DATA

6.4.1 Censored Data Linear Rank Statistics

Consider now a censored sample from (6.1). Let $w_{(1)} < \cdots < w_{(k)}$ represent the distinct ordered residuals about the hypothesized model

$\boldsymbol{\beta} = \boldsymbol{\beta}_0$ so that as before $w = y - \mathbf{z}\boldsymbol{\beta}_0$. Further, let $w_{i1}, \ldots, w_{im_i}$ be right censored residuals in the interval $[w_{(i)}, w_{(i+1)})$, $i = 0, \ldots, k$, where $w_{(0)} = -\infty$ and $w_{(k+1)} = \infty$. Also let $\mathbf{z}_{(i)}$ and $\mathbf{z}_{ij}$ represent the corresponding regression vectors. Note that it is being assumed for the moment that there are no ties among the uncensored residuals.

In order to generalize the previous results to this case it is necessary to consider generalizations of the rank vector to the censored data case. One possible extension (Peto, 1972a) takes the rank vector as the labels corresponding to the ordered sample of all censored and uncensored w values along with the associated censoring indicators. This statistic is the maximal invariant statistic under monotone increasing transformations on the observed w values, but its sampling distribution depends in a complicated way on the censoring mechanism (Crowley, 1974). As a consequence, its use will usually not yield simple rank statistics for censored samples. An alternative approach, as discussed in Section 4.2.2 and by Kalbfleisch and Prentice (1973), views the rank vector of the underlying w values to be of primary interest. This rank statistic is only partially observed owing to the censoring, and the censored data rank vector is taken to be the whole set of possible underlying rank vectors given the data. It should be noted that this definition of a censored rank vector does not utilize the ordering of censored w values between adjacent uncensored values. That is, no use is made of the ordering among $w_{i1}, \ldots, w_{im_i}$ in the rank vector probabilities.

The calculations follow very closely those of Section 4.2.2 that lead to the marginal likelihood for $\boldsymbol{\beta}$ in the proportional hazards model. In evaluating the total probability that the uncensored rank vector should be one of those possible on the sample, we calculate first the probability of the event $w_{ij} \geq w_{(i)}$, $j = 1, \ldots, m_i$, $i = 1, \ldots, k$, given the uncensored residuals $w_{(1)} < \cdots < w_{(k)}$. This gives

$$\prod_{i=0}^{k} \prod_{j=1}^{m_i} F(\tau_{(i)} - \mathbf{z}_{ij}\boldsymbol{\gamma})$$

where as previously, $F(\tau) = \int_\tau^\infty f(u)\,du$, $\boldsymbol{\gamma} = \sigma^{-1}(\boldsymbol{\beta} - \boldsymbol{\beta}_0)$, and $\tau_{(i)} = (w_{(i)} - \alpha)\sigma^{-1}$, $i = 1, \ldots, k$. Note that the term in $i = 0$ can be dropped since $\tau_{(0)} = -\infty$. The total accumulated probability of possible underlying rank vectors $\{\mathbf{r}\}$ is then

$$P(\{r\}) = \int_{\tau_{(1)} < \cdots < \tau_{(k)}} \cdots \int \prod_{1}^{k} \left[f(\tau_{(i)} - \mathbf{z}_{(i)}\boldsymbol{\gamma}) \prod_{j=1}^{m_i} F(\tau_{(i)} - \mathbf{z}_{ij}\boldsymbol{\gamma})\, d\tau_{(i)} \right]. \tag{6.15}$$

Again, since the τ's enter (6.15) as dummy variables, $P(\{r\})$ is completely independent of α and σ. Note that at $\boldsymbol{\gamma} = \mathbf{0}$ (that is, $\boldsymbol{\beta} = \boldsymbol{\beta}_0$), (6.15) can be integrated directly, giving

$$P(\{r\})|_{\boldsymbol{\gamma}=\mathbf{0}} = \sum_1^k n_i^{-1},$$

where

$$n_i = \sum_{j=i}^{k} (1 + m_j)$$

is the number of individuals with w values known to be equal or greater than $w_{(i)}$.

As with uncensored data, a score test for $\boldsymbol{\beta} = \boldsymbol{\beta}_0$ may be based on (6.15) giving

$$\mathbf{v} = \left.\frac{d \log P(\{r\})}{d\boldsymbol{\gamma}}\right|_{\boldsymbol{\gamma}=\mathbf{0}}$$

$$= \sum_{i=1}^{k} (\mathbf{z}'_{(i)} c_i + \mathbf{s}'_{(i)} C_i), \tag{6.16}$$

where $\mathbf{s}_{(i)} = \sum_{j=1}^{m_i} \mathbf{z}_{ij}$, c_i is a score corresponding to $w_{(i)}$, and C_i is a score corresponding to each of $w_{i1}, \ldots, w_{im_i}$. This statistic is of the type discussed in Section 6.2.1. The scores are explicitly

$$c_i = \int_{\tau_{(1)}<\cdots<\tau_{(k)}} \!\!\cdots \int \left(\frac{-d \log f(\tau_{(i)})}{d\tau_{(i)}}\right) \prod_{j=1}^{k} [n_j F^{m_j}(\tau_{(j)}) f(\tau_{(j)})\, d\tau_{(j)}]$$

and

$$C_i = \int_{\tau_{(1)}<\cdots<\tau_{(k)}} \!\!\cdots \int \left(\frac{-d \log F(\tau_{(i)})}{d\tau_{(i)}}\right) \prod_{j=1}^{k} [n_j F^{m_j}(\tau_{(j)}) f(\tau_{(j)})\, d\tau_{(j)}].$$

As in (6.11) these scores can be expressed in terms of functions on (0, 1). Set $u_j = 1 - F(\tau_{(j)})$, $j = 1, \ldots, k$, and define for $0 < u < 1$

$$\phi(u) = \frac{-f'(F^{-1}(1-u))}{f(F^{-1}(1-u))}$$

$$\Phi(u) = (1-u)^{-1} f(F^{-1}(1-u)).$$

The scoring system can now be written

$$\begin{aligned} c_i &= \int_{u_1<\cdots<u_k} \!\!\cdots \int \phi(u_i) \prod_1^k [n_j(1-u_j)^{m_j}\, du_j] \\ C_i &= \int_{u_1<\cdots<u_k} \!\!\cdots \int \Phi(u_i) \prod_1^k [n_j(1-u_j)^{m_j}\, du_j]. \end{aligned} \tag{6.17}$$

The test statistic (6.16) is simply the average of uncensored test statistics (Section 6.3.1) over possible underlying rank vectors. This, along with the assumption that the censoring mechanism and the mechanism (6.1) are independent, shows that the expectation of v is identically zero, under $\boldsymbol{\beta} = \boldsymbol{\beta}_0$, regardless of the actual sampling distribution f_0. Note also that the sum of scores given by any realization of (6.16) is zero, that is,

$$\sum_1^k (c_i + m_i C_i) = 0. \tag{6.18}$$

This occurs since (6.18) is the average of sums of scores corresponding to possible underlying (uncensored) rank vectors, each of which, from Section 6.3 has value zero.

In order to list some specific scoring schemes let

$$J(g(u_i)) = \int \cdots \int_{u_1 < \cdots < u_k} g(u_i) \prod_{j=1}^{k} [n_j (1 - u_j)^{m_j} \, du_j],$$

for an arbitrary function g. A simple calculation gives

$$J((1 - u_i)^l) = \prod_{j=1}^{i} \left(\frac{n_j}{n_j + l} \right), \qquad l = 1, 2, \ldots. \tag{6.19}$$

A logistic score generating density $f(\tau) = e^{\tau}(1 + e^{\tau})^{-2}$ gives $\phi(u) = 2u - 1$, $\Phi(u) = u$ so that, from (6.17) and (6.19) the corresponding scores are those given by (6.9). This test is discussed in Section 6.2 and is a censored data generalization of the Wilcoxon test similar to that proposed by Peto and Peto (1972).

An extreme value density $f(\tau) = \exp(\tau - e^{\tau})$ yields $\phi(u) = -\log(1 - u) - 1$, $\Phi(u) = -\log(1 - u)$. Direct integration of (6.17) gives

$$J(\log(1 - u_i)) = -\sum_{j=1}^{i} n_j^{-1}$$

so that the corresponding scores are given in (6.7) and discussed in Section 6.2.

6.4.2 Variance Estimation

A permutation approach is commonly used to calculate the variance of a linear rank statistic. This approach does not extend in a convenient way to arbitrarily censored data, because the expectation involved is a complicated function of the censoring mechanism. If censoring does not depend on $\mathbf{z}$, however, a permutation approach is still appropriate, though the reference set consists only of realizations generating the same set of scores (i.e., is conditional on $m_0, m_1, \ldots, m_k$). As in Chapters 3 and 4, however, the "observed" Fisher information matrix

$$V_0 = \frac{-d^2 \log P(\{r\})}{d\gamma \, d\gamma}\bigg|_{\gamma=0} \tag{6.20}$$

provides a variance estimator that is generally appropriate (see Prentice, 1978, for further discussion of this point). After straightforward differentiation of (6.15), (6.20) can be written

$$V_0 = \sum_{i=1}^{k} \left[\mathbf{z}'_{(i)}\mathbf{z}_{(i)} J(\psi_1(u_i)) + \sum_{j=1}^{m_i} \mathbf{z}'_{ij}\mathbf{z}_{ij} J(\psi_2(u_i))\right] - [J(\mathbf{b}'\mathbf{b}) - \mathbf{v}\mathbf{v}'],$$

where

$$\psi_1(u) = \frac{-d^2 \log f(\tau)}{d\tau^2}\bigg|_{\tau = F^{-1}(1-u)}$$

$$\psi_2(u) = \frac{-d^2 \log F(\tau)}{d\tau^2}\bigg|_{\tau = F^{-1}(1-u)}$$

and

$$\mathbf{b} = \sum_{i=1}^{k} [\mathbf{z}_{(i)}\phi(u_i) + \mathbf{s}_{(i)}\Phi(u_i)].$$

V_0 can be calculated explicitly for logistic and extreme value score generating densities. A logistic density gives a test statistic (6.16) with scores (6.9) and having variance estimator

$$V_0 = \sum_{i=1}^{k} \left[a_i(1 - a_i^*)\left(2\mathbf{z}'_{(i)}\mathbf{z}_{(i)} + \sum_{j=1}^{m_i} \mathbf{z}'_{ij}\mathbf{z}_{ij}\right) - (a_i^* - a_i)\mathbf{x}'_{(i)}\left(a_i\mathbf{x}_{(i)} + 2\sum_{j=i+1}^{k} a_j\mathbf{x}_{(j)}\right)\right] \tag{6.21}$$

where

$$a_i = \prod_{j=1}^{i}\left(\frac{n_j}{n_j+1}\right), \quad a_i^* = \prod_{j=1}^{i}\left(\frac{n_j+1}{n_j+2}\right) \quad \text{and} \quad \mathbf{x}_{(i)} = 2\mathbf{z}_{(i)} + \mathbf{s}_{(i)},$$
$$i = 1, \ldots, k.$$

An extreme minimum value distribution gives scores (6.7) and a test statistic with variance estimator (6.8).

Conditions on the regression vector $\mathbf{z}$ for the asymptotic normality of (6.16) will be similar to those given in Hájek and Sĭdák (1967, p. 152). Essentially it is required that each observation have an asymptotically trivial contribution to (6.16). The adequacy of an asymptotic normal approximation would then be suspect if the sample contains extreme and isolated $\mathbf{z}$ values that individually have a dominating effect on the test statistic. Similarly, conditions for asymptotic normality would exclude situations in which censoring times converge rapidly to zero with increasing sample size, for this may leave a few observations with an asymptotically nonnegligible effect on (6.16). Under $\boldsymbol{\beta} = \boldsymbol{\beta}_0$ and asymptotic normality of $\mathbf{v}$ the statistic

$$\mathbf{v}' V_0^{-1} \mathbf{v} \tag{6.22}$$

has an asymptotic χ^2 distribution with degrees of freedom the dimension of $\mathbf{z}$, assuming V_0 nonsingular.

6.4.3 More General Inference

As in Section 6.3.2 the approximate χ^2 statistic (6.22) can form the basis for more general inference on $\boldsymbol{\beta}$. An approximate confidence region for $\boldsymbol{\beta}$ is given by those $\boldsymbol{\beta}_0$ values for which (6.22) does not exceed specified percentage points of the χ^2 distribution. An estimator $\tilde{\boldsymbol{\beta}}$ may be defined as the $\boldsymbol{\beta}_0$ value, or values, that minimize (6.22).

Partial tests on $\boldsymbol{\beta}$ are frequently important. For example, one may wish to compare several survival curves, or evaluate the prognostic importance of some factor, while adjusting for certain nuisance factors. Suppose $\mathbf{z}\boldsymbol{\beta} = \mathbf{z}_1\boldsymbol{\beta}_1 + \mathbf{z}_2\boldsymbol{\beta}_2$ and it is desired to test $\boldsymbol{\beta}_1 = \boldsymbol{\beta}_1^0$. Such a test may be based on the statistic (6.16) as applied to residuals

$$w = y - \mathbf{z}_1\boldsymbol{\beta}_1^0 - \mathbf{z}_2\tilde{\boldsymbol{\beta}}_2,$$

where $\tilde{\boldsymbol{\beta}}_2$ maximizes (6.22) subject to $\boldsymbol{\beta}_1 = \boldsymbol{\beta}_1^0$. Once again, numerical aspects have not been sufficiently addressed that this approach can currently be advocated. A related procedure that involves stratification on nuisance factors and calculation of statistics of the type (6.16) within strata, is, however, convenient and widely used.

Suppose the sample is stratified on the basis of values of the nuisance regressor variable $\mathbf{z}_2$ in such a manner that it is reasonable to assume homogeneity with respect to $\mathbf{z}_2$ within strata. Within the jth stratum let $\mathbf{v}_j$, V_{0j} represent the test statistic (6.15) and variance estimator (6.20) corresponding to the hypothesis $\boldsymbol{\beta}_1 = \boldsymbol{\beta}_1^0$. An overall test for $\boldsymbol{\beta}_1 = \boldsymbol{\beta}_1^0$ can then be based on an approximate χ^2 distribution for

$$\left(\sum \mathbf{v}_j'\right)\left(\sum V_{0j}\right)^{-1}\left(\sum \mathbf{v}_j\right), \tag{6.23}$$

where the summation is over strata and the number of degrees of freedom, assuming $\sum V_{0j}$ nonsingular, is the dimension of $\mathbf{z}_1$. If log-rank scores are being used, this is the Mantel Haensel or stratified log-rank test of Section 4.4.

6.5 PROPERTIES AND DISCUSSION

An important aspect of rank tests is their invariance under monotone increasing transformations on the response variable. For example, if w were replaced by some $w' = h(w)$, where h is strictly monotone increasing, the rank vector and scores (6.15) would be unchanged. This feature is most easily appreciated in terms of a test of $\boldsymbol{\beta} = \mathbf{0}$. Monotone transformations on y itself, then, do not affect the test procedure so that equivalent tests arise for example from taking y to be normal or log-normal or indeed from taking any monotone function of y to be normal. As a second example, the proportional hazards model of Chapter 4 can be thought of as arising from a model which specifies that some monotone increasing function of $y = \log t$ has an extreme value error. For this reason, with uncensored data the log-rank scores are (locally) optimum within the whole proportional hazards class. See Kalbfleisch (1978b) for further discussion of this.

Since the Weibull model is a special case of both the proportional hazards and accelerated failure time models, we may compare the methods of Section 4.2 with those of this chapter by supposing the error density f in (6.1) to have the extreme value form. It is convenient to consider the exponential special case $(\lambda_0(t) = \lambda)$ since $\boldsymbol{\beta}$ then has the same meaning in both models. Tests for $\boldsymbol{\beta} = \mathbf{0}$ are equivalent from the two approaches as was just noted. More generally, however, the proportional hazards test of $\boldsymbol{\beta} = \boldsymbol{\beta}_0 \neq 0$ is based on the generalized rank vector $\{r(\mathbf{t})\}$ of the observed survival times t, whereas the log-rank methods of Section 6.4 are based on the generalized rank vector $\{r(\mathbf{w})\}$ based on "centered" values $w = \log t - \mathbf{z}\boldsymbol{\beta}_0$. As $\boldsymbol{\beta}_0$ departs more radically from zero the generalized rank vector $\{r(\mathbf{t})\}$ becomes increasingly predictable under the hypothesis and so less informative against local alternatives (values of $\boldsymbol{\beta}$ near $\boldsymbol{\beta}_0$). If there is no censoring and a single covariate, Kalbfleisch (1974) shows that the proportional hazards test has efficiency approximately of the form $\exp(-\beta_0^2\mu_2)$ where μ_2 is the variance of the finite population of z values. In contrast, the log-rank test based on residuals about $\boldsymbol{\beta}_0$ is fully efficient. It is important to note, however, that these are local efficiencies and the efficiency against distant alternatives may or may not be similar. The local efficiency comparison is somewhat unfair in that the proportional hazards test has the same efficiency when compared with any specific $\lambda_0(t)$ in the proportional hazards class whereas the efficiency of this particular log-rank procedure changes if hazards are proportional but outside the Weibull class. Both procedures would be inefficient if (6.1) were appropriate but the error density departs from the extreme value form. For the log-rank test based on $r(\mathbf{w})$, however, the reduction in efficiency is independent of $\boldsymbol{\beta}_0$.

Efficiency properties of the censored data rank tests are generally not known. From their construction, however, it would be expected that they would have better local power or efficiency properties than other scoring procedures based on the same generalized rank vector as used here. It is possible, of course, that tests which take account of the ordering of successive censored w values would improve the local power, though any increase would be expected to be small.

Gehan (1965a) proposed a censored generalization of the two sample Wilcoxon statistic that differs from (6.9). Gehan's statistic and the multiple sample generalization of Breslow (1970) have been widely used in the clinical trials setting. The censored and uncensored scores are, respectively,

$$c_i = i - n_i$$

$$C_i = i.$$

The scores are simply the number of residuals known to be smaller than the residual being scored minus the number known to be larger. This statistic will provide a close approximation to (6.9) if the censoring is slight but can be very misleading with heavy censorship (see Prentice and Marek, 1979 and problems 66 and 67 in Appendix 2). It therefore cannot be advocated for general use.

The methods of this chapter have assumed that uncensored w values are distinct. Tied uncensored values may be handled in the manner suggested in Chapter 4: tied values at the same $\mathbf{z}$ may be ordered arbitrarily because their "actual" ordering does not affect the statistic (6.15). More generally, if the fraction of tied values is small, tied values may be assigned the average of their possible scores with a suitable adjustment to the variance matrix. Perhaps more appropriate, but computationally difficult, the rank vector may be taken, as in Section 4.2.2 to consist of all possible underlying rank vectors that are consistent with the observed tied and censored data. The same idea allows the methods of this chapter to be extended to doubly censored data.

6.6 ILLUSTRATION

For illustration, we apply the methods of this chapter to the carcinogenesis data of Section 1.1.1. The results may be compared with the parametric methods of Chapter 3 and the proportional hazards methods of Section 4.2.6.

To begin, suppose as in Chapter 3 that $y = \log(t - 100)$ adheres to a linear model (6.1) where t represents time to carcinogenesis as given in

Table 1.1. As previously, suppose the scalar regression variable z has value zero for group 1 and value 1 for group 2. A test of $\beta = \beta_0$ can be based on the generalized rank vector $r(\mathbf{w})$ of (6.6), where $w = y - z\beta_0$. To test equality of the two survival curves the hypothesis $\beta = 0$ is considered. In this case $w = y$ and the generalized rank vector based on w coincides with that based on t itself, since $y = \log(t - 100)$ is strictly monotone increasing. Monotonicity considerations also show the rank test for $\beta = 0$ to be independent of the guarantee time so that such a time need not be specified. The test (6.23) involves n_j, $j = 1, \ldots, k$, which under $\beta = 0$ are simply the number of rats alive just before the jth smallest (not necessarily distinct) failure time. Thus $k = 36$, $n_1 = 40$, $n_2 = 39, \ldots, n_{10} = 31$, $n_{11} = 29, \ldots, n_{36} = 2$. Note that since tied failure times occur only at the same z value (same sample) they may be ordered arbitrarily. The generalized Wilcoxon scores (6.9) give a test statistic $v = 2.887$ and a variance estimator $V_0 = 3.066$ using (6.16). This leads to an approximate $\chi^2_{(1)}$ statistic (6.23) of value 2.72. In comparison, the log-rank scores (6.7) give $v = 4.584$, $V_0 = 7.653$ and an approximate $\chi^2_{(1)}$ statistic of nearly identical value 2.75 as was obtained in Section 4.2.6 and compared there with other methods (e.g., maximum likelihood) in the proportional hazards model. From Section 3.8.2 parametric procedures using a Weibull regression model, that is, an extreme value error in (6.1), yield $\hat{\beta} = .213$ and a $\chi^2_{(1)}$ value of 3.90. The parametric procedure suggests somewhat stronger evidence for a difference between the survival curves than do the rank regression tests (.05 level versus .10 level of significance). The parametric method, however, is much more sensitive to the error modeling. As previously noted, for example, a parametric analysis using a log-normal model, that is, normal error in (6.1), gives $\hat{\beta} = .154$ and a $\chi^2_{(1)}$ value of 1.32 for testing $\beta = 0$. Such a statistic is not at all suggestive of a survival difference between groups.

For estimation of β, still with $y = \log(t - 100)$, the statistic (6.23) was scanned over a fine grid of β_0 values. The test statistic is a monotone decreasing function of β_0 for either generalized Wilcoxon or log-rank scores. These score functions are plotted in Fig. 6.1 and indicate approximate linearity over plausible values of β. An approximate 95% confidence interval for β is given by those β_0 values for which (6.18) is less than 3.84, the upper 5% end point of a $\chi^2_{(1)}$ distribution. Using generalized Wilcoxon scores (6.9) one obtains $(-.025, .414)$ as an approximate 95% interval for β whereas log-rank scores give $(-.038, .424)$. Values of (6.23) are plotted in Fig. 6.2. The two tests nearly coincide over plausible β values. The dashed line determines an approximate 95% confidence interval for β. The corresponding estimators $\tilde{\beta}$ at which (6.23) is minimized are .180 in each case, intermediate to those values from the two parametric analyses mentioned above.

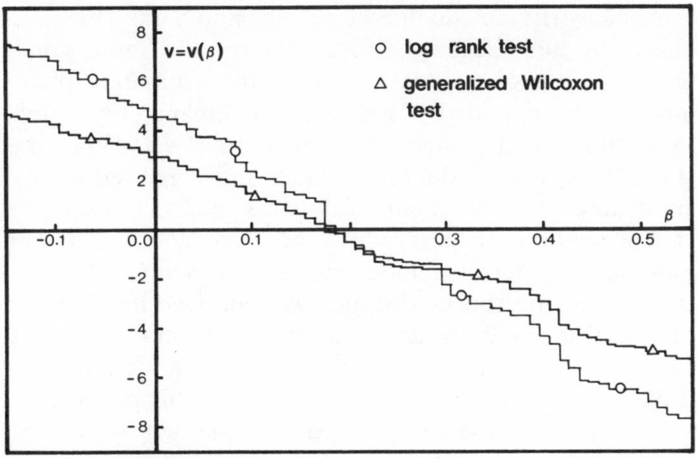

Figure 6.1 Values of certain rank statistics over plausible values of a two sample regression coefficient β.

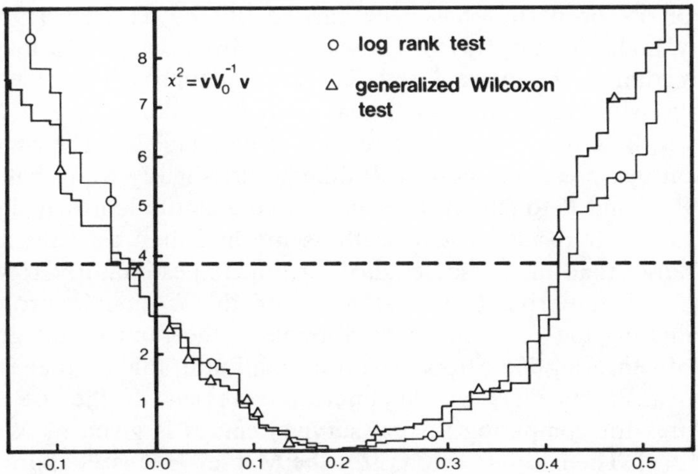

Figure 6.2 Values of approximate $\chi^2_{(1)}$ statistics, corresponding to certain rank tests, over plausible values of a two sample regression coefficient β.

BIBLIOGRAPHIC NOTES

There are now many books wholly or partially devoted to univariate methods of statistical inference based on ranks. Some relatively recent introductory books with extensive bibliographies are those by Lehmann (1975), Hájek (1969), Hollander and Wolfe (1973), and Conover (1971).

More advanced works include those by Hájek and Sĭdák (1967), Lehmann (1959), and Cox and Hinkley (1974, Chapter 6). Some notable contributions to the use of residual ranks for the estimation of regression coefficients have been made by Hodges and Lehmann (1963) and Adichie (1967), in addition to the papers by Jurečková (1969, 1971), mentioned above. Hogg (1974) reviewed these methods and suggested some adaptive scoring procedures. Further theoretical works related to rank procedures include a collection of papers edited by Puri (1970) and a book on multivariate nonparametric methods by Puri and Sen (1971).

Rank tests with type II censored data were derived in a similar manner to that of Section 6.4.2 by Johnson and Mehrotra (1972). A rather general, but somewhat ad hoc, means of generating rank tests with censored data is given by Peto and Peto (1972). This paper and another by Peto (1972a) may usefully be read in parallel with the current development though some claims in regard to asymptotic properties of their estimators need further restriction. The log-rank test (6.7) can be regarded as an outgrowth of a series of papers by Mantel and collaborators (Mantel and Haenszel, 1959; Mantel, 1963; Mantel, 1966). Work on its properties has been carried out by Thomas (1969), and Crowley and Thomas (1975). Crowley (1974b) discusses various likelihoods leading to the log-rank test. The general approach considered here for ranks with censored data is that used by Kalbfleisch and Prentice (1973) which, in turn, amplified a remark of Peto (1972b). The generalized Wilcoxon test presented here (6.7) differs only slightly from that presented by Peto and Peto (1972). They have used quantile scores analogous to (6.12) and their results would be the same had they used the Kaplan–Meier rather than the Altshuler survivor function estimator. Crowley and Thomas (1975) derive (6.7) and (6.8) for the k sample problem by approximating the conditional distribution of the rank vector given the vector of censoring indicators. The presentation in this chapter is similar to that of Prentice (1978). An interesting review of the use of rank procedures for comparing several survival curves is given by Ware and Byar (1979). Their report emphasizes the Mantel–Haenszel approach and Gehan's (1965a,b) generalizations of the Wilcoxon test. A recent review article by several statisticians involved with clinical trials (Peto et al., 1977) presents the use of the log-rank test, and other features of design and analysis of clinical trial data, in a nontechnical manner.

CHAPTER 7

Multivariate Failure Time Data and Competing Risks

7.1 SOME GENERALIZATIONS

The preceding chapters have presented methods for data analyses when there is a single, possibly censored, failure time on each study subject. Failure data may be more general in two important respects. First, the failure on an individual may be one of several distinct types or causes. Problems arising in the analysis of such data are commonly referred to as competing risk problems. Second, there may be more than one failure time on each study subject. Such multivariate failure times may correspond to repeated occurrences of some similar event or to the occurrence of events of entirely different natures. In general, study subjects may experience a variable number of failures each with its own type or cause.

The hazard function representation of failure time is generalized in this chapter to permit the inclusion of competing risks and multiple failures. The methods of preceding chapters are generalized in order to study the relationship between covariates and certain "cause"-specific hazard functions. The use of time dependent covariates gives a promising approach to other competing risk problems. In Section 7.3 the hazard function modeling for an individual study subject is continued beyond the first failure time in order to provide a framework for the analysis of multivariate failure time data. This formulation provides a link to the literature on point processes and suggests topics for further research.

7.2 COMPETING RISKS

7.2.1 Competing Risks Problems

Much of the material in this section is similar to and extends that given by Prentice et al. (1978). In fact, their paper arose largely through attempts to consolidate a rather diffuse competing risks literature for inclusion in this book. Nearly all of the competing risks literature has presumed the existence of "latent" or "potential" failure times for an individual corresponding to each of the m failure types. In contrast, the methods presented below are based on statistical models for observable quantities only. Some discussion of the latent failure time approach is given in Section 7.2.5.

Suppose as before that each study subject has an underlying failure time T that may be subject to censoring, and a covariate vector $\mathbf{z}$, or more generally a covariate function $Z = \{\mathbf{z}(u): u \geq 0\}$. Suppose also that when failure occurs it may be of one of m distinct types or causes denoted by $J \in \{1, 2, \ldots, m\}$. For example in the mouse radiation study mentioned in Section 1.1.1 (data listed in Appendix 1) the failure types are death from thymic lymphoma, death from reticulum cell sarcoma, or death from other causes. In demographic mortality studies failure types are often broad categories such as cancer, heart disease, or accidents. In industrial life testing a certain piece of equipment may develop one of several faults causing its failure.

Three distinct problems arise in the analysis of such data:

1. The estimation of the relationship between covariates and the rate of occurrence of failures of specific types.
2. The study of the interrelation between failure types under a specific set of study conditions.
3. The estimation of failure rates for certain types of failure given the "removal" of some or all other failure types.

For instance, problem 1 arose in the analysis of data from the University Group Diabetes Program (e.g., Cornfield, 1971; Gilbert et al., 1975). The primary end point for treatment evaluation was patient survival time. Since the differential effects of treatment may depend markedly on cause of death, and since distinct causes of death may relate to different prognostic factors, an analysis of overall survival time may well be inadequate. In the UGDP study, significant treatment differences in respect to overall survival time were not found. An analysis of cause-specific treatment differences led, however, to the finding of primary interest and controversy in the study; namely, that one treatment, tolbutamide, appeared to give rise to a higher risk of cardiovascular death. Similarly the Hoel data (Appendix 1) may be used to study the effect of a germ-free environment on specific causes of death in irradiated male mice. Methods for the estimation of cause-specific failure rates are considered in Section 7.2.3.

Problem 2 is also of interest in a number of contexts. For example, Thomas et al. (1975a,b) summarized their experience with marrow transplantation in the treatment of acute leukemia. A little background is given here since this application serves to illustrate and clarify a number of points in relation to problems 2 and 3: end stage leukemia patients in relapse are given ordinarily lethal doses of radiation and chemotherapy in an attempt to eradicate their leukemia. Following such conditioning and immunosuppression, bone marrow cells from an HLA (human lymphocyte antigen) matched sibling donor are infused into the patient's bloodstream. These cells lodge in "spacings" in the patient marrow, repopulate, and give rise to a new hemopoietic and immunologic system for the patient. Major causes of death relate to (1) recurrence of leukemia and (2) graft versus host disease (GVHD). The latter occurs as an immunologic reaction of the new marrow graft against the patient. GVHD is evident primarily through effects on the skin, liver, and gut.

A question of significant biologic implication concerns the relationship between GVHD and relapse due to recurrent leukemia. For example, it may happen that a graft versus host (GVH) reaction of a certain degree of severity is useful in preventing recurrence. The immunologic reaction of the marrow graft on the host may even selectively kill residual or new leukemia cells. On the other hand, a severe GVH reaction may simply weaken the patient or be an indication of a poorly functioning marrow graft which, in turn, may be associated with an increased risk of leukemic relapse. Knowledge of the interrelation between the two failure modes would be valuable in defining the required degree of tissue similarity between marrow donor and patient and could be used to predict the usefulness of autologous marrow replacement following treatment in nonhematologic diseases. Note that this question is posed in terms of a patient group with well defined eligibility criteria under a specific treatment program. The question does not involve, for example, inferences about recurrent leukemia rates if mortality associated with graft versus host disease could somehow be obviated.

A second and more general application involves the relationship between censoring due to withdrawal or removal from a study and failure, in a follow-up study. The methods of the preceding chapters all assume an "independent censoring mechanism." The regression coefficients and survivor function estimators may be seriously affected if study subjects are selectively censored when, for example, they are thought to be at a high risk of failure, relative to a comparable group of study subjects at risk at the same follow-up time. In the formulation of this chapter we may regard censoring as an additional failure mode and attempt to study the relationship between times to censoring and the corresponding times to "actual" failure.

Several authors (e.g., Cox, 1959; Tsiatis, 1975; Peterson, 1975) have indicated that data of the type (T, J) are not adequate to study the interrelation between failure types. These results extend directly to data $(T, J, \mathbf{z})$ with $\mathbf{z}$ time independent. Section 7.2.5 presents some suggestions for the use of time dependent covariates to study such relationships.

Problem 3 has often been regarded as *the* problem of competing risk analysis. In order to put forward solutions to problems of this type one must assume that data under one set of study conditions in which m failure types are operative are somehow relevant to a different set of study conditions in which only certain of the failure modes can occur. Of course the failure rate function for a specific failure type may be affected in a variety of ways by "removal" of other failure types. "Removal" itself may involve a variety of mechanisms, with corresponding different structures for the remaining failure types. For these reasons, problem 3 is largely nonstatistical. It is unrealistic to think that general statistical methods can be put forward to estimate failure rates under the removal of other causes. A good deal of knowledge of the physical or biological mechanisms giving rise to the failures as well as knowledge of the mechanism giving rise to the removal of certain failure types is necessary before reasonable methods can be proposed in any given setting.

To illustrate difficulties associated with problem 3 consider again the marrow transplantation example. One can envisage two mechanisms that could potentially give rise to the removal of mortality associated with graft versus host disease. First, this could occur via the use of a treatment that reduced the severity of the GVH reaction allowing the patient to survive through the acute GVHD time period (approximately 6 months) and be at subsequent risk only for the remaining types of failure. This mechanism would in no way alter the donor–recipient matching, the conditioning treatments prior to transplantation, or the marrow grafting procedure itself. On the other hand, it could substantially alter the relation between recurrent leukemia and the incidence rates of the other causes of failure. As a matter of fact, there is a treatment called antithymocyte globulin (ATG) that appears to be effective in reducing the severity of the GVH reaction. Unfortunately its use may also give rise to an increased risk of fatal interstitial pneumonia. The effect of ATG on the recurrence of leukemia is unknown. It is clear, however, that the introduction of an agent, such as ATG, that affects the immune system as well as other body systems may substantially alter recurrent leukemia rates.

A second conceivable mechanism to "remove" GVHD as a cause of death would involve a protocol change in the degree of genetic similarity (tissue matching) between donor and recipient before a patient undergoes transplantation. As an extreme special case leukemia patients receiving marrow grafts from an identical twin donor do not experience graft versus host disease. Such a protocol change could give rise to a completely different immunologic response of the marrow graft against residual or new leukemia cells. Times to recurrent leukemia following such a change in matching criteria would not be expected to be closely related to recurrent leukemia times that would have been experienced under previous criteria.

7.2.2 Representation of Competing Risk Failure Rates

Suppose that failure time is continuous and, for the moment, that the covariate $Z = \{\mathbf{z}(u): u \geq 0\}$ is fixed, defined, or ancillary (see Section 5.3 for definitions). That is, we are excluding from consideration "internal" covariate processes that require survival of the study subject for their existence. As previously the overall failure rate or hazard function with such covariates is defined by

$$\lambda(t; Z) = \lim_{\Delta t \to 0} \frac{P(t \leq T < t + \Delta t | T \geq t, Z)}{\Delta t}.$$

One can similarly define a type or cause-specific hazard function

$$\lambda_j(t; Z) = \lim_{\Delta t \to 0} \frac{P(t \leq T < t + \Delta t, J = j | T \geq t, Z)}{\Delta t}, \tag{7.1}$$

for $j = 1, \ldots, m$. In words, $\lambda_j(t; Z)$ is the instantaneous rate for failure of type j at time t given Z and in the presence of the other failure types. Assuming that failure type j must be a unique element of $\{1, 2, \ldots, m\}$ gives

$$\lambda(t; Z) = \sum_1^m \lambda_j(t; Z). \tag{7.2}$$

In some studies two or more causes of failure may occur simultaneously. In the current framework such joint events may be taken as defining additional failure types.

With the covariate types under consideration the survivor function, given Z, has clear meaning and is written

$$F(t; Z) = \exp\left[-\int_0^t \lambda(u; Z)\, du\right] \tag{7.3}$$

The failure time (sub)density function for failure type j, given Z, is

$$f_j(t; Z) = \lim_{\Delta t \to 0} \frac{P(t \leq T < t + \Delta t, J = j | Z)}{\Delta t}$$

$$= \lambda_j(t; Z) F(t; Z), \quad j = 1, \ldots, m. \tag{7.4}$$

Note that (7.2) to (7.4) show that the likelihood function can be written entirely in terms of the cause-specific hazard functions. Further, these functions are identifiable; that is, they can be estimated from data of the type $(t, j; Z)$ without further assumptions.

7.2.3 Estimation of Cause-Specific Hazard Functions

Suppose n study subjects give rise to data $(t_i, \delta_i, j_i; Z_i)$, $i = 1, \ldots, n$ where t_i is the observed survival time, δ_i is the censoring indicator, j_i is the failure type (does not enter the likelihood if $\delta_i = 0$) and Z_i is the covariate or covariate function for the ith individual. Under an independent noninformative censoring mechanism the likelihood function is proportional to

$$\prod_1^n \{[\lambda_{j_i}(t_i; Z_i)]^{\delta_i} F(t_i; Z_i)\} = \prod_1^n \left\{[\lambda_{j_i}(t_i; Z_i)]^{\delta_i} \prod_{j=1}^m \exp\left[-\int_0^{t_i} \lambda_j(u; Z_i)\, du\right]\right\}. \tag{7.5}$$

Arguments leading to this parallel those in Section 5.2. Upon rearrangement the likelihood factors into a separate component for each failure type $j = 1, \ldots, m$. Moreover the likelihood factor involving a specific $\lambda_j(t; Z)$ is precisely that which would be obtained by regarding all failures of types other than j as censored at the individual's failure time. It follows that any of the methods of preceding chapters can be used for inference on the $\lambda_j(t; Z)$'s. Some results for inference on the cause-specific hazard functions are given in the ensuing paragraphs. It is convenient to denote

$$F_j(t; Z) = \exp\left[-\int_0^t \lambda_j(u; Z)\, du\right], \qquad j = 1, \ldots, m \tag{7.6}$$

although these functions will not, in general, have any survivor function interpretation for $m > 1$.

Consider first homogeneous (no covariates) competing risk failure time data. Two useful graphical estimators of the distribution of (T, J) are (1) cumulative "incidence" plots and (2) cumulative hazard plots. Cumulative incidence plots provide estimators of the subdistribution functions

$$I_j(t) = P(T < t, J = j) = \int_0^t f_j(u)\, du, \qquad j = 1, \ldots, m \tag{7.7}$$

and cumulative hazard plots provide estimators of $-\log F_j(t)$ with F_j as defined in (7.6).

The nonparametric estimation technique of Kaplan and Meier (Section 1.3) is readily generalized to include competing risks as follows: Let $t_{j1} < t_{j2} < \cdots < t_{jk_j}$ denote the k_j failure times for failures of type j,

$j = 1, \ldots, m$. Suppose failure type j occurs with multiplicity d_{ji} at time t_{ji}. The likelihood function can be written

$$L = \prod_{j=1}^{m} \left(\prod_{i=1}^{k_j} \left\{ [F_j(t_{ji}) - F_j(t_{ji} + 0)] \prod_{\substack{k=1 \\ k \neq j}}^{m} F_k(t_{ji}) \right\}^{d_{ji}} \prod_{l=1}^{C_{ji}} F_j(t_{jil+0}) \right) \tag{7.8}$$

where $t_{ji1}, \ldots, t_{jiC_{ji}}$ denote the C_{ji} censored times between t_{ji} and the next larger failure time in the combined sequence of failure times of all types. Note that failure times of different types are allowed to tie with each other in (7.8) though a specific individual must have a unique failure type. The likelihood (7.8) factors into a component for each failure type. As one would expect the likelihood factor for the jth failure type is exactly that which would be obtained by regarding failure times for failures of types other than j as censored. It follows that the "nonparametric" maximum likelihood estimator of F_j is

$$\hat{F}_j(t) = \prod_{\{i | t_{ji} < t\}} \left(\frac{n_{ji} - d_{ji}}{n_{ji}} \right) \tag{7.9}$$

where n_{ji} is the number of study subjects at risk just prior to t_{ji}. One can readily show that, provided there are no ties with different failure types,

$$\hat{F}(t) = \prod_{j=1}^{m} \hat{F}_j(t)$$

is the overall Kaplan–Meier survivor function estimator. The corresponding estimator of the cause-specific hazard function $\lambda_j(t)$ has value d_{ji}/n_{ji} at t_{ji}, $i = 1, \ldots, k_j$, and value zero elsewhere. The cumulative "incidence" function (7.7) is then estimated by

$$\hat{I}_j(t) = \sum_{\{i | t_{ji} < t\}} d_{ji} n_{ji}^{-1} \hat{F}(t_{ji}), \qquad \text{for } j = 1, \ldots, m. \tag{7.10}$$

Note that (7.10) reduces to $1 - \hat{F}(t)$ when there is a single failure mode ($m = 1$). A plot of $\hat{I}_j(t)$ versus t gives estimates of the probability that failure type j will occur before any specified time t within the range of observation. The cumulative hazard plot, $-\log \hat{F}_j(t)$ versus t, provides some supplementary information. Specifically, the plot gives a visual impression, by means of its "slope', over specific time intervals of the rate of occurrence of failures of type j.

Consider now inference on the relationship between cause-specific hazard functions and regression vectors or functions Z. Any of the modeling approaches of preceding chapters may be utilized. For example proportional hazards modeling in which the cause-specific hazard function at time t depends on Z only in terms of the concurrent value $\mathbf{z}(t)$ gives

$$\lambda_j(t; Z) = \lambda_{0j}(t) \exp[\mathbf{z}(t)\boldsymbol{\beta}_j], \qquad j = 1, \ldots, m. \tag{7.11}$$

Note that both the "shape" functions λ_{0j} and the regression coefficients $\boldsymbol{\beta}_j$ have been permitted to vary arbitrarily over the m failure types.

Let $t_{ji} < \cdots < t_{jk_j}$ denote the k_j failures of type j, $j = 1, \ldots, m$ and let Z_{ji} be the regression function for the individual that fails at t_{ji}. The method of partial likelihood then gives

$$L(\boldsymbol{\beta}_1, \ldots, \boldsymbol{\beta}_m) = \prod_{j=1}^{m} \prod_{i=1}^{k_j} \left(\frac{\exp[\mathbf{z}_{ji}(t_{ji})\boldsymbol{\beta}_j]}{\sum_{l \in R(t_{ji})} \exp[\mathbf{z}_l(t_{ji})\boldsymbol{\beta}_j]} \right). \tag{7.12}$$

Estimation and comparison of the $\boldsymbol{\beta}_j$'s can be conducted by applying standard asymptotic likelihood techniques individually to the m factors in (7.12) exactly as in Chapters 4 and 5. Expression (7.12) accommodates tied failure times on different failure modes whereas a modification to the product of terms of the type (4.7) or (4.8) would be required if tied failure times occur on the same failure mode. The functions $F_j(t; Z)$ in (7.6) can be estimated at specified Z, using the methods of Chapters 4 and 5, upon inserting the maximum likelihood estimators $\hat{\boldsymbol{\beta}}_1, \ldots, \hat{\boldsymbol{\beta}}_m$ from (7.12). The corresponding estimators of the cumulative incidence

$$I_j(t; Z) = P(T < t, J = j; Z) = \int_0^t \lambda_j(u; Z) F(u; Z) du, \qquad j = 1, \ldots, m$$

can be obtained simply by inserting the appropriate estimators for F and the λ_j functions.

The cause-specific hazard functions could similarly be modeled using an "accelerated failure time" model

$$\lambda_j(t; Z) = \lambda_{0j}\{t \exp[\mathbf{z}(t)\boldsymbol{\beta}_j]\} \exp[\mathbf{z}(t)\boldsymbol{\beta}_j], \qquad j = 1, \ldots, m. \tag{7.13}$$

It would be necessary to restrict the covariate to be fixed or a step function in order to preserve the multiplicative relationship between covariates and failure time. Again because of the factorization (7.5) inference on a particular $\boldsymbol{\beta}_j$ can be conducted using the methods of Chapter 3 or 6 upon treating failure times of types other than j as censored.

A specialization of (7.11) that is expected to give rise to more efficient $\boldsymbol{\beta}$ estimators, if applicable, is that given by

$$\lambda_j(t; Z) = \lambda_0(t) e^{\gamma_j} \exp[\mathbf{z}(t)\boldsymbol{\beta}_j], \qquad j = 1, \ldots, m. \tag{7.14}$$

This time, the cause-specific hazards are assumed to be proportional to each other with proportionality factors e^{γ_j} (for uniqueness set $\gamma_1 = 0$). This proportionality assumption can be checked graphically on the basis of plots of $\log[-\log \hat{F}_j(t; Z_0)]$ versus t, for specified Z_0, where the $\hat{F}_j$ are estimated from an analysis of model (7.11). Such plots should be separated by approximately a constant difference for various values of j

under (7.14). Estimation of the γ_j's and $\boldsymbol{\beta}_j$'s can be carried out using methods similar to those of Sections 4.2 and 4.3. A partial likelihood for these parameters can be obtained as the product, over successive failure times $t_{(i)}$, of conditional probabilities. At time $t_{(i)}$, the contribution is the conditional probability that (i) fails of cause $j_{(i)}$ given the risk set at $t_{(i)}$ and the information that an item fails of some unspecified cause at $t_{(i)}$. The partial likelihood can then be written

$$\frac{\prod_{j=1}^{m} \exp(\gamma_j k_j + \boldsymbol{\beta}_j \mathbf{s}_j)}{\prod_{i=1}^{k} \{\sum_{j=1}^{m} \sum_{l \in R(t_{ji})} \exp[\gamma_j + \boldsymbol{\beta}_j \mathbf{z}_l(t_{ji})]\}}. \tag{7.15}$$

In (7.15) k_j is the number of failures of type j, $k = \sum k_j$, $t_1 < \cdots < t_k$ are ordered combined failure times and $\mathbf{s}_j$ is the sum of $\mathbf{z}(t)$ values at their time of failure for individuals with failure type j. Again, tied failure times can be handled as in (4.7) or (4.8).

The proportional risk model (7.14) has some convenient properties even though it would often be more restrictive than is desirable. For instance, the probability that an individual with fixed covariate $\mathbf{z}$ fails of cause j is simply

$$\begin{aligned} P(J = j; \mathbf{z}) &= \int_0^\infty f_j(t; \mathbf{z})\, dt \\ &= \frac{\exp(\gamma_j + \mathbf{z}\boldsymbol{\beta}_j)}{\sum_{k=1}^{m} \exp(\gamma_k + \mathbf{z}\boldsymbol{\beta}_k)}, \qquad j = 1, \ldots, m \end{aligned} \tag{7.16}$$

regardless of $\lambda_0(\cdot)$. It follows that T and J are statistically independent. Note that in the case of no regression variable the likelihood (7.15) reduces to the multinomial likelihood based on (7.16). The corresponding maximum likelihood estimators of the proportionality factors e^{γ_j}, subject to $\gamma_1 = 0$ are

$$e^{\hat{\gamma}_j} = \frac{k_j}{k_1}.$$

7.2.4 Cause-Specific Hazard Function Estimation with Internal Covariates

As discussed in Section 5.3 the hazard function definition can be extended to include "internal" time dependent covariates which are defined as covariates that require survival of the study subject for their existence. Denote by $X(t) = \{\mathbf{x}(u), u \leq t\}$ such a covariate function from time 0 up to t. Also denote by $Y = \{\mathbf{y}(u), u \geq 0\}$ a covariate function defined over the entire study period for all fixed, defined or ancillary covariates. Let $Z(t) = \{Y, X(t)\}$, so that $Z(t) = Y$ for all t if no internal covariates are present. The overall hazard function is defined as

$$\lambda(t; Z(t)) = \lim_{\Delta t \to 0} \frac{P[t \leq T < t + \Delta t | T \geq t, Z(t)]}{\Delta t}.$$

Cause-specific hazard functions can be defined analogously to (7.1) as

$$\lambda_j(t; Z(t)) = \lim_{\Delta t \to 0} \frac{P[t \leq T < t + \Delta t, J = j | T \geq t, Z(t)]}{\Delta t}, \qquad j = 1, \ldots, m. \tag{7.17}$$

The overall likelihood, conditional on Y, can be obtained by arguments similar to those in Section 5.3. Essentially, the contribution to the likelihood in $[t, t + dt)$ is partitioned as in (5.11). The term in the product integral arising from the failure information is

$$\mathop{\mathscr{P}}_{0}^{t} P[D_t(dt)|H(t)] \tag{7.18}$$

where $H(t)$ is, as before, the complete history of the process to time t and $D_t(dt)$ specifies the items that fail and the failure types in $[t, t + dt)$. This term (7.18) again gives rise to (7.5) but it should be noted that estimation of the survivor function $P(T \geq t; Y)$ would require an integration over the process $X(t)$.

A partial likelihood, useful in the proportional hazards case, can be obtained as the product over all failure times t_{ji}, $j = 1, \ldots, m$, $i = 1, \ldots, k_j$ of the probability that failure occurs on the individual with corresponding covariate function $Z_{ji}(t_{ji})$ given the risk set $R(t_{ji})$, the covariate functions $Z_l(t_{ji})$, $l \in R(t_{ji})$, and the occurrence of a failure of type j at t_{ji}. The partial likelihood can be written

$$\prod_{j=1}^{m} \prod_{i=1}^{k_j} \left(\frac{\lambda_j(t_{ji}; Z_{ji}(t_{ji}))}{\sum_{l \in R(t_{ji})} \lambda_j(t_{ji}; Z_l(t_{ji}))} \right)$$

which reduces precisely to (7.12) under a proportional hazards model,

$$\lambda_j(t; Z(t)) = \lambda_{0j}(t) \exp[\mathbf{z}(t)\boldsymbol{\beta}_j], \qquad j = 1, \ldots, m.$$

Standard asymptotic likelihood techniques applied to (7.12) then provide a viable means of estimating the relative risk parameters $\boldsymbol{\beta}$, even when the covariate process includes internal components.

7.2.5 Identifiability and the Multiple Decrement Function

Most of the competing risk literature approaches problems of types 1–3 above (Section 7.2.1) by defining latent or conceptual failure times corresponding to each failure type. Let $\tilde{T}_1, \ldots, \tilde{T}_m$ denote such conceptual times. The actual failure time is defined to be

$$T = \min(\tilde{T}_1, \ldots, \tilde{T}_m),$$

and the corresponding failure type is

$$J = \{j \mid \tilde{T}_j \leq \tilde{T}_k, \qquad k = 1, \ldots, m\}.$$

Problems of types 1–3 are then posed in terms of a multiple decrement function. Since such a specification has only been given for fixed covariates we will restrict ourselves to a covariate vector $\mathbf{z}$ that is independent of t in this section. The multiple decrement function, or joint survivor function, is specified by

$$Q(t_1, \ldots, t_m; \mathbf{z}) = P(\tilde{T}_1 > t_1, \tilde{T}_2 > t_2, \ldots, \tilde{T}_m > t_m; \mathbf{z}). \tag{7.19}$$

All estimable quantities can be written in terms of Q. For example,

$$F(t; \mathbf{z}) = Q(t, t, \ldots, t; \mathbf{z}),$$

and

$$\begin{aligned} \lambda_j(t; \mathbf{z}) &= \lim_{\Delta t \to 0} \frac{P(t \leq T_j < t + \Delta t \mid T \geq t; \mathbf{z})}{\Delta t} \\ &= \left. \frac{-d \log Q(t_1, \ldots, t_m; \mathbf{z})}{dt_j} \right|_{t_1 = \cdots = t_m = t}, \qquad j = 1, \ldots, m. \end{aligned} \tag{7.20}$$

Since the likelihood function (7.6) can be written entirely in terms of the $\lambda_j(t, \mathbf{z})$'s it follows that functions of Q other than those given in (7.20) cannot be estimated without further assumption. Such functions are therefore termed "nonidentifiable."

For example, the marginal distributions of the latent failure times are generally not identifiable. Let

$$Q_j(t_j; \mathbf{z}) = Q(0, \ldots, 0, t_j, 0, \ldots, 0)$$

represent the marginal "survivor" function for T_j. Q_j is in 1–1 correspondence with the corresponding "hazard" function

$$h_j(t; \mathbf{z}) = \frac{-d \log Q_j(t; \mathbf{z})}{dt}. \tag{7.21}$$

Because these quantities cannot without further assumption be expressed as functions of (7.20) the marginal functions are nonidentifiable. One strong assumption would assert that $\tilde{T}_1, \ldots, \tilde{T}_m$ are statistically independent. That is,

$$Q(t_1, \ldots, t_m; \mathbf{z}) = \prod_{j=1}^{m} Q_j(t_j; \mathbf{z}). \tag{7.22}$$

It follows easily under (7.22), that

$$\lambda_j(t; \mathbf{z}) = h_j(t; \mathbf{z}). \tag{7.23}$$

Actually (7.22) is more restrictive than (7.23) in that for (7.23) to hold we require only

$$Q(t, \ldots, t; \mathbf{z}) = \prod_{j=1}^{m} Q_j(t; \mathbf{z}),$$

assuming the derivatives (7.20) and (7.21) exist.

Without further assumption, however, a hypothesis of independence is wholly untestable. Independence would require the factorization (7.22) to hold, but both the marginal and joint survivor functions are non-identifiable. Data of the type $(T, J; \mathbf{z})$ do not allow one, for example, to distinguish between an independent competing risk model and an infinitude of dependent models giving rise to the same λ_j's. Even if a parametric model with dependent risks is assumed for Q, and all parameters are estimable, it will be impossible to distinguish between the assumed model and one with independent risks but the same cause-specific hazard functions. The following example illustrates this point: suppose it is assumed that

$$Q(t_1, t_2) = \exp\{1 - \alpha_1 t_1 - \alpha_2 t_2 - \exp[\alpha_{12}(\alpha_1 t_1 + \alpha_2 t_2)]\} \tag{7.24}$$

where $\alpha_1, \alpha_2 > 0$ and $\alpha_{12} > -1$. The parameter α_{12} measures the dependence between T_1 and T_2 assuming (7.24). The cause-specific hazard functions are

$$\lambda_j(t) = \alpha_j\{1 + \alpha_{12} \exp[\alpha_{12}(\alpha_1 + \alpha_2)t]\}, \qquad j = 1, 2. \tag{7.25}$$

The likelihood function can be written in terms of the $\lambda_j(t)$'s and it is immediately clear that all three parameters are estimable. The likelihood obtained from (7.24), however, is identical to that arising from the independent risks model with cause-specific hazard functions (7.25) and joint survivor function

$$Q^*(t_1, t_2) = \exp\left(1 - \alpha_1 t_1 - \alpha_2 t_2 - \frac{\alpha_1 e^{\alpha_{12}(\alpha_1+\alpha_2)t_1} + \alpha_2 e^{\alpha_{12}(\alpha_1+\alpha_2)t_2}}{\alpha_1 + \alpha_2}\right). \tag{7.26}$$

It follows that the estimated value of α_{12} should not be taken as an indication of dependence unless there is external evidence to support (7.24). To put it another way, the estimate of the degree of dependence between T_1 and T_2 arises from a model assumption that cannot be checked from the data.

The "marginal" hazard functions from Q and Q^* are

$$h_j(t) = \alpha_j[1 + \alpha_{12} \exp(\alpha_j \alpha_{12} t)]$$

and (7.25), respectively. It is of interest to note that Q and Q^* give rise to proportional cause-specific hazard functions with proportionality factor

$$\frac{\lambda_1(t)}{\lambda_2(t)} = \frac{\alpha_1}{\alpha_2},$$

which in the notation of Section 7.2.3 has maximum likelihood estimator k_1/k_2. The value of k_1/k_2 does not, however, lead to any direct statement about $h_1(t)$ and $h_2(t)$.

The latent failure time formulation is also frequently used to study the association between specific failure types and fixed covariates (e.g. David and Moeschberger, 1978). Such an approach involves the specification and estimation of a regression model $Q(t_1, \ldots, t_m; \mathbf{z})$. Usually, for tractability, the latent failure times are assumed independent. Otherwise, a parametric form must be imposed so that the multiple decrement function is estimable. Since only the derivatives (7.20) enter the likelihood function, it is easier and less restrictive to model only the cause-specific hazard functions as in Section 7.2.3. No assumption then needs to be placed on the interrelation among failure types.

7.2.6 Estimation of the Interrelation between Failure Types

As indicated above, data of the type $(T, J; \mathbf{z})$, with $\mathbf{z}$ time independent do not, without further assumptions, allow one to study the interrelation, or even test for independence, among competing failure modes. More comprehensive data are needed. One possibility with human or animal data is to attempt to use the frequency with which multiple pathologic entities are present at failure (e.g., autopsy) to provide some information on the way such entities are related throughout the individual's life time. The difficulty, of course, is that one may be observing only local phenomena just prior to death, in which the presence of one pathologic entity substantially changes the risk of occurrence of others. Also diagnostic procedures related to one pathologic entity may increase the probability of discovering certain other abnormalities. Similar statements could be made in relation to studies of equipment failure in which a number of different faults may be simultaneously observed at breakdown.

A second and more promising approach involves the use of time dependent covariates. In some studies each, or at least some, failure types will have corresponding risk indicators that can be regularly measured over the course of the study. Such risk indicators can be used to define time dependent regression variables. A test for no association between values of such a risk indicator and the cause-specific hazard function (7.17) for another failure type would provide indirect evidence for or against a hypothesis of no association between the two failure modes.

The illustrations mentioned in Section 7.2.1 can illuminate these ideas. Consider first the marrow transplantation example. Regular measurements of the presence and severity of a graft versus host reaction on the skin, gut, and liver are taken over the patients' post-transplantation course. Such measurements provide rather direct information on the risk of a GVHD death. These data may be used to define one or more time dependent covariates. A test for no association between values of these covariates and the recurrent leukemia relapse rates then provides evidence for or against an association between GVHD and recurrent leukemia as causes of death. In addition, estimation of the coefficients for the time dependent covariates may provide valuable information on characteristics of any suggested interrelation between failure types. As an illustration of this approach (also given in Prentice et al., 1978) a proportional hazards model of the form (7.11) was specified for recurrent leukemia relapse rates (failure type $J = 1$). This model was then applied to data on 135 marrow transplant recipients with the following covariates: (1) an indicator time dependent covariate that takes value zero from the time of transplant to the date of diagnosis of GVHD and value one, thereafter (this is an "internal" time dependent covariate), (2) a fixed covariate giving the patients age at transplant, and (3) an indicator variable indicating whether the marrow donor was a matched sibling (allogeneic transplant) or an identical twin (syngeneic transplant). Table 7.1 gives maximum likelihood estimates and corresponding standard errors for the corresponding regression coefficients.

This analysis indicates a reduction in recurrent leukemia rates by an estimated factor $\exp(-.764) = .47$ upon the onset of GVHD. Asymptotic likelihood theory applied to the GVHD indicator coefficient indicates that this reduction is significant at the .05 level. The analysis suggests a negative relationship between the two failure types. When the risk of

Table 7.1 PROPORTIONAL HAZARDS ANALYSIS OF LEUKEMIA MARROW TRANSPLANT DATA

Covariate	Coefficient (β)	Standard Error	Normal Deviate
GVHD risk indicator	−.764	.370	−2.06
Syngeneic (0) vs Allogeneic (1)	.054	.340	0.16
Age in years/10	.127	.098	1.29

7.3 MULTIVARIATE FAILURE TIME DATA WITH COMPETING RISKS

7.3.1 Hazard Function Representation

In this section, situations are considered in which at least some of the failure types under consideration have the property that the survival of the study subject does not end at the time of their occurrence. The study subject may then be followed beyond the time of first failure for second and subsequent failures. Denote by (T_k, J_k) the time and type of the kth failure on a specific individual $k = 1, 2, \ldots$. The failure times are measured from the same initial time origin so that $T_1 \le T_2 \le \cdots$. The cause-specific hazard functions for the first failure time may be specified exactly as in Section 7.2. The hazard function modeling may be continued to the second and subsequent failure times. In its most general form the cause-specific hazard function for the kth failure on an individual can be permitted to be arbitrary as a function of time and to depend arbitrarily upon the time and type of previous failures, as well as upon covariates. We therefore define the cause-specific hazard functions for the kth failure on a study subject as

$$\lambda_j^k(t; Z_k(t)) \\ = \lim_{\Delta t \to 0} \frac{P[t \le T_k < t + \Delta t, J_k = j \mid T_k \ge t; (t_i, j_i), i = 1, \ldots, k-1; Z(t)]}{\Delta t}, \tag{7.27}$$

for $j = 1, \ldots, m$, $k = 1, 2, \ldots$. Here we have denoted $Z_k(t) = \{Z(t): (t_i, j_i), i = 1, \ldots, k-1\}$ for $k > 1$, with $Z_1(t) = Z(t)$, where the covariate process $Z(t)$ may include fixed, defined, or ancillary covariates $Y = \{\mathbf{y}(u); u \ge 0\}$ as well as "internal" covariates $X(t) = \{\mathbf{x}(u), u \le t\}$. Note that $\lambda_j^k(t; Z_k(t))$ will be identically zero for $t < t_{k-1}$. The possibility that some failure types become impossible following the occurrence of other failure types (or that the set of possible failure types varies with k in a more general manner) can be accommodated in (7.27) simply by restricting (7.27) to be identically zero for the appropriate (k, j) values. A few examples will serve to illustrate the notation.

Consider first a clinical study in early breast cancer which is designed to compare conservative and radical surgery of the primary tumor. Suppose that following mastectomy, patients are followed for disease recurrence and subsequently until death or censoring, or in some cases until death or censoring without recurrence. One can define "failure" as either recurrence of disease ($J = 1$) or death ($J = 2$). The first failure time on an individual then represents disease free survival time while the second failure time, defined only for patients with recurrence, gives time to death ($T_2 \geq T_1$). Questions of interest may concern treatment comparisons in respect to (1) time to recurrence, (2) time to death without recurrence, (3) disease free survival, (4) mortality following recurrence, or (5) overall mortality. Such questions can be addressed directly in terms of models for the hazard functions: (1) $\lambda_1^1(t; Z(t))$; (2) $\lambda_2^1(t; Z(t))$; (3) $\lambda_1^1(t; Z(t)) + \lambda_2^1(t; Z(t))$, (4) $\lambda_2^2(t; Z_2(t))$; and (5) $\lambda_2^1(t; Z(t)) + \lambda_2^2(t; Z_2(t))$, respectively. For example, for a person that dies without recurrence this last expression (5) takes value $\lambda_2^1(t; Z(t))$ from time zero to the time of death, whereas for an individual that experiences disease recurrence (5) has value $\lambda_2^1(t; Z(t))$ from time zero to the time t_1 of disease recurrence and value $\lambda_2^2(t; Z_2(t))$ thereafter.

The heart transplant data discussed in Chapter 5 can be thought of in the notation of this section by defining "failure" to be either transplantation ($J = 1$) or death ($J = 2$). The first failure time T_1 is then the time to transplantation or death without transplantation, and T_2, defined only for transplant patients, is time to death. Of primary interest in this example is a comparison of the mortality rate function $\lambda_2^1(t; Z(t))$ for patients that are still awaiting a heart to $\lambda_2^2(t; Z_2(t))$ for patients, at the same time t (from entry into the program), who have already received a heart. This formulation permits the effect of transplantation to depend on initial prognostic factors and characteristics of the transplantation procedure (e.g., tissue similarity of donor and recipient), as well as on the waiting time for the transplant. The latter possibility is through the inclusion of $(t_1, j_1 = 1)$ in $Z_2(t) = \{Z(t), (t_1, j_1)\}$.

As a rather different illustration suppose there is only a single failure type ($m = 1$) that can occur repeatedly on each study subject. Possible applications include times to asthmatic attacks, epileptic seizures, infection episodes in patients undergoing immunosuppressive therapy, bleeding incidents among hemophiliacs, or times to breakdown in electronic computers. In such circumstances the times to failure $T_1, T_2, \ldots$ on an individual study subject may be thought of as a realization on a point process (e.g., Cox and Lewis, 1966). Models and statistical methods for point processes have been considerably studied but usually within the framework of a single study subject (process). It has then been necessary to place rather severe assumptions such as stationarity, or stronger renewal-type assumptions, on the process in order to develop practical estimation techniques. In most of the applications envisaged here there would be a reasonable number of study subjects, each of which gives rise to a point process of information. It may then not be necessary to restrict the hazard functions (7.27) in such a severe manner. Some restrictions of (7.27) corresponding to models of the type used for point processes are discussed in Section 7.3.4. Note that in the examples mentioned earlier in this paragraph it may be of value to refine the definition of failure, for example to distinguish bacterial, viral, and fungal infections, and to include additional failure types corresponding to entirely different events.

7.3.2 The Likelihood Function

Set

$$\lambda^k(t; Z_k(t)) = \sum_{j=1}^{m} \lambda_j^k(t; Z_k(t)), \qquad k = 1, 2, \ldots$$

so that $\lambda^k(t; Z_k(t))$ is the overall (all causes) failure rate for the kth failure. It takes value zero up to t_{k-1}. As in Section 7.2 the likelihood function will have a particularly convenient expression if there are no "internal" time dependent covariates, so that $Z(t) = Y$, all t. One can then informally build up the likelihood contribution for an individual study subject as follows: proceeding from time 0 the probability of no failure in $[0, t_1)$ is $\exp[-\int_0^{t_1} \lambda^1(t; Z_1(t))\, dt]$. Given that there are no failures in $[0, t_1)$ the probability element for a failure of type j_1 in $[t_1, t_1 + dt)$ is $\lambda_{j_1}^1(t_1; Z_1(t_1))\, dt_1$. Similarly given (t_1, j_1) the probability of no failure in (t_1, t_2) is $\exp[-\int_{t_1}^{t_2} \lambda^2(t; Z_2(t))\, dt]$ after which the probability element for a failure of type j_2 in $[t_2, t_2 + dt_2)$ is $\lambda_{j_2}^2(t_2; Z_2(t_2))\, dt_2$. This procedure is continued until the final time of failure or censoring. General independent censoring schemes can be incorporated by adapting the arguments of Section 5.2. The likelihood contribution is then

$$\prod_{k=1}^{r} \left\{ \exp\left[-\int_{t_{k-1}}^{t_k} \lambda^k(t; Z_k(t)) \right] [\lambda_{j_k}^k(t_k; Z_k(t_k))]^{\delta_k} \right\} \tag{7.28}$$

where $t_0 = 0$, t_r is the total observation period for the study subject and $\delta_k = 1$, $k = 1, \ldots, r-1$, and $\delta_r = 0$ if t_r is a censored failure time and $\delta_r = 1$ otherwise. The overall likelihood (conditional on Y), based on data $\{(t_{ki}, j_{ki}),\ k = 1, \ldots, r_i, \delta_{ri}; Y_i\}$ on study subject i, $i = 1, \ldots, n$ is the product of terms (7.28). After some rearrangement it can be written

$$L = \prod_{i=1}^{n} \left[\prod_{k=1}^{r_i} \left([\lambda_{j_{ki}}^k(t_{ki}; Z_k(t_{ki}))]^{\delta_{ki}} \prod_{j=1}^{m} \exp\left[-\int_{t_{(k-1)i}}^{t_{ki}} \lambda_j^k(t; Z_k(t))\, dt \right] \right) \right]. \tag{7.29}$$

This likelihood function factors into a separate component for each $\lambda_j^k(t; Z_k(t))$; $j = 1, \ldots, m$, $k = 1, 2, \ldots, \tilde{r}$ where $\tilde{r}$ is the maximum of $r_1, \ldots, r_n$. The likelihood factor for a particular $\lambda_j^k(t; Z_k(t))$ is precisely the same as would be obtained from a sample with a single failure time and single failure type in which study subject i begins observation at $t_{(k-1)i}$ and fails at t_{ki} if $J_{ki} = j$, or is censored at t_{ki} if $J_{ki} \neq j$ or $\delta_{ki} = 0$. Many of the methods of the preceding chapters can therefore be readily adapted to the estimation of $\lambda_j^k(t; Z_k(t))$ for any particular (j, k). Sample size limitations, however, usually necessitate restrictions among the λ_j^k functions for feasible estimation.

When $Z(t)$ is relaxed to include internal time dependent covariates, the likelihood can be written by a method similar to that used in Section 5.3. The contribution of the failure time information is (7.29), which arises as a product integral.

$$\mathscr{P}_0^\infty P[D_t(dt)|H(t)] = \prod_{\text{all } k} \mathscr{P}_0^\infty P[D_t^k(dt)|R_k(t), Z_k(t)] \tag{7.30}$$

where $H(t)$ is the history of the process as before including all failure times and types, $D_t(dt)$ gives failure times and types in $[t, t+dt)$, and $D_t^k(dt)$ denotes the set of individuals that experience their kth failure in $[t, t+dt)$. The risk set $R_k(t)$ is the set of individuals at risk of their kth failure at time t ($k-1$ but not k failures have occurred prior to t) and $Z_k(t)$ gives full information on items in $R_k(t)$ up to time t. The full likelihood would involve, as before, additional contributions from the covariate processes.

7.3.3 Estimation of Cause and Failure Specific Hazard Functions

Consider first a homogeneous population with no covariates. The likelihood function (7.29) applies to a random sample from such a population with $Z_k(t) = \{(t_i, j_i),\ i = 1, \ldots, k-1\}$ consisting of the times and types of the preceding failures on an individual, for $k \geq 2$. A nonparametric maximization of the hazard, or cumulative hazard, functions proceeds exactly as in Section 7.2.3 for the hazard functions $\lambda_j^1(t)$, $j = 1, \ldots, m$ pertaining to the failure time T_1. The maximization is more complex thereafter, however, since the λ_j^k functions for $k \geq 2$ are conditional on the preceding failure times and types. Some assumptions concerning the effect of these conditioning variables on the corresponding failure rates will generally be required for estimation. For example, the conditioning data (t_i, j_i), $i = 1, \ldots, k-1$, could be used to divide the study subjects at risk into strata $s = 1, 2, \ldots$. One might then assume a common hazard function for a failure of type j as a function of $t - t_{k-1}$ for all study subjects in stratum s. That is, one might assume in stratum s that

$$\lambda_j^k(t; (t_i, j_i), i = 1, \ldots, k-1) = h_j^k(t - t_{k-1}; s) \tag{7.31}$$

for some hazard function $h_j^k(u; s)$. The hazard functions $h_j^k(u; s)$ would then be estimated exactly as in Section 7.2.3.

Alternatively one might assume a model

$$\lambda_j^k(t;(t_i,j_i), i=1,\ldots,k-1)=\lambda_{0j}^k(t)\exp(\mathbf{z}_k\boldsymbol{\beta}_{kj}) \tag{7.32}$$

where the regression vector $\mathbf{z}_k$ would include functions of the conditioning data (t_i, j_i), $i = 1, \ldots, k-1$. The methods of Section 7.2 could then be used to estimate $\{\lambda_{0j}^k(\cdot), \boldsymbol{\beta}_{kj}\}$ with the risk sets $R_k(t)$ defined as indicated in Section 7.3.2. Further work is needed on methodology for nonparametric estimation and graphical presentation of multivariate failure time data.

Consider now inference on the relationship between the cause and failure specific hazard functions $\lambda_j^k(t; Z_k(t))$ and the covariate functions $Z_k(t)$. For any of the covariate types a partial likelihood for the $\lambda_j^k(t; Z_k(t))$ functions may be written

$$\prod_k \prod_{j=1}^m \prod_{i=1}^{d_{kj}} \left(\frac{\lambda_j^k(t_{kji}; Z_{kji}(t_{kji}))}{\sum_{l\in R_k(t_{kji})} \lambda_j^k(t_{kji}; Z_{kl}(t_{kji}))}\right) \tag{7.33}$$

where $t_{kj1}, t_{kj2}, \ldots$ are the d_{kj} failure times in which the kth failure on an individual is of type j. The (k, j, i) contribution to (7.33) derives from the probability that failure occurs on the study subject with covariate $Z_{kji}(t_{kji})$ at time t_{kji} given the risk set $R_k(t_{kji})$ and given the occurrence at time t_{kji} of the kth failure on a study subject, the failure being of type j.

Consider a proportional hazards model

$$\lambda_j^k(t, Z_k(t)) = \lambda_{0j}^k(t)\exp[\mathbf{z}_k(t)\boldsymbol{\beta}_{kj}], \tag{7.34}$$

all (k, j), where $\mathbf{z}_k(t)$ consists of the value of the function $Z(t)$ at time t as well as additional components that are functions of (t_i, j_i), $i = 1, \ldots, k-1$. The partial likelihood (7.33) then reduces to

$$\prod_k \prod_{j=1}^m \prod_{i=1}^{d_{kj}} \left(\frac{\exp[\mathbf{z}_{kji}(t_{kji})\boldsymbol{\beta}_{kj}]}{\sum_{l\in R_k(t_{kji})} \exp[\mathbf{z}_{kl}(t_{kji})\boldsymbol{\beta}_{kj}]}\right). \tag{7.35}$$

It follows that the relative risk parameters $\boldsymbol{\beta}_{kj}$ can be estimated individually in the same manner as in Chapters 4 and 5. The likelihood factorization (7.29) implies that the corresponding $\lambda_{0j}^k(\cdot)$ functions can also be estimated using the techniques of Chapters 4 and 5 upon inserting the corresponding $\hat{\boldsymbol{\beta}}_{kj}$ from (7.35), provided covariates are not of the internal time dependent type.

The discussion of Section 7.2 on the use of time dependent covariates to study the relationship between failure types as well as the comments on extrapolation to estimate failure rates for some failure types given the removal of other failure types apply also to multivariate failure time data with competing risks.

The next section considers a variety of other hazard models.

7.3.4 Other Hazard Function Models

1. Special cases of (7.34)

The model (7.34) includes an arbitrary shape function $\lambda_{0j}^{k}(t)$ for each failure type $j = 1, 2, \ldots$ and each failure time $k = 1, 2, \ldots$. There are a variety of restrictions that could be placed on these functions or upon the corresponding regression coefficients. A modulated nonhomogeneous Poisson model requires that failure type j has a common shape function $\lambda_{0j}^{1}(t) = \lambda_{0j}^{2}(t) = \cdots = \lambda_{0j}(t)$ which is modulated by the covariate vector $\mathbf{z}_k(t)$ to give the hazard functions (7.27). The hazard functions can then be written

$$\lambda_j^k(t; Z_k(t)) = \lambda_{0j}(t) \exp[\mathbf{z}_k(t)\boldsymbol{\beta}_{kj}], \qquad \text{all } (k, j). \tag{7.36}$$

Recall that $\mathbf{z}_k(t)$ includes the value of fixed and time dependent covariates at time t as well as specific functions of the previous failure times and types (t_i, j_i), $i = 1, \ldots, k-1$. The model (7.36) presumes that the effect of previous failures on the cause-specific hazard functions can be adequately modeled, with suitable definition of $\mathbf{z}_k(t)$, by means of the multiplicative factor $\exp[\mathbf{z}_k(t)\boldsymbol{\beta}_{kj}]$. Such an assumption would likely be adequate in many situations. For example, the model (7.36) would be an appealing one for the study of bacterial infection incidence in the marrow transplantation setting discussed in Section 7.2. The time from marrow transplantation t is the crucial variable affecting a patient's immune competence and thereby the patient's ability to fight infections. All patients experience severe immunodeficiency in the first few weeks following transplantation after which immunity gradually returns at a rate depending on patient and treatment characteristics and particularly depending on whether or not the patient is experiencing graft versus host disease. If, for example, T_k was defined as the time from transplantation to the kth bacterial infection for a patient, the model (7.36) would permit this infection incidence at a particular time to depend in a multiplicative manner on the number and timing of previous infections as well as upon patient and treatment characteristics. The inclusion of interaction terms between t and the previous infection times would allow even greater flexibility.

A "fuller" partial likelihood than (7.35) can be formed from (7.36). It can be written in the above notation as

$$\prod_k \prod_{j=1}^{m} \prod_{i=1}^{d_{kj}} \left(\frac{\exp[\mathbf{z}_{kji}(t_{kji})\boldsymbol{\beta}_{kj}]}{\sum_{\text{all } \kappa} \sum_{l \in R_\kappa(t_{kji})} \exp[\mathbf{z}_{\kappa l}(t_{kji})\boldsymbol{\beta}_{\kappa j}]} \right). \tag{7.37}$$

This likelihood factors into a separate component for each j, but the coefficients $\boldsymbol{\beta}_{kj}$, $k = 1, 2, \ldots$ for any particular j will need to be estimated simultaneously in an application of asymptotic maximum likelihood procedures to (7.37). The contribution to (7.37) from the failure at t_{kji} is based on the probability that the particular individual with covariate $\mathbf{z}_{kji}(t_{kji})$ fails given the risk sets $R_k(t_{kji})$, $k = 1, 2, \ldots$, and the occurrence of a failure of type j at t_{kji}. In some circumstances it may be reasonable to restrict the coefficients to have some common components as k ranges over $1, 2, \ldots$.

A modulated Poisson process (Cox, 1973) would be given by (7.36) with $\lambda_{0j}(t)$ restricted to be a constant for each $j = 1, \ldots, m$. Likelihood methods could easily be developed for such a model.

2. *Modulated Renewal Process*

This model assumes that the hazard functions for the time intervals between successive failures on a study subject have a shape function that depends arbitrarily on the failure number and type and that is modulated by the covariate process. The hazard function is

$$\lambda_j^k(t; Z(t)) = \lambda_{0j}^k(t - t_{k-1}) \exp[\mathbf{z}_k(t)\boldsymbol{\beta}_{kj}], \qquad \text{all } (k, j). \tag{7.38}$$

Such a model would be suitable when the process of repair, recovery, or renewal following a failure leads to hazard functions that can be more parsimoniously expressed in terms of gaps between failure times rather than in terms of the total observation time. With a suitable allowance, in the definition of $\mathbf{z}_k(t)$, for dependence on previous failure times and types this model may also be sufficiently general for many situations. Cox (1973) considers (7.38) with a single failure type. His model does not permit the regression coefficient or the shape function λ_{0j}^k to depend on k. The modulated Poisson process is also a special case of (7.38) upon setting $\lambda_{0j}^k(u)$ equal to a constant for each $j, k = 1, \ldots, m$.

Once again a partial likelihood can be formed from (7.38). For this purpose it is convenient to denote the interfailure times by $U_k = T_k - T_{k-1}$, $k = 1, 2, \ldots$, with $T_0 = 0$. One can also define $\tilde{R}_k(u)$ to be the set of individuals that are observed beyond their $(k-1)$st failure for which $U_k \geq u$. Denote by u_{kji} the value of U_k corresponding to the failure time t_{kji}. A partial likelihood for the coefficients $\boldsymbol{\beta}_{kj}$, $j = 1, \ldots, m$, is then

$$\prod_{j=1}^{m} \prod_{i=1}^{d_{kj}} \left(\frac{\exp[\mathbf{z}_{kji}(t_{kji})\boldsymbol{\beta}_{kj}]}{\sum_{l \in \tilde{R}_k(u_{kji})} \exp[\mathbf{z}_{kl}(t_{kji})\boldsymbol{\beta}_{kj}]} \right). \tag{7.39}$$

The jth component of this likelihood derives from the probability that the failure at interfailure time u_{kji} occurs on individual with corresponding covariate value $\mathbf{z}_{kji}(t_{kji})$ given the interfailure risk sets $\tilde{R}_k(u_{kji})$, and the fact that at interfailure time u_{kji} the kth failure on a study subject occurs and the failure is of type j. Again equality of certain components of $\boldsymbol{\beta}_{kj}$, $k = 1, 2, \ldots$, will be in order in some situations.

The model (7.38) can be restricted by setting $\lambda^k{}_{0j}(u) = \lambda_{0j}(u)$, $k = 1, 2, \ldots$. This gives rise to a partial likelihood that is the product of terms (7.39) over k but with the denominator term summed over all k. In addition, (7.38) can be relaxed to allow the $\lambda_{0j}(u)$ functions to be specific to each individual study subject for the study of the effects of time dependent covariates. The overall likelihood then is the product of terms (7.39) over individuals. Such a model should provide a very thorough accommodation of correlations among failure times on the same study subject and could be used when there is only a single study subject or single point process of information. Whittemore and Keller (1978) have applied this type of model to discrete failure time data relating asthmatic attacks to air pollution levels.

3. Other Models

The two classes of models (7.34) and (7.38) can both be relaxed by allowing the shape function and the regression coefficients to depend on a (time dependent) stratification variable, $s = s(z_k(t))$ in place of the dependence on the total number of failures, k. For example, s may be defined in terms of the number of preceding failures, of each particular failure type or upon the number, type and times of preceding failures. A partial likelihood function is readily obtained with a general definition of strata.

7.3.5 Joint Survivor Function Representation of Multivariate Failure Time Data

In a subset of the situations considered above one could consider an alternative representation of multivariate failure time data. In particular, if there are no covariates or only fixed covariates $\mathbf{z}$ and if each failure type $J = 1, \ldots, m$ can occur at most once then one can specify a joint survivor or multiple decrement function

$$Q(y_1, \ldots, y_m; \mathbf{z}) = P(Y_1 \geq y_1, \ldots, Y_m \geq y_m; \mathbf{z}) \tag{7.40}$$

as in Section 7.2.5, where Y_j represents the time to failure of type j. In contrast T_k, defined above, is the time to the kth failure, regardless of the failure type. With genuinely multivariate data the identifiability problems discussed in Section 7.2.5 are lessened to the extent that if all m failure times are observed on at least some of the study subjects the whole function (7.40) is identifiable.

Since the hazard function representation (7.27) completely specifies the distribution of the observed failure times (conditional on $\mathbf{z}$) it is clear that (7.40) must be expressible in terms of these hazard functions. Let $q(y_1, \ldots, y_m; \mathbf{z})$ represent the density function corresponding to (7.40). Within any of the $m!$ regions corresponding to a specific rank vector for $y_1, \ldots, y_m$ the density $q(y_1, \ldots, y_m; \mathbf{z})$ admits a simple expression in terms of the hazard functions (7.27). For example, within the region $y_1 < y_2 < \cdots < y_m$ one has $Y_i = T_i$, $i = 1, \ldots, m$ so that

$$q(y_1, \ldots, y_m) = \prod_{i=1}^{m} \left\{ \exp\left[- \int_{y_{i-1}}^{y_i} \lambda^i(t; \mathbf{z})\, dt \right] \lambda_i^i(y_i; \mathbf{z}) \right\} \tag{7.41}$$

with $y_0 = 0$. A similar expression holds in each of the other regions.

The use of (7.40) would seem to have little advantage over (7.27) for the study of the effects of regression variables on the occurrence of certain types of failures. Modeling and estimation based on (7.27), rather than (7.40), would seem to give more direct information on the failure process. As has been seen above, some flexible specifications of (7.27) accommodate censoring readily and permit feasible inference procedures even when time dependent covariates are present.

BIBLIOGRAPHIC NOTES

The actuarial approach to competing risk problems, in the absence of regressor variables, has been discussed by many authors. For example, Seal (1954), Cornfield (1957), Elveback (1958), Kimball (1958, 1969), Berkson and Elveback (1960), Chiang (1961; 1968, Chapter II; 1970), Pike (1970), and Hoel (1972) discuss models and statistical methods for the estimation of crude, partial crude, and net probabilities. An assumption of independence between latent failure times pervades most of this work though Kimball (1958, 1969) considers a somewhat different, but equally strong assumption. Chiang's work mostly involves the additional convenient assumption that cause-specific hazard functions are proportional within specified failure time intervals. As pointed out by Makeham (1874) and Cornfield (1957), the assumption that the elimination of certain causes of failure only nullifies the corresponding arguments in the multiple decrement function Q, has no prior validity. Gail (1975) has provided an excellent review of the above literature. Aalen (1976) has considered some formal properties of nonparametric estimators for the multiple decrement model.

The fact that data of the form (T, J) do not allow one to discriminate between an independent risk model and an infinitude of dependent risk models has been pointed out by Cox (1959; 1962, p. 112) and more recently by Tsiatis (1975) and Peterson (1975). Peterson (1976) provides bounds on Q and the Q_j functions corresponding to specified λ_j's.

Many authors have considered parametric models for competing risk data. Only some recent references are listed here: Marshall and Olkin (1967), Moeschberger and David (1971), and Moeschberger (1974) utilize exponential and Weibull models. The Marshall and Olkin bivariate models include a positive probability that the latent failure time (T_1, T_2) are equal. Nádas (1971) has used a bivariate normal model and Hoel (1972) fitted an independent risks Makeham–Gompertz model. The reader is referred to the monograph by David and Moeschberger (1978) for a more detailed discussion of estimation from the multiple decrement formulation. A historical perspective is given by Seal (1977).

The proportional hazards cause-specific hazard function model (7.11) has been utilized by Holt (1978) and Prentice and Breslow (1978). The discussion of Section 7.2 generalizes its use by Prentice et al. (1978).

Fisher and Kanarek (1974) and Williams and Lagakos (1977) consider procedures for investigating the consequences of a lack of "independent" censoring on the survivor function and regression coefficient estimators. Fisher and Kanarek suppose that censored individuals possess a hazard rate that is proportional to that for the remainder of the sample. The effect of varying this proportionality parameter on estimators of $F(t; \mathbf{z})$ can be readily investigated, though such a parameter is, of course, nonidentifiable.

Rather little seems to have been written on the analysis of multivariate failure time data from the perspective of Section 7.3. Cox (1972) gave hazard function definitions for bivariate survival data. Cox (1973) considered a less general form of the modulated renewal process of Section 7.3.4. Other approaches to the modeling of bivariate and multivariate failure time data have been considered by Barlow and Proschan (1976), Basu (1971) and Johnson and Kotz (1975). Lagakos, Sommer, and Zelen (1978) consider a Markov renewal model that is a special case of (7.38). The monographs Cox (1962b) and Cox and Lewis (1966) and the book Snyder (1975) provide a good deal of additional material on point processes, mostly under assumptions that give more structure to the process than was assumed above. Some work related to Section 7.3 has recently been carried out by Barbara Williams and Arthur Peterson at the University of Washington.

CHAPTER 8

Miscellaneous Topics

8.1 ANALYSIS OF PAIRED FAILURE TIMES

8.1.1 Background

Suppose again that there is a single possibly censored failure time for each study subject, and a single failure type. Failure times may occur in pairs either naturally or by design. Natural pairing arises, for example, in twin studies and in studies in which two failure times are recorded on each individual or piece of equipment. In a matched pair study, individuals sharing certain characteristics are assigned to a pair. Table 8.1, from Batchelor and Hackett (1970), provides an example of naturally paired failure times. The survival times of closely and poorly HLA (human lymphocyte antigen) matched skin grafts on the same burned individual are given along with the percentage of body surface of full thickness burn.

Typically, certain primary regressor variables, $\mathbf{z}$ (such as quality of the match in Table 8.1) are to be related to failure time and other factors (such as characteristics of the individual receiving the graft) form the basis for pairing. One possibility for the analysis of such data is to extend the regression vector to include not only the primary variables but also variables that describe the matching. The methods of preceding chapters can then be used directly. Such a procedure may be quite appropriate in certain matched pair situations in which the matching depends on a small number of explicit quantitative criteria. With natural pairing, however, the characteristics common to a pair that are relevant to failure time may be too detailed and extensive to permit quantification and modeling. Estimation of the primary regression coefficient in this case can be handled by allowing each pair to define a new stratum, leading to an analysis based on within pair data. Such an analysis is also valid even if matching factors can be explicitly modeled, but with some associated reduction in efficiency. The remainder of this section is devoted to models in which each pair includes a parameter or parameters specifically in-

Table 8.1 DAYS OF SURVIVAL OF CLOSELY AND POORLY MATCHED SKIN GRAFTS ON THE SAME PERSON, FROM BATCHELOR AND HACKETT (1970)*

Case number	4	5	7	8	9	10	11	12	13	15	16
Survival of close match	37	19	57^+	93	16	21–23	20	18	63	29	60^+
Graft			(57^+)						(77, 29)		
Survival of poor match	29	13	15	26	11	15–18	26	19–23	43	15	38–42
Graft										(18)	
Amount of burn	30	20	25	45	20	18	35	25	50	30	30

*Amount of burn is percentage of body surface of full thickness burn. A (+) indicates censoring, and the numbers in parentheses are survival times of other grafts on the same person.

tended to describe failure time properties of the stratum defined by that pair.

Any of the parametric or partially parametric methods of preceding chapters may be adapted to paired failure times. Proportional hazards methods are considered in Section 8.1.2, log-linear methods are considered in Section 8.1.3, and competing risks problems are considered in Section 8.1.4.

8.1.2 Proportional Hazards Methods

Let $\{t_{si}; Z_{si}(t_{si})\}$ represent the failure time and regression function for the ith individual in the sth pair. Let $\mathbf{z}_{si}(t_{si})$ be the value of $Z_{si}(t_{si})$ at time t_{si}. Recall that a covariate function $Z(t)$ is defined to include the covariate values over the entire study period for fixed, defined, or ancillary covariates as well as the values for internal time dependent covariates up to time t (see Section 5.3 for definitions).

Under a proportional hazards model the hazard function for the sth pair may be written

$$\lambda(t; Z(t), s) = \lambda_{0s}(t) \exp[\mathbf{z}(t)\boldsymbol{\beta}]. \tag{8.1}$$

A partial (also conditional) likelihood can be formed as the product over pairs of the conditional probability for the "pair rank" given the smallest failure time $t_{s(1)}$ in the sth pair. With continuous failure times the pair rank has only two possible values, depending on which member of the pair fails first. Denote

$$r_s = \begin{cases} 0, & t_{s1} < t_{s2} \\ 1, & t_{s2} < t_{s1} \end{cases}$$

and set $\mathbf{d}_s = \mathbf{z}_{s2}\{t_{s(1)}\} - \mathbf{z}_{s1}\{t_{s(1)}\}$. The contribution to the conditional likelihood from pair s is then

$$P(r_s | t_{s(1)}; \mathbf{d}_s) = \frac{\exp(r_s \mathbf{d}_s \boldsymbol{\beta})}{1 + \exp(\mathbf{d}_s \boldsymbol{\beta})}. \tag{8.2}$$

The conditional likelihood

$$\prod_{s \in S} \left(\frac{\exp(r_s \mathbf{d}_s \boldsymbol{\beta})}{1 + \exp(\mathbf{d}_s \boldsymbol{\beta})} \right), \tag{8.3}$$

where S consists of all pairs for which r_s can be specified, is precisely a binary logistic likelihood. Standard asymptotic likelihood methods can be applied to (8.3). The score statistic is

$$\mathbf{u}(\boldsymbol{\beta}) = \sum_{s \in S} \mathbf{d}_s' \left(r_s - \frac{\exp(\mathbf{d}_s \boldsymbol{\beta})}{1 + \exp(\mathbf{d}_s \boldsymbol{\beta})} \right) \tag{8.4}$$

while the information matrix is

$$I(\boldsymbol{\beta}) = \sum_{s \in S} \mathbf{d}_s' \mathbf{d}_s \left(\frac{\exp(\mathbf{d}_s \boldsymbol{\beta})}{[1 + \exp(\mathbf{d}_s \boldsymbol{\beta})]^2} \right). \tag{8.5}$$

Further detail on estimation from (8.3) is given by Cox (1970).

A possible difficulty arises from the absence of any contribution to (8.3) from pairs for which r_s cannot be specified. To investigate this, suppose there are fixed potential censoring times and let t_s^0 represent the minimum of the potential censoring times for the sth pair. The pair rank r_s will be observed if and only if $t_{s(1)} \leq t_s^0$. One can calculate

$$P(r_s | t_{s(1)} \text{ and } t_{s(1)} < t_s^0; \mathbf{d}_s) = \frac{\exp(r_s \mathbf{d}_s \boldsymbol{\beta})}{1 + \exp(\mathbf{d}_s \boldsymbol{\beta})}$$

as before. It follows that (8.3) is a proper conditional likelihood and that no systematic bias results from the omission of pairs for which $t_{s(1)} > t_s^0$.

A distinct, but related, concern arises in follow-up studies in twin populations, for example. Pairs in which one or both members have died before the initiation of the study may well be excluded from study because of difficulty in the identification of such pairs or necessity to interview study subjects. Failure time, on the other hand, may refer to time since birth or since some other time prior to the beginning of the study. For a pair to be included in the study it is then required that both t_{s1} and t_{s2} exceed a certain time t_s^0 (e.g., from birth to initiation of study). A simple calculation gives

$$\dot{P}(r_s|t_{s1}, t_{s2} > t_s^0; \mathbf{d}_s) = \frac{\exp(r_s\mathbf{d}_s\boldsymbol{\beta})}{1+\exp(\mathbf{d}_s\boldsymbol{\beta})} \tag{8.6}$$

as before, provided the regressor variable is now restricted to be time independent. This means that, subject to the adequacy of the proportional hazards model, there is no bias involved in restricting the study to pairs in which both members are currently alive.

Table 8.1 gives $r_s = 1$ for nine pairs and $r_s = 0$ for two pairs. A single indicator regressor variable for quality of the match (0—close match; 1—poor match) leads to $d_s \equiv 1$, all s, and a likelihood proportional to

$$\frac{\exp(9\beta)}{(1+\exp\beta)^{11}},$$

so that $\hat{\beta} = \log\frac{9}{2} = 1.504$ with asymptotic variance $11/[(2)(9)] = .611$. This yields an approximate 95% confidence interval for β of $[1.504 \pm 1.96(.611)^{1/2}] = (-.028, 3.036)$. Correspondingly, an approximate interval for e^{β}, the ratio of hazard rates for poor and well matched grafts, is (0.97, 20.83). In good agreement, an exact test for $\beta = 0$ based on the binomial parameter $p = e^{\beta}/(1+e^{\beta})$ yields a significance level of $2(1+11+55)/2^{11} = .065$. An additional regressor variable defined as the product of the above indicator variable and amount of burn can be used to test whether the importance of the HLA tissue typing depends on the severity of the burn.

8.1.3 Log-Linear Models

We may also consider log-linear failure time models for paired data. Suppose $y_{si} = \log t_{si}$ follows a model

$$y_{si} = \alpha_s + \mathbf{z}_{si}\boldsymbol{\beta} + \sigma_s v_{si} \tag{8.7}$$

with error p.d.f. $f_s(v_{si})$ and $\mathbf{z}_{si}$ time independent. Expression (8.7) is overparametrized, having more unrelated parameters than failure times. The assumption of common error distribution ($\sigma_s = \sigma$, $f_s = f$; all s) relieves this problem and leads to

$$x_s = y_{s2} - y_{s1} = \mathbf{d}_s\boldsymbol{\beta} + \sigma w_s, \tag{8.8}$$

where the error w_s, corresponding to the pairwise difference in log failure times, has symmetric p.d.f.

$$g(w) = \int_{-\infty}^{\infty} f(v)f(w+v)\,dv. \tag{8.9}$$

For example, a Weibull density for t_{si} leads to an extreme minimum density for v_{si} which in turn gives a logistic density for w_s. Log-normal t_{si}'s, on the other hand, give a normal regression model for x_s.

With uncensored data the methods for Chapter 3 can be used for inference on $\boldsymbol{\beta}$. These methods have somewhat greater efficiency than the partially parametric method of Section 8.1.2. For example, assuming a two parameter Weibull model the relative efficiency of the maximum likelihood estimator based on (8.3) to that based on (8.8) is 75% at $\boldsymbol{\beta} = \mathbf{0}$ and less than 75% elsewhere.

In the presence of censoring, however, a special problem arises with pairs in which both members are censored. Such pairs clearly can convey some information on $\boldsymbol{\beta}$ even though the range of possible x_s values is unrestricted. In terms of (8.7), such pairs provide data consistent with an arbitrarily large α_s. Some authors (e.g., Sampford and Taylor, 1959) have simply ignored such pairs even though there is potential for significant bias (Holt and Prentice, 1974).

A proper conditional likelihood may be based on the distribution of x_s given that at least one survival time is uncensored. This likelihood is generally complicated and depends explicitly on the potential censoring times for individuals that fail. Such censoring times would often be unavailable.

Following Chapter 6, analysis from (8.7) may also be based on pseudo-ranks for each pair. For simplicity assume a common error distribution ($\sigma_s = \sigma$, $f_s = f$). Now in order to test $\boldsymbol{\beta} = \boldsymbol{\beta}_0$, form $w_{si} = y_{si} - \mathbf{z}_{si}\boldsymbol{\beta}_0$ and let

$$r_s = r_s(\boldsymbol{\beta}_0) = \begin{cases} -1, & w_{s1} < w_{s2} \\ \;\;\,1, & w_{s2} < w_{s1} \end{cases}.$$

Then

$$P(r_s = -1; \mathbf{z}_{s1}, \mathbf{z}_{s2}) = \int_{\infty}^{\infty} \int_{\tau_{s1}}^{\infty} \prod_{1}^{2} [f(\tau_{si} - \mathbf{z}_{si}\boldsymbol{\gamma})\, d\tau_{si}],$$

where

$$\tau_{si} = \frac{w_{si} - \alpha_s}{\sigma}, \qquad \boldsymbol{\gamma} = \frac{\boldsymbol{\beta} - \boldsymbol{\beta}_0}{\sigma},$$

and

$$P(r_s = 1; \mathbf{z}_{s1}, \mathbf{z}_{s2}) = 1 - P(r_s = -1; \mathbf{z}_{s1}, \mathbf{z}_{s2}).$$

The contribution from pair s to the score statistic for testing $\boldsymbol{\beta} = \boldsymbol{\beta}_0$ ($\boldsymbol{\gamma} = \mathbf{0}$) is simply

$$\begin{aligned} \mathbf{u}_s &= \frac{d \log P(r_s; \mathbf{z}_{s1}, \mathbf{z}_{s2})}{d\gamma} \\ &= c r_s \mathbf{d}_s' \end{aligned}$$

where $\mathbf{d}_s = \mathbf{z}_{s2} - \mathbf{z}_{s1}$, as before, and

$$c = 2\int_{-\infty}^{\infty}\int_{\tau_1}^{\infty}\left(\frac{-d\log f(\tau_1)}{d\tau_1}\right)\prod_1^2 [f(\tau_i)\,d\tau_i].$$

Under $\boldsymbol{\beta} = \boldsymbol{\beta}_0$, r_s takes values -1 or $+1$, each with probability $\frac{1}{2}$. Hence, the variance of $\mathbf{u}_s$ is

$$U_s = c^2\mathbf{d}_s'\mathbf{d}_s.$$

Let S denote the set of pairs for which $r_s = r_s(\boldsymbol{\beta}_0)$ can be specified and set

$$\mathbf{u} = \sum_S \mathbf{u}_s, \qquad U = \sum_S U_s.$$

A test for $\boldsymbol{\beta} = \boldsymbol{\beta}_0$ can then be based on an approximate χ^2 distribution (degrees of freedom the dimension of $\boldsymbol{\beta}$, assuming U nonsingular) for

$$\mathbf{u}'U^{-1}\mathbf{u} = \sum_S \mathbf{d}_s r_s\left(\sum_S \mathbf{d}_s'\mathbf{d}_s\right)^{-1}\sum_S \mathbf{d}_s' r_s. \tag{8.10}$$

Note that (8.10) is independent of the choice of f.

The data of Table 8.1 lead to a $\chi^2_{(1)}$ statistic, for testing the hypothesis of no association between HLA match and survival, of value

$$\mathbf{u}'U^{-1}\mathbf{u} = (9-2)(11)^{-1}(9-2) = 4.45$$

which is significant at the .05 level. Such a test is suitable regardless of the error distribution f.

The omission of pairs from (8.10) for which r_s cannot be specified (censored w_{si} less than first uncensored w_{si}) does not bias the test since r_s takes value ± 1 with probability $\frac{1}{2}$, under $\boldsymbol{\beta} = \boldsymbol{\beta}_0$, even conditional on r_s being observed. Interval estimation and partial tests for $\boldsymbol{\beta}$ can be based on (8.10) as discussed in Chapter 6. Note that, unlike inference based on (8.3), the pairs utilized in testing $\boldsymbol{\beta} = \boldsymbol{\beta}_0$ via (8.10) depend on the hypothesized value $\boldsymbol{\beta}_0$.

8.1.4 Competing Risks

Methods for the analysis of paired failure times may be extended to allow for competing causes of failure and multivariate failure times. This simply involves specialization of the results of Chapter 7 to paired data. For example, with a single failure time for each study subject we may suppose that the cause-specific hazard function for cause j in pair s is of the form

$$\lambda_j(t; Z(t), s) = \lambda_{0js}(t)\exp[\mathbf{z}(t)\boldsymbol{\beta}_j]. \tag{8.11}$$

Setting

$$r_{js} = \begin{cases} 0 \text{ if first failure in pair } s \text{ is of cause } j \text{ and } t_{s1} < t_{s2} \\ 1 \text{ if first failure in pair } s \text{ is of cause } j \text{ and } t_{s2} < t_{s1} \end{cases}$$

leads to a conditional likelihood for the cause-specific regression coefficients $\boldsymbol{\beta}_j$ of the form (8.3)

$$\prod_j \prod_{s \in S_j} \left(\frac{\exp(\mathbf{d}_s \boldsymbol{\beta}_j r_{js})}{1 + \exp(\mathbf{d}_s \boldsymbol{\beta}_j)} \right),$$

where S_j consists of pairs for which r_{js} can be specified for some j. The more restrictive model in which the shape functions $\lambda_{0js}(\cdot)$, $j = 1, \ldots, m$, are required to be proportional to each other can be written

$$\lambda_j(t; Z(t), s) = \lambda_{0s}(t) \exp(\gamma_j + \mathbf{z}\boldsymbol{\beta}_j). \tag{8.12}$$

This model gives a partial likelihood of the form (7.15). Let $j(1)$ represent the cause of failure for the first individual to fail in pair s and let $\mathbf{z}_{s(1)}$ be the corresponding regressor variable at the time of failure. Let $j(2)$ and $\mathbf{z}_{s(2)}$ be the corresponding quantities for the second failure. The partial likelihood for $(\boldsymbol{\beta}_j, \gamma_j)$, $j = 1, \ldots, m$, can be written

$$\prod_{s \in S_1} \left(\frac{\exp(\gamma_{j(1)} + \mathbf{z}_{s(1)} \boldsymbol{\beta}_{j(1)})}{\sum_{j=1}^m [\exp(\gamma_j + \mathbf{z}_{s1} \boldsymbol{\beta}_j) + \exp(\gamma_j + \mathbf{z}_{s2} \boldsymbol{\beta}_j)]} \right) \prod_{s \in S_2} \left(\frac{\exp(\gamma_{j(2)} + \mathbf{z}_{s(2)} \boldsymbol{\beta}_{j(s)})}{\sum_{j=1}^m \exp(\gamma_j + \mathbf{z}_{s(2)} \beta_j)} \right) \tag{8.13}$$

where S_1 consists of pairs in which a failure occurs before any censoring, and S_2 consists of pairs in which both members are observed to failure.

8.2 FIXED STUDY PERIOD SURVIVAL AND BIOLOGICAL ASSAY

In some studies the only failure time data recorded are whether or not each individual survives some study period common to all individuals. Such studies can be viewed as special cases of classical quantal response bioassay. For example, Table 8.2 presents data on death or survival of insects, after 5 hours of exposure to various levels of gaseous carbon disulfide.

Each of the parametric and semiparametric survival models of preceding chapters may be specialized to this type of data. Such specialization delivers the usual binary response regression models. Consider first the log-linear models of Chapter 3. Log failure time y is presumed to arise from a model

$$y = \mu + \mathbf{z}\boldsymbol{\gamma} + \sigma v,$$

Table 8.2 MORTALITY OF ADULT BEETLE AFTER FIVE HOURS EXPOSURE TO GASEOUS CARBON DISULFIDE (FROM BLISS [1935])

Dosage ($\log_{10} CS_2$mg/liter)	1.6907	1.7242	1.7552	1.7842	1.8113	1.8369	1.8610	1.8839
Insects	59	60	62	56	63	59	62	60
Killed	6	13	18	28	52	53	61	60
Probit fit ($\hat{\mu} = 1.771$, $\hat{\sigma} = .051$)	3.27	10.89	23.65	33.88	49.60	53.28	59.63	59.21
Logit fit ($\hat{\mu} = 1.772$, $\hat{\sigma} = .029$)	3.45	9.84	22.45	33.89	50.10	53.29	59.22	58.74
Log F ($m_2 = 1$) ($\hat{\mu} = 1.818$, $\hat{\sigma} = .016$, $\hat{m}_1 = 2.79$)	6.07	11.23	20.10	29.69	48.57	54.84	60.93	59.75

with error p.d.f. $f(v)$ and fixed covariate $\mathbf{z}$. The probability that failure time t is less than a study period T may be written

$$P(\mathbf{z}) = \int_{-\infty}^{\alpha+\mathbf{z}\boldsymbol{\beta}} f(v)\, dv,$$

where $\alpha = (\log T - \mu)/\sigma$ and $\boldsymbol{\beta} = \boldsymbol{\gamma}/\sigma$. Weibull, log-logistic, and log-normal models for T then yield the so-called "multi-hit" model

$$P(\mathbf{z}) = 1 - \exp(-e^{\alpha+\mathbf{z}\boldsymbol{\beta}}),$$

the logit model (8.2),

$$P(\mathbf{z}) = \frac{\exp(\alpha + \mathbf{z}\boldsymbol{\beta})}{1 + \exp(\alpha + \mathbf{z}\boldsymbol{\beta})}$$

and the probit model

$$P(\mathbf{z}) = \int_{-\infty}^{\alpha+\mathbf{z}\boldsymbol{\beta}} (2\pi)^{-1/2} \exp(-\tfrac{1}{2} v^2)\, dv$$

respectively. The terminology multi-hit, logit, and probit usually apply to situations in which there is a scalar regression variable z that gives the dosage of some toxic substance. The present notation includes the possibility that the regression variable $\mathbf{z}$ may include more complicated functions of such a dosage variable.

The proportional hazards model (4.1) may also be used to induce a binary response model. A hazard function $\lambda_0(t)e^{\mathbf{z}\boldsymbol{\beta}}$ gives once again the multi-hit model

$$P(\mathbf{z}) = 1 - \exp(-e^{\alpha+\mathbf{z}\boldsymbol{\beta}}),$$

where $\alpha = \alpha(T) = \log \int_0^T \lambda_0(u)\, du$. Upon relaxing the proportional hazards model to include "defined" time dependent covariates (see Section 5.1) $\mathbf{x}(t)$ with coefficient $\boldsymbol{\gamma}$ the probability of failure before time T becomes

$$P(\mathbf{z}) = 1 - \exp[-e^{\mathbf{z}\boldsymbol{\beta}} g(T;\gamma)]$$

where $g(T;\gamma) = \int_0^T \lambda_0(u) \exp[\mathbf{x}(u)\gamma]\, du$.

The generalized F regression model (Chapters 2 and 3) offers additional special cases. For example, the subclass with $m_2 = 1$ gives

$$P(\mathbf{z}) = \left(\frac{\exp(\alpha + \mathbf{z}\boldsymbol{\beta})}{1 + \exp(\alpha + \mathbf{z}\boldsymbol{\beta})}\right)^{m_1}$$

whereas $m_1 = 1$ leads to

$$P(\mathbf{z}) = 1 - \left(\frac{\exp(-\alpha - \mathbf{z}\boldsymbol{\beta})}{1 + \exp(-\alpha - \mathbf{z}\boldsymbol{\beta})}\right)^{m_2}.$$

Model fitting and estimation from such binary response models will be very briefly considered. For individual i with regression vector $\mathbf{z}_i$, let

$$r_i = \begin{cases} 1, & t_i < T \\ 0, & \text{otherwise.} \end{cases}$$

The likelihood is simply

$$\prod_1^n P(\mathbf{z}_i)^{r_i} Q(\mathbf{z}_i)^{1-r_i},$$

where $Q = 1 - P$. The score statistic in regard to a parameter $\boldsymbol{\theta} = (\theta_1, \theta_2, \ldots)$ that includes α, $\boldsymbol{\beta}$ and possibly other shape parameters has jth component

$$s_j = \sum_{i=1}^n [r_i P(\mathbf{z}_i)^{-1} - (1 - r_i) Q(\mathbf{z}_i)^{-1}] \left(\frac{dP(\mathbf{z}_i)}{d\theta_j}\right), \tag{8.14}$$

and the (j, k) element of the information matrix, Σ, is

$$\left(\Sigma\right)_{jk} = \sum_{i=1}^n [P(\mathbf{z}_i) Q(\mathbf{z}_i)]^{-1} \left(\frac{dP(\mathbf{z}_i)}{d\theta_j}\right)\left(\frac{dP(\mathbf{z}_i)}{d\theta_k}\right). \tag{8.15}$$

Further,

$$\frac{dP(\mathbf{z})}{d\alpha} = f(\alpha + \mathbf{z}\boldsymbol{\beta}), \qquad \frac{dP(\mathbf{z})}{d\boldsymbol{\beta}} = \mathbf{z}' f(\alpha + \mathbf{z}\boldsymbol{\beta}) \tag{8.16}$$

so that model fitting with any specific f is straightforward provided $P(\mathbf{z}_i)$ can be conveniently calculated. Score tests for log-normal (probit) and log-logistic (logit) failure time models relative to the generalized F regression model are also straightforward (Prentice, 1976b). Let

$$\boldsymbol{\theta} = \begin{pmatrix} \boldsymbol{\theta}_1 \\ \boldsymbol{\theta}_2 \end{pmatrix},$$

where $\boldsymbol{\theta}_1 = \begin{pmatrix} \alpha \\ \boldsymbol{\beta} \end{pmatrix}$ and $\boldsymbol{\theta}_2 = \begin{pmatrix} m_1 \\ m_2 \end{pmatrix}$. Similarly partition the score statistic (8.14) into $\mathbf{s} = \begin{pmatrix} s_1 \\ s_2 \end{pmatrix}$ and covariance matrix (8.15) into

$$\Sigma = \begin{pmatrix} \Sigma_{11} & \Sigma_{12} \\ \Sigma_{21} & \Sigma_{22} \end{pmatrix}.$$

Consider now a score test for $\boldsymbol{\theta}_2 = \boldsymbol{\theta}_2^0$. A Newton–Raphson iteration based on (8.14)–(8.16) gives $\hat{\alpha} = \hat{\alpha}(\boldsymbol{\theta}_2^0)$ and $\hat{\boldsymbol{\beta}} = \hat{\boldsymbol{\beta}}(\boldsymbol{\theta}_2^0)$. Under $\boldsymbol{\theta}_2 = \boldsymbol{\theta}_2^0$, the asymptotic distribution of the test statistic

$$\mathbf{s}_2'\left\{\Sigma_{22} - \Sigma_{21}\Sigma_{11}^{-1}\Sigma_{12}\right\}^{-1}\mathbf{s}_2 \tag{8.17}$$

is χ_2^2 with all quantities in (8.17) evaluated at $\hat{\alpha}$, $\hat{\beta}$. For example, to test a log-logistic hypothesis we need only note that

$$\frac{dP(\mathbf{z})}{dm_j} = \frac{(-1)^j \log\{1 + \exp[(-1)^j(\alpha + \mathbf{z}\boldsymbol{\beta})]\}}{1 + \exp[(-1)^j(\alpha + \mathbf{z}\boldsymbol{\beta})]}, \qquad j = 1, 2.$$

With the data of Table 8.2 and $z = \log CS_2$/milligram/liter, (8.17) has value 7.95 which has significance level .019 on the basis of a $\chi_{(2)}^2$ (exponential) distribution. The log-normal model is most conveniently tested in terms of the parameters (q, p) of Chapter 3. Now take

$$\boldsymbol{\theta}_2 = \begin{pmatrix} q \\ p \end{pmatrix}$$

and note that

$$\frac{d \log P(\mathbf{z})}{dq} = \frac{[(\alpha + \mathbf{z}\boldsymbol{\beta})^2 + 2]\phi(\alpha + \mathbf{z}\boldsymbol{\beta})}{6}$$

$$\frac{d \log P(\mathbf{z})}{dp} = \frac{[(\alpha + \mathbf{z}\boldsymbol{\beta})^3 + 3(\alpha + \mathbf{z}\boldsymbol{\beta})]\phi(\alpha + \mathbf{z}\boldsymbol{\beta})}{24},$$

where $\phi(w) = (2\pi)^{-1/2} \exp(-\frac{1}{2}w^2)$.

The statistic (8.17), with the data of Table 8.2, gives a $\chi_{(2)}^2$ value of 7.40 which occurs at the .025 upper end point of the $\chi_{(2)}^2$ distribution. Table 8.2 also gives parameter estimates for the fit of probit and logit models as well as the generalized F model with $m_2 = 1$.

8.3 RETROSPECTIVE STUDIES

8.3.1 General

Preceding chapters have assumed that failure time is being monitored in a prospective fashion. That is, it is assumed that individuals are selected randomly (conditional on covariates) from the population under study and are followed to observe their time and type of failure or failures.

In some settings the prospective type of study is often too expensive and time consuming. This is particularly true in the study of risk factors for rare diseases. The retrospective approach attempts to circumvent these difficulties by selecting individuals on the basis of failure time and failure type, after which their covariate data are ascertained. The covariate function would typically include certain exposures that are to be related to the disease under study as well as confounding factors. As a refinement, cases (with the disease or diseases under study) and controls (without such diseases) are sometimes matched on certain factors as part of the sampling procedure.

In the notation of Section 7.2, the prospective study involves sampling (T, J) given $Z(T)$, where T is failure time, $J \in \{1, 2, \ldots, m\}$ is cause of failure, and $Z(T)$ is a covariate function to be related to (T, J). The retrospective study, on the other hand, involves sampling $Z(T)$ given (T, J). Alternatively, if matching takes place on factors $\mathbf{x} = \mathbf{x}(t)$ the retrospective study can be viewed as sampling $Z(T)$ given $\{T, J, \mathbf{x}(T)\}$.

In this section it is shown that the proportional hazards prospective model can be used to induce a model for the corresponding retrospective study, and further that the prospective regression coefficient can be estimated from retrospective data. We do not attempt to discuss the many other possible approaches to the analysis of case control studies.

8.3.2 Relative Risk Estimation

As usual T denotes time from some well defined event. For instance, in an epidemiologic study T may represent patient age, time since menopause, time since exposure to a suspected carcinogen, or duration of employment in a certain industry. Suppose a particular failure type or disease ($J = 1$) is of primary interest. It is convenient to denote control study subjects, that is, study subjects that have not failed, by $J = 0$. The regression function $Z(T)$ represents study subject characteristics, including exposure factors that are to be related to disease incidence.

Suppose now that a cause-specific hazard function of the type discussed in Section 7.2 is specified for the study disease; that is,

$$\lambda_1(t; Z(t)) = \lambda_{01}(t) \exp[\mathbf{z}(t)\boldsymbol{\beta}] \tag{8.18}$$

where $\exp[\mathbf{z}(t)\boldsymbol{\beta}]$ ($\boldsymbol{\beta}_1$ in the notation of Section 7.2) is the relative risk function relating the regression function $Z(t)$ to the hazard function for the study disease. The "instantaneous" odds ratio for disease ($J = 1$) at time t for an individual with regressor vector $Z(t)$ versus an individual with standard regressor function $Z_0(t)$ is

$$\frac{\lambda_1(t; Z(t))}{\lambda_1(t; Z_0(t))}.$$

Upon substitution from (8.18) and, for notational simplicity, assuming $\mathbf{z}_0(t) = 0$, this reduces immediately to $\exp[\mathbf{z}(t)\boldsymbol{\beta}]$. Further, in a similar manner to Cornfield (1951) or Prentice and Breslow (1978), this odds ratio can be "inverted" to give

$$\frac{P[Z(t)|(t, 1)]/P[Z(t)|(t, 0)]}{P[Z_0(t)|(t, 1)]/P[Z_0(t)|(t, 0)]} = \exp[\mathbf{z}(t)\boldsymbol{\beta}]. \tag{8.19}$$

Since each of the quantities on the left side of (8.19) can be estimated from a case control study it follows that the proportional hazards regression coefficient $\boldsymbol{\beta}$ can be also estimated from such a study.

Suppose, at a specific failure time t, m cases in the study population contact the disease (fail of cause $J = 1$). Suppose also that n controls are selected randomly from the subset of the population known to be disease free at t. Let $Z_1(t), \ldots, Z_m(t), Z_{m+1}(t), \ldots, Z_{m+n}(t)$ be the respective regression functions that are sampled. In an epidemiologic study these data are typically obtained by means of an individual interview or by review of hospital, pharmacy, or employment records. The probability that the covariate functions $Z_1(t), \ldots, Z_m(t)$ correspond to cases ($J = 1$) given the set of $(m + n)$ covariate functions is

$$\frac{\prod_{i=1}^{m} P[Z_i(t)|(t, 1)] \prod_{m+1}^{m+n} P[Z_i(t)|(t, 0)]}{\sum_{l \in R(m, n)} \prod_{i \in l} P[Z_i(t)|(t, 1)] \prod_{i \notin l} P[Z_i(t)|(t, 0)]} \tag{8.20}$$

where $R(m, n)$ is the set of $\binom{m+n}{n}$ subsets of size m from $\{1, 2, \ldots, m + n\}$ and $l = \{l1, \ldots, lm\}$. From (8.19) one can write

$$P[Z_i(t)|(t, 1)] = \exp[\mathbf{z}_i(t)\boldsymbol{\beta}] \frac{P[Z_i(t)|(t, 0)]P\{Z_0(t)|(t, 1)\}}{P[Z_0(t)|(t, 0)]}.$$

Substitution into (8.20) gives

$$\left(\frac{\exp[\mathbf{s}_i(t_i)\boldsymbol{\beta}]}{\sum_{l \in R(m_i, n_i)} \exp[\mathbf{s}_l(t_i)\boldsymbol{\beta}]}\right). \tag{8.21}$$

where $\mathbf{s}(t) = \mathbf{z}_1(t) + \cdots + \mathbf{z}_m(t)$ and $\mathbf{s}_l(t) = \mathbf{z}_{l1}(t) + \cdots + \mathbf{z}_{lm}(t)$. Note that under (8.18) the covariate functions need be known only at the time t of case or control ascertainment in order to calculate (8.21).

A conditional likelihood for $\boldsymbol{\beta}$ based on case and control samples of respective sizes m_i and n_i at times t_i, $i = 1, \ldots, k$ is simply the product of terms (8.21) over the k distinct ascertainment times

$$\prod_{i=1}^{k} \frac{\exp[\mathbf{s}_i(t_i)\boldsymbol{\beta}]}{\sum_{l \in R(m_i, n_i)} \exp[\mathbf{s}_l(t_i)\boldsymbol{\beta}]}.$$

This likelihood is of the same form as the partial likelihood based on prospective data (Chapter 4). Note however that the "risk sets" $R(m_i, n_i)$ include only those cases and controls with failure or survival times equal to t_i. Computations will be practical if the m_i, $i = 1, \ldots, k$ are small (e.g., all equal to 1). Prentice and Breslow (1978) suggest an asymptotically equivalent estimator that is practical even if both the m_i's and the n_i's are large. The same paper also generalizes the methods of this section to include studies in which cases and controls are matched on certain characteristics and to include the possibility that several case groups (diseases) are simultaneously under study.

8.4 A BAYESIAN ANALYSIS OF THE PROPORTIONAL HAZARDS MODEL

This section gives a brief outline of a nonparametric Bayesian analysis of survival time data arising from the proportional hazards model of Chapter 4. This requires a brief discussion of some of the recently proposed methods of nonparametric Bayesian inference, contained in Section 8.4. Section 8.4.2 considers application to the proportional hazards model.

8.4.1 Nonparametric Bayesian Procedures

We consider here specific application of some nonparametric Bayesian procedures to survival distributions. A more detailed and mathematically complete discussion of these methods can be found in Ferguson (1973) and Doksum (1974).

Suppose that the (conditional) survivor function of the random variable T is

$$\exp[-\Lambda_0(t)] = P[T \geq t | \Lambda_0].$$

The probability statement is shown conditional on $\Lambda_0(\cdot)$ since Λ_0, the parameter in the model, is the realization of a stochastic process to be defined. We consider a partition of $[0, \infty)$ into a finite number k of disjoint intervals

$$[a_0 = 0, a_1)[a_1, a_2), \ldots, [a_{k-1}, a_k = \infty)$$

and define the hazard contribution of the ith interval as

$$q_i = P\{T \in [a_{i-1}, a_i) | T \geq a_{i-1}, \Lambda_0\} \tag{8.22}$$

if $P(T \geq a_{i-1} | \Lambda_0) > 0$ and otherwise $q_i = 1$ $(i = 1, \ldots, k)$. Clearly then we have

$$\Lambda_0(a_i) = \sum_{j=1}^{i} -\log(1 - q_j) = \sum_{j=1}^{i} r_j, \qquad i = 1, \ldots, k. \tag{8.23}$$

Doksum (1974) has considered this situation and has shown that a probability distribution can be specified on the space $\{\Lambda_0(t)\}$ by specifying the finite dimensional distributions of $q_1, \ldots, q_k$ for each partition $[a_{i-1}, a_i)$, $i = 1, \ldots, k$. Accordingly, independent prior probability densities can be specified for $q_1, \ldots, q_k$ subject to some consistency conditions and the resulting processes are called tailfree or neutral to the right by Doksum. Examination of (8.23), however, makes it clear that $\Lambda_0(t)$ is by this construction a nondecreasing independent increments process. This process in $\Lambda_0(t)$ is called a subordinator by Kingman (1975).

The problem then reduces to the specification of a nondecreasing independent increments process for $\Lambda_0(t)$. To do this, one need only specify independent priors for the r_i's (or for the q_i's) subject to the condition that the distribution of $r_i + r_{i+1}$ must be the same as would be obtained by direct application of the rules to the combined interval $[a_{i-1}, a_{i+1})$.

The Dirichlet process (Ferguson, 1973) was the first random probability measure used in a Bayesian context. The process can be defined in very general probability spaces. But in the above context, where the observable variate T has a natural ordering, it is convenient to think of the process as unfolding sequentially in time as above.

Let c be a positive real number and $\exp[-\Lambda^*(t)]$ be a completely specified survivor function. Suppose that $q_1, \ldots, q_k$ are independent with

$$q_i \sim B_e[\gamma_{i-1} - \gamma_i, \gamma_i], \qquad i = 1, \ldots, k,$$

where $X \sim B_e\{a, b\}$ means that X has a beta distribution with parameters a, b and $\gamma_i = c \exp[-\Lambda^*(a_i)]$, $i = 1, \ldots, k$. We are adopting here the conventions that $B_e(0, b)$ is the distribution with unit mass at 0 $(b > 0)$ and $B_e(a, 0)$ is the distribution with unit mass at 1 $(a \geq 0)$. That this is a valid assignment is easily verified by considering the hazard contribution of $[a_{i-1}, a_{i+1})$,

$$q_c = 1 - (1 - q_i)(1 - q_{i+1}) \sim B_e(\gamma_{i-1} - \gamma_{i+1}, \gamma_{i+1})$$

which is the result obtained by combining the intervals or by applying the definition directly.

The Dirichlet process takes its name from the fact that the random probabilities

$$P_i = P\{T \in [a_{i-1}, a_i)|\Lambda_0\}, \qquad i = 1, \ldots, k,$$

have a k variate Dirichlet distribution, a fact easily deduced from the distribution of $q_1, \ldots, q_k$. In fact if $A_1, \ldots, A_k$ is any set of mutually exclusive and exhausitve events, the variates

$$P(t \in A_i|\Lambda_0), \qquad i = 1, \ldots, k,$$

have also a k variate Dirichlet distribution with parameters

$$cP^*(A_1), \ldots, cP^*(A_k)$$

where $P^*(A_i)$ is the probability attached to A_i by the probability distribution with survivor function $\exp[-\Lambda^*(t)]$. Details and properties of the Dirichlet process have been considered by Ferguson (1973, 1974) and Doksum (1974). More recently Susarla and Van Ryzin (1976) have considered applications of the Dirichlet process to the nonparametric estimation of the survivor function from censored data.

A second process, similar to the Dirichlet process, is obtained if $\Lambda_0(t)$ is a gamma process similar to that used by Moran (1959) in modeling dam inputs. More specifically, let $r_i = -\log(1 - q_i)$ have the independent gamma distributions

$$r_i \sim G(\alpha_i - \alpha_{i-1}, c), \qquad i = 1, \ldots, k, \tag{8.24}$$

where $\alpha_i = c\Lambda^*(a_i)$, $i = 1, \ldots, k$ and c and Λ^* have the same meanings as before. The notation $X \sim G(a, b)$ indicates that X has the gamma distribution with parameters a (shape) and b (scale). The conventions adopted are that $G(0, c)$ is the distribution with unit mass at 0 and $G(\infty - \infty, c)$ or $G(\infty, c)$ is the distribution with unit mass at ∞. We shall write $\Lambda_0 \sim G(c\Lambda^*, c)$ to describe this process for $\Lambda_0(t)$.

Ferguson (1973) gives an interpretation of the parameters c and Λ^* of the Dirichlet process. Similar interpretations are available for the gamma hazard process. If we consider a partition $[0, t)$, $[t, \infty)$ and let

$$q = P\{T \in [0, t)|\Lambda_0\} = 1 - \Lambda_0(t),$$

then

$$r = \Lambda_0(t) \sim G(\Lambda^*(c), c)$$

and

$$E[\Lambda_0(t)] = \Lambda^*(t)$$

$$\text{var}[\Lambda_0(t)] = \frac{\Lambda^*(t)}{c}.$$

The specified Λ^* can be viewed as an initial guess at Λ_0 and c is a specification of the weight attached to that guess.

In the following sections, the gamma hazard process is used as a prior distribution in analyzing data arising from the proportional hazards model. Calculations could equally have been carried out using the Dirichlet process although most results are then more complicated.

8.4.2 The Estimation of $\boldsymbol{\beta}$ in the Proportional Hazards Model

Let T_i be a random variable with conditional survivor function

$$P(T_i \geq t_i | z_i, \Lambda_0) = \exp[-\Lambda_0(t_i) e^{\mathbf{z}_i \boldsymbol{\beta}}] \qquad (i = 1, \ldots, n) \tag{8.25}$$

independently. This is the proportional hazards model with time independent covariates $\mathbf{z}_i$. For the moment, we consider the case of no censoring. We suppose that $\Lambda_0 \sim \mathbf{G}(c\Lambda^*, c)$ and consider the problem of estimating $\boldsymbol{\beta}$ on the basis of the data $(t_1, \mathbf{z}_1), \ldots, (t_n, \mathbf{z}_n)$. One way to proceed is to calculate the probability density of $t_1, \ldots, t_n$ conditional on the $\mathbf{z}_i$'s, Λ_0 having been eliminated, and to interpret this as a likelihood function for $\boldsymbol{\beta}$. Conditional on Λ_0,

$$P(T_1 \geq t_1, \ldots, T_n \geq t_n | \boldsymbol{\beta}, \mathbf{Z}, \Lambda_0) = \exp\left[-\sum \Lambda_0(t_i) e^{\mathbf{z}_i \boldsymbol{\beta}}\right] \tag{8.26}$$

where $\mathbf{Z}$ is the design matrix with ith column $\mathbf{z}_i$. Without loss of generality, we consider $t_1 \leq t_2 \leq \cdots \leq t_n$ and define $r_i = \Lambda_0(t_i) - \Lambda_0(t_{i-1})$ $(i = 1, \ldots, n+1)$ where $t_0 = 0$ and $t_{n+1} = \infty$. The t_i's are now playing the role of the a_i's in the preceding section. We have further that

$$r_i \sim G(c\Lambda^*(t_i) - c\Lambda^*(t_{i-1}), c), \tag{8.27}$$

$(i = 1, \ldots, n+1)$ independently, and since $\Lambda_0(t_i) = \Sigma_1^i r_j$, $i = 1, \ldots, n+1$, (8.26) implies

$$P(T_1 \geq t_1, \ldots, T_n \geq t_n | \boldsymbol{\beta}, \mathbf{Z}, r_1 \cdots r_{n+1}) = \exp\left(-\sum_1^n r_j A_j\right) \tag{8.28}$$

where

$$A_j = \sum_{l \in R(t_j)} e^{\mathbf{z}_l \boldsymbol{\beta}} \qquad (j = 1, \ldots, n) \tag{8.29}$$

and $R(t_j)$ is the set of individuals at risk at time $t_j - 0$. Integrating (8.28) with respect to the distribution (8.27) of $r_1, \ldots, r_n$ gives

$$P(T_1 \geq t_1, \ldots, T_n \geq t_n | \boldsymbol{\beta}, \mathbf{Z}) = \exp\left[-\sum c B_j \Lambda^*(t_j)\right] \tag{8.30}$$

where $B_j = -\log[1 - \exp(\mathbf{z}_j \beta)/(c + A_j)]$.

The expression (8.30) is valid for any cumulative hazard $\Lambda^*(t)$. In order to avoid problems with fixed discontinuities, however, we assume that $\Lambda^*(t)$ is absolutely continuous. The multiple decrement function (8.29) is then absolutely continuous except along any hyperplane with $t_i = t_j$ for some $i \neq j$. Thus if there are no ties in the data, $(t_1 < t_2 < \cdots < t_n)$ the p.d.f. of $T_1, \ldots, T_n$ is computed by differentiation and yields

$$L(\boldsymbol{\beta}) = c^n \exp\left[-\sum cB_j\Lambda^*(t_j)\right] \prod_1^n [\lambda^*(t_i)B_i] \tag{8.31}$$

where $\lambda^*(t) = (d/dt)\Lambda^*(t)$. The expression (8.31) can be interpreted as a likelihood function for β on the data $T_1 = t_1, \ldots, T_n = t_n$. Right censoring is easily accommodated since the appropriate likelihood is obtained by differentiating (8.30) only with respect to the observed failure times. This gives

$$\exp\left[-\sum cB_j\Lambda^*(t_j)\right] \prod_1^n [c^*(t_i)B_i]^{\delta_i} \tag{8.32}$$

where $\delta_j = 0$ or 1 for censored or failure times t_j, respectively. The standard convention is adopted here that censored times tied with failure times are adjusted an infinitesimal amount to the right.

Two cases are of particular interest. If c is near 0, then to a first order approximation

$$L(\boldsymbol{\beta}) \cong K \prod_1^n \left[-\log\left(\frac{1-e^{z_i\boldsymbol{\beta}}}{c+A_i}\right)\right] \cong K \prod_1^n \left(\frac{e^{z_i\boldsymbol{\beta}}}{A_i}\right). \tag{8.33}$$

The last term in (8.33) is proportional to the partial likelihood or the marginal likelihood of $\boldsymbol{\beta}$. Small values of c correspond to placing little faith in the prior estimate $\Lambda^*(t)$ of $\Lambda_0(t)$. On the other hand,

$$\lim_{c\to\infty} L(\boldsymbol{\beta}) = \exp\left[-\sum \Lambda^*(t_j)e^{z_j\boldsymbol{\beta}}\right] \prod \lambda^*(t_i)e^{z_i\boldsymbol{\beta}},$$

which is the appropriate likelihood if it is assumed that $\Lambda_0(t) = \Lambda^*(t)$ at the outset.

In effect, (8.33) gives a spectrum of likelihoods ranging from truly nonparametric situations (c near 0) to situations where $\Lambda_0(t)$ is assumed completely known. By allowing $\Lambda^*(t)$ to depend on one or more unknown parameters, for example, $\Lambda^*(t) = \lambda t$, the likelihood (8.32) corresponds to a generalization of the usual parametric analysis ($c \to \infty$) with exponential survivals. An examination of the likelihood for varying c can lead to an evaluation of how assumption dependent is the analysis.

The leukemia data of Gehan (1965) reproduced in Table 8.3 can be used to illustrate this point. The covariate is treatment group specified by the indicator variable $z = -.5, .5$. We select $\Lambda^*(t) = \lambda t$ where λ is left

Table 8.3 TIMES OF REMISSION IN WEEKS OF LEUKEMIA PATIENTS

Group 1 ($Z=.5$)	1, 1, 2, 2, 3, 4, 4, 5, 5, 8, 8, 8, 8, 11, 11, 12, 12, 15, 17, $22+\epsilon$,† $23+\epsilon$†
Group 0 ($Z=-.5$)	6, 6, 6, 6,* 7, 9,* 10, 10,* 11,* 13, 16, 17,* 19,* 20,* 22, 23, 25,* 32,* 32,* 34,* 35*

*Censored.

†These data were 22, 23 but were adjusted slightly in the positive direction to break the ties with items in group 0.

Table 8.4 ESTIMATION OF β

	Proportional Hazards	$c=1$	$c=5$	$c=25$	$c=125$	$c=\infty$
$\hat{\beta}$	1.652	1.606	1.512	1.461	1.490	1.58
$\text{var}(\hat{\beta})$	.148	.134	.124	.130	.146	.158
$\hat{\beta}/\sqrt{\text{var}(\hat{\beta})}$	4.29	4.39	4.29	4.05	3.90	3.97

unspecified. The censored result (8.32) then gives a joint likelihood for λ and β for each specified c:

$$L(\beta, \lambda)\exp\Big(-c\lambda \sum t_j B_j\Big) \prod (c\lambda B_i)^{\delta_i}.$$

Maximum likelihood equations are easily formed and the maximization is carried out using a Newton–Raphson iterative technique. For the data in Table 8.3, there are ties present and these were broken at random in order to apply the results above. The exact analysis taking account of the ties is more difficult and the interested reader is referred to Kalbfleisch (1978a).

Table 8.4 summarizes the estimation of β for various c values and gives also the results for the exponential regression model ($c\to\infty$) and for the proportional hazards model. The entries $\text{var}(\hat{\beta})$ give the asymptotic variance of $\hat{\beta}$ based on the entry in the inverse of the information matrix for β, λ. The estimation of β is very stable over the range $0<c<\infty$.

8.4.3 The Posterior Distribution of Λ_0

In this section, the posterior distribution of the underlying process Λ_0 is obtained when a sample $(t_1, \mathbf{z}_1), \ldots, (t_n, \mathbf{z}_n)$ is obtained from the model

(8.25). The prior distribution of Λ_0 is the gamma process with parameters c and Λ^* as before.

Consider again a partition of $[0, \infty)$ into disjoint intervals $[a_0 = 0, a_1), \ldots, [a_{k-1}, a_k = \infty)$ and suppose that the data is $(t_1, \mathbf{z}_1)$ so that $n = 1$. The extension to other n values follows easily from the results for $n = 1$. As before $r_j = -\log(1 - q_j)$ where q_j is the hazard contribution of the jth interval $j = 1, \ldots, k$. Assume that $a_{i-1} \le t_1 < a_i$ and let $r_{i1} = -\log(1 - q_{i1})$ and $r_{i2} = -\log(1 - q_{i2})$, where q_{i1} and q_{i2} are the hazard contributions of the intervals $[a_{i-1}, t_1)$ and $[t_1, a_i)$, respectively. Then

$$P(T_1 \ge t_1 | \mathbf{r}, r_{i1}, \mathbf{z}_1) = \exp[-(t_1 + \cdots + r_{i-1} + r_{i1})e^{\mathbf{z}_1\boldsymbol{\beta}}], \tag{8.34}$$

and

$$P[T_1 \ge t_1, r_j < r_{0j} \; (j = 1, \ldots, k) | \mathbf{z}_1] = H(t_1, r_{01}, \ldots, r_{0k}, \mathbf{z}_1)$$

is obtained by integrating (8.34) with respect to the independent gamma prior distributions of $r_1, \ldots, r_{i-1}, r_{i1}, r_{i2}, r_{i+1}, \ldots, r_k$ over the appropriate range. The posterior distribution of $\mathbf{r}$ given $T_1 = t_1$ is then specified by

$$P[r_j < r_{0j} \; (j = 1, \ldots, k) | T_1 = t_1, z_1] = \frac{\dfrac{\partial}{\partial t_1} H(t_1, r_{01}, \ldots, r_{0k}, z_1)}{\dfrac{\partial}{\partial t_1} H(t_1, \infty, \ldots, \infty, \mathbf{z}_1)} .$$

There does not appear to be a simple closed form expression for this but the probability laws can be simply characterized using moment generating functions. Straightforward calculation shows that

$$M_{\mathbf{r}}(\boldsymbol{\theta} | T_1 = t_1) = E(e^{\theta_1 r_1 + \cdots + \theta_k r_k} | T_1 = t_1) = \prod_{j=1}^{k} M_{r_j}(\theta_j | T_1 = t_1)$$

where

$$M_{r_j}(\theta_j | T_1 = t_1) = \left(\frac{c_1}{c_1 - \theta_j}\right)^{\alpha_j - \alpha_{j-1}}, \qquad j = 1, \ldots, i-1$$

$$M_{r_j}(\theta_j | T_1 = t_1) = \left(\frac{c}{c - \theta_j}\right)^{\alpha_j - \alpha_{j-1}}, \qquad j = i+1, \ldots, k.$$

for $c_1 = c + e^{\mathbf{z}_1\boldsymbol{\beta}}$. The moment generating function of r_i is

$$\left(\frac{c_1}{c_1 - \theta_i}\right)^{\alpha(t_1) - \alpha_{i-1}} \left(\frac{c}{c - \theta_i}\right)^{\alpha_i - \alpha(t_1)} \frac{\log[(c_1 - \theta_i)/(c - \theta_i)]}{\log(c_1/c)} . \tag{8.35}$$

Here as before $\alpha_j = c\Lambda^*(a_j)$, $j = 1, \ldots, k$, and $\alpha(t_1) = c\Lambda^*(t_1)$. Thus r_i is distributed as the sum of three independent random variables X, Y, U where X and Y are gamma variables and $U \sim A(c_1, c)$ with density

$$\frac{(1/u)(e^{-cu} - e^{-c_1 u})}{\log(c_1/c)}$$

whose moment generating function is the last factor in (8.35). All finite dimensional distributions of the posterior process have now been obtained and characterization of the process is straightforward. This special case ($n = 1$) is covered in the next paragraph.

The generalization of these results to obtain the posterior distribution of Λ_0 given $(t_1, \mathbf{z}_1), \ldots, (t_n, \mathbf{z}_n)$ is straightforward. Given $t_1 < t_2 < \cdots < t_n$, $\Lambda_0(t)$ is an independent increments process. At t_i the increment is

$$U_i \sim A(c + A_i, c + A_{i+1}), \qquad i = 1, \ldots, n,$$

and between $[t_{i-1}, t_i)$ increments occur as for the gamma process $\mathbf{G}(c\Lambda^*, c + A_i)$. This result is easily seen by inserting t_n first then $t_{n-1}, \ldots, t_1$. Insertion of t_i affects the process only at points $t \leq t_i$.

The estimation of the survivor function can be performed in several ways. For example, $E[\Lambda_0(t)|\mathbf{t}, \mathbf{Z}, \boldsymbol{\beta}]$ would yield an optimum estimate if losses were squared error loss in $\Lambda_0(t)$. If $t_1 < t_2 < \cdots < t_n$ and $t_{i-1} \leq t < t_i$, then the posterior distribution of $\Lambda_0(t)$ is that of the sum of independent variables $X_1 + U_1 + \cdots + X_{i-1} + U_{i-1} + D_i$ where $X_j \sim G(c\Lambda^*(t_{j-1}) - c\Lambda^*(t_j),\ c + A_j)$, $u_j \sim A(c + A_j,\ c + A_{j+1})$ $(j = 1, \ldots, i-1)$, and $D_i \sim G(c\Lambda^*(t) - c\Lambda^*(t_{i-1}),\ c + A_i)$. Now easy calculation shows that

$$E(U_j) = \frac{\exp(\mathbf{z}_i\boldsymbol{\beta})}{(c + A_j)(c + A_{j+1})} \log\left(\frac{c + A_j}{c + A_{j+1}}\right)$$

and

$$E(\Lambda_0(t)|\mathbf{t}, \mathbf{Z}, \beta\} = \sum_1^{i-1} [E(X_j) + E(U_j)] + E(D_i).$$

For c small, $E(U_j) \cong 1/A_j$ $(j = 1, \ldots, n-1)$ and $E(X_j) \cong E(D_i) \cong 0$ so

$$E(\Lambda_0(t)|\mathbf{t}, \mathbf{Z}, \beta) \cong \sum_{j|t_j<t} \frac{1}{A_j}. \tag{8.36}$$

In Section 4.3 the maximum likelihood estimate of $\Lambda_0(t)$ is obtained [see expression (4.21)] as

$$\begin{aligned} \hat{\Lambda}_0(t) &= \sum_{j|t_j<t} \left(\frac{1 - e^{\mathbf{z}_j\boldsymbol{\beta}}}{A_j}\right)^{\exp(-\mathbf{z}_j\boldsymbol{\beta})} \\ &\cong \sum_{j|t_j<t} \frac{1}{A_j}. \end{aligned} \tag{8.37}$$

Again, the two results are similar.

Table 8.5 LOG-RANK AND GENERALIZED WILCOXON SIGNIFICANCE TESTS TO IDENTIFY ASSOCIATIONS BETWEEN VARIOUS FACTORS AND MORTALITY CAUSED BY THYMIC LEUKEMIA, NONTHYMIC LEUKEMIA, AND ALL NATURAL CAUSES

			Thymic Leukemia		Nonthymic Leukemia		All Natural Causes	
Factor	*Class*	*No. Mice*	*O**	*O/E*†	*O*	*O/E*	*O*	*O/E*
MHC	*k*	110	39	1.10	9	1.46	67	1.13
	b	93	28	.88	3	.51	47	.86
Log-rank significance level			.36		.10		.15	
Generalized Wilcoxon			.38		.21		.23	
Antibody	<0.5	101	45	1.64	9	1.82	70	1.52
(% gp70	0.5–5.0	27	4	.38	2	.94	9	.48
ppt.)	5.0–20.0	53	12	.63	1	.29	25	.77
	>20.0	19	4	.50	0‡	.00	7	.50
Log-rank significance level			.0002		.09		<.0001	
Generalized Wilcoxon			.0001		.08		<.0001	
Virus	$<10^{1.6}$	17	3	.47	1	.78	8	.75
(PFU/ml)	$10^{1.6}$–$10^{3.0}$	35	1	.07	2	.71	9	.39
	$10^{3.0}$–$10^{4.0}$	33	2	.16	3	1.08	11	.51
	$>10^{4.0}$	90	50	2.12	5	1.21	63	1.75
Log-rank significance level			<.0001		.92		<.0001	
Generalized Wilcoxon			<.0001		.93		<.0001	
Sex	Male	96	34	1.09	7	1.28	65	1.23
	Female	108	33	.92	5	.77	50	.80
Log-rank significance level			.49		.38		.02	
Generalized Wilcoxon			.79		.43		.07	
Albino	*c/c*	92	29	.91	5	.84	46	.83
	c/+	112	38	1.08	7	1.15	69	1.15
Log-rank significance level			.49		.60		.08	
Generalized Wilcoxon			.46		.68		.16	
Gpd-1	*b/b*	30	14	3.01	1	2.86	20	2.60
	b/a	70	5	.35	1	.61	15	.55
Log-rank significance level			<.0001		.23		<.0001	
Generalized Wilcoxon			<.0001		.16		<.0001	

**O*, Number of deaths.

†*O/E*, Ratio of the number of deaths to the "expected number" of deaths (calculated under a hypothesis of equality of mortality).

‡This group excluded from the comparison.

tests as applied to strata formed from each of the six factors mentioned above. Such tests were carried out for each of the mortality categories of thymic leukemia, nonthymic leukemia, and all natural causes. Corresponding to each factor (covariate) are the stratum definitions along with the number of mice known to belong to that stratum. The tabular entries are the number of deaths, the ratio of observed to "expected" deaths, the log-rank significance level, and the generalized Wilcoxon (Gehan scores) significance level. As noted in Section 6.5, however, it would be better to use the generalized Wilcoxon test of Section 6.2. The Gehan generalization is satisfactory only if the censoring is not severe.

A total of 67 mice died of thymic leukemia. The rank tests indicate strong associations between thymic leukemia mortality and each of the level of virus, the level of viral antibody, and the Gpd-1 phenotype. Only 12 animals died of nonthymic leukemia so that there is limited ability to detect associations between this mortality type and the genetic and viral factors. It does seem, however, that the strong association noted between virus levels and thymic leukemia mortality is not evident for nonthymic leukemia, suggesting different etiologies for the two diseases. A total of 115 animals died of natural causes. Not surprisingly, in view of the fact that more than half of these deaths were attributed to thymic leukemia, the factors virus level, antibody level, and Gpd-1 phenotype appear strongly related to all natural causes of mortality. In addition the female mice may experience lower natural mortality rates than do the males.

The following statistical questions arise in the analyses of Table 8.5:

1. How should classes be formed?
2. Do the multiple significance tests affect the interpretation of the suggested associations?
3. What should be done if test statistics such as the log-rank and generalized Wilcoxon tests are in qualitative disagreement?
4. Are sample sizes adequate to permit the use of the asymptotic likelihood theory?
5. How are these or other inferences affected by the substantial fraction missing data on Gpd-1 phenotype?

As usual such practical problems as 1–5 are more difficult to answer precisely than are many of the associated theoretical questions. We provide here some discussion of these problems along with possible approaches to their resolution.

In regard to question 1, the four genetic factors are all binary so that no question of class formation arises. Antibody level and virus levels are at least partially quantitative. Antibody levels less than 5% (gp70 precipitate) could not be detected, as was the case with virus levels less than $10^{1.6}$(PFU/ml). Virus levels in excess of 10^4(PFU/ml) also could not be further specified. Beyond these restrictions grouping could be done as desired. The formation of three or four classes on the basis of a quantitative factor would often provide adequate resolution without unduly compromising efficiency (the χ^2 degrees of freedom are generally one fewer than the number of classes formed). Roughly equal sample sizes among classes is a sensible criterion, though natural division points may be present and some limited unbalance in the initial groups may be preferable if censoring depends markedly on the factor under consideration. Of course formation of groups on the basis of the observed mortality data itself would invalidate the corresponding tests. Table 8.5 defines four classes for each of the antibody and virus levels. Some collapsing of these classes particularly for the nonthymic leukemia mortality may be preferable.

Six significance tests are carried out for each of the three mortality categories in Table 8.5. If significance levels were based on k *independent* test statistics the probability that one or more would show significance at the α% level is $1-(1-\alpha)^k$ which for $k=6$ and $\alpha=.05$, .01, .001, and .0001 has values .26, .06, .01, and .001, respectively. This point should be kept in mind if a large number of factors are simultaneously studied in an exploratory manner for association with failure time. Frequently, however, a study is designed to examine the prognostic value of one or a small number of factors (e.g., MHC and Gdp-1 in the current data set), and other variables are known or suspected risk factors that are included for model building purposes or in order to avoid confounding. The multiple testing problem is then not severe in respect to the primary factors. The related multiple comparison problem involving tests to compare all pairs of mortality curves when the data are divided into $s>2$ groups has been discussed by Koziol and Reid (1977). They consider both the log-rank and (Gehan) generalized Wilcoxon tests.

The log-rank and (Gehan) generalized Wilcoxon significance levels in Table 8.5 are generally in good agreement. Some relatively minor differences can be noted. For example, the comparison of male and female mortality from all natural causes has significance level .02 under the log-rank test and significance level .07 under the generalized Wilcoxon test. It is possible, however, that large differences in these significance levels occur [e.g., Prentice and Marek (1979)].

There are several possibilities for attempting to resolve a discrepant result among rank tests. One approach (e.g., Tarone and Ware, 1977) would define and utilize a new statistic with scores intermediate to the two being compared. A more promising, but little studied, method would permit the rank test to adapt to the data, in the event of such discrepancy. For example, any of the censored data rank tests may be generalized by *stratifying on the time axis itself*. The time axis may be divided into s (e.g., 2 or 3) strata and separate χ^2_{r-1} statistics may be calculated in each stratum, where r is the number of groups being compared (see Peto et al., 1977 for the log-rank generalization). Specifically, the χ^2 statistic in a stratum is based on failure time data in which study subjects at risk during the stratum that do not fail or are not censored in that particular stratum have their failure or censoring times replaced by censored times at the upper end point of the stratum. An overall $\chi^2_{s(r-1)}$ statistic is then formed by adding the individual x^2_{r-1} statistics over strata. This generalization would permit the log-rank test, for example, to be more sensitive to hazard ratios that vary among the failure time strata than is the case with the ordinary log-rank test. This generalization of the log-rank test is precisely the score statistic based on a global test for $\boldsymbol{\beta} = \mathbf{0}$ in a proportional hazards model when time dependent indicator covariates are defined in each stratum for any $r-1$ of the groups. An alternative statistic that would usually be more efficient, but also more difficult computationally, is the score statistic based on a test for $\boldsymbol{\beta} = \mathbf{0}$ with indicator variables for $r-1$ of the groups along with time dependent covariates that are products of these indicator variables and simple functions of time, such as t or $\log t$. These latter tests would be sensitive for the detection of hazard ratios different from unity when such ratios are monotone and relatively smooth functions of time. Of course, for formal testing, the subdivision must not be based on previous examination of the failure times.

Problem 4 is concerned with sample sizes for the use of asymptotic results. This topic is of particular concern for the tests on nonthymic leukemia mortality because of the small number of deaths. In such situations it may be advisable to estimate the actual sampling distribution of the test statistic (under the null hypothesis) by simulation, rather than by relying solely on asymptotic results. Very briefly, if censoring is independent of the covariates under consideration a permutation approach to estimating the actual distribution of the test statistic is valid. From Chapter 6 the rank statistic can be written

$$\mathbf{v} = \sum_{i=1}^{k} \left(\mathbf{z}_{(i)} c_i + \sum_{j=1}^{m_i} \mathbf{z}_{ij} C_i \right)$$

where $\mathbf{z}_{(1)}, \ldots, \mathbf{z}_{(k)}$ are covariate values corresponding to the uncensored failure times $t_{(1)}, \ldots, t_{(k)}$, and $\mathbf{z}_{i1}, \ldots, \mathbf{z}_{im_i}$ are covariate values corresponding to the m_i censored times in $[t_{(i)}, t_{(i+1)})$ with $t_{(0)} = 0$, $t_{(k+1)} = \infty$. The scores (c_i, C_i), $i = 1, \ldots, k$, are known for any particular rank test. Under the hypothesis $\boldsymbol{\beta} = \mathbf{0}$ and the assumption of censorship independent of $\mathbf{z}$ the permutation distribution of $\mathbf{v}$ may be generated. Each step in the simulation simply involves a random assignment of the scores $\{c_i, C_i, \ldots, C_i; i = 1, \ldots, k\}$ to the $k + (m_1 + \cdots + m_k)$ z values and the calculation of $\mathbf{v}$. A more complex simulation is required if the censoring depends on $\mathbf{z}$ in an important manner. One possibility in this regard is to generate failure times at each specific $\mathbf{z}$ from an exponential distribution with the actual censoring scheme at each particular $\mathbf{z}$ approximated by a progressive type II censoring procedure.

Before problem 5 mentioned above is addressed, some regression analyses for the mouse leukemia data are presented. Problems relating to missing covariate data are then discussed for both the rank tests and the regression methods.

The analyses presented in Table 8.5 provide no insight into the joint association between two or more covariates and mortality and only limited insight into the magnitude of relative risks associated with the covariates. Table 8.6 gives results of four applications of the proportional hazards model (4.1) to these data. Proportional hazards models were assumed for the cause-specific mortality rate (7.1) for thymic leukemia as well as for all natural causes. There were too few nonthymic leukemia deaths to support such an analysis.

The left side of Table 8.6 gives the codes for seven indicator covariates defined from the six factors under study, and the tabular entries are the coefficients and (asymptotic) standard errors corresponding to covariates included in a particular run. The first column under either mortality category indicates that the production of detectable viral antibody is associated with reduced mortality, particularly thymic leukemia mortality, after adjusting mortality rates for possible influences of MHC, sex, and coat color. The second column shows that this association is no longer close to significant when the mortality rates are permitted to vary among three virus level categories. The antibody–virus association is largely explained by the suppressed virus levels that occur in antibody producing animals. Virus and antibody levels could be entered into these regression analyses in a quantitative manner with equal ease.

The initial analyses using log-rank and Wilcoxon tests provide valuable guidance for more refined model building. In addition, they are themselves useful summaries for many purposes. In particular, they are valuable aids in presenting the results of an analysis to nonstatisticians since these tests (log-rank and Wilcoxon) can be described and explained at a very intuitive level. Because of the close relationship between the log-rank and the proportional hazards model (see Section 4.2.5), the main results of an analysis based on the latter can often be described with reasonable accuracy using log-rank or stratified log-rank tests. For example, the second log-rank test in Table 8.5 illustrates the strong association

Table 8.6 PROPORTIONAL HAZARDS REGRESSION ANALYSIS OF FACTORS RELATED TO MORTALITY IN AKR × (B6 × AKR)F_1 MICE

	Thymic Leukemia Mortality				All Natural Mortality			
Factor (Code)	Analysis 1 Coef. (S.E.)	Analysis 2 Coef. (S.E.)	Analysis 3 Coef. (S.E.)	Analysis 4 Coef. (S.E.)	Analysis 5 Coef. (S.E.)	Analysis 6 Coef. (S.E.)	Analysis 7 Coef. (S.E.)	Analysis 8 Coef. (S.E.)
MHC	0.166	−1.179	n.i.	−0.122	−0.055	−0.333	n.i.	−0.206
(0 = *k*/*k*; 1 = *k*/*b*)	(.294)	(.299)		(.551)	(.232)	(.237)		(.433)
Antibody (% gp70 ppt)	−0.996†	−0.405	n.i.	−0.896	−0.607*	−0.225	n.i.	−0.441
(0 < 0.5; 1 ≥ 0.5)	(.324)	(.388)		(.582)	(.249)	(.263)		(.425)
Virus (PFU/ml)	n.i	1.559	1.226	1.914	n.i.	0.776	0.503	1.236
($<10^{1.6}$; $0 \geq 10^{1.6}$)		(.825)	(1.238)	(1.313)		(.881)	(.659)	(.732)
Virus	n.i.	−2.891‡	−1.672*	−1.701*	n.i.	−1.349‡	−0.132	−0.373
($1 < 10^4$; $0 > 10^4$)		(.603)	(.824)	(.824)		(.272)	(.437)	(.447)
Sex	−0.097	0.058	n.i.	−0.279	−0.392	−0.345	n.i.	−0.848*
(0 = *M*; 1 = *F*)	(.290)	(.301)		(.504)	(.231)	(.238)		(.380)
Albino	0.121	0.275	n.i.	0.676	0.301	0.444*	n.i.	0.413
(0 = *c*/+; 1 = *c*/*c*)	(.274)	(.277)		(.481)	(.217)	(.221)		(.361)
Gpd-1	n.i.	n.i.	−1.510*	−1.977†	n.i.	n.i.	−1.598‡	−1.792‡
(0 = *b*/*b*; 1 = *b*/*a*)			(.626)	(.636)			(.439)	(.414)

Key: n.i., not included in analysis; S.E., standard error.
*$p < .05$.
†$p < .01$.
‡$p < .001$.

between detectable viral antibody and mortality. A stratified log-rank test of this same variable with strata defined by the four virus level categories would illustrate the extent to which virus level accounts for this association. For the statistician, however, the proportional hazards presentation has the advantage of exhibiting these facts and others quickly and in an easily understood manner.

The final two analyses in Table 8.6, under either mortality category, are restricted to the 100 animals with known Gpd-1 phenotype. Analyses 3 and 8 indicate a much weaker virus-mortality association when taking account of Gpd-1 phenotype. This reflects a strong association between Gpd-1 and virus level. The significant Gpd-1 coefficients in the presence of virus level covariates may also reflect a role for Gpd-1 in relation to mortality, beyond its association with virus level.

Useful supplementary information may be obtained by studying the relation between the covariates themselves. For example, a binary logistic model applied to whether or not virus levels exceed 10^4 PFU/ml shows strong ($p < .001$) associations between virus levels and both antibody production and Gpd-1. The same type of analysis with the presence or absence of detectable antibody as dependent variables showed simultaneous associations between antibody and each of MHC, virus level, and Gpd-1 phenotype.

A troublesome point in the analysis of Tables 8.5 and 8.6 is that one of the most important prognostic factors, Gpd-1 phenotype, is available on only about one-half of the experimental animals. Moreover, the probability of a missing Gpd-1 phenotype is related to the survival time of the animal.

Let us first consider the problem of missing covariate data in more general terms. Suppose $\mathbf{z} = (\mathbf{z}_1, \mathbf{z}_2)$ and suppose that $\mathbf{z}_2$ may be missing. Define an indicator variable C for each animal that takes values 0 or 1 according to whether or not $\mathbf{z}_2$ is missing. One possible likelihood for the estimation of parameters in a failure time model $f(t|\mathbf{z})$ would be that based only on data points for which $\mathbf{z}_2$ is not missing ($C = 1$). The conditional likelihood contribution from a study subject with $C = 1$ is then

$$f(t|\mathbf{z}, C = 1) = \frac{P(C = 1|t, \mathbf{z})}{P(C = 1|\mathbf{z})} f(t|\mathbf{z}). \tag{8.38}$$

Suppose $P(C = 1|t, \mathbf{z})$ is independent of t. The conditional likelihood (8.38) is then equal to that obtained by simply applying $f(t|\mathbf{z})$ to all study subjects that do not have missing data, and no bias is associated with such a procedure.

If a substantial fraction of sample points have missing covariate values it is important to somehow include study subjects with $C = 0$ in the analysis to better examine the relationship between t and $\mathbf{z}_1$. The natural framework in this regard is to consider the fuller conditional likelihood with contributions (8.38) as before from study subjects with $C = 1$ and contributions

$$f(t|\mathbf{z}_1, C = 0) = \frac{P(C = 0|t, \mathbf{z}_1)}{P(C = 0|\mathbf{z}_1)} \frac{\int_{\mathbf{z}_2} f(t|\mathbf{z})p(\mathbf{z})\, d\mathbf{z}_2}{\int_{\mathbf{z}_2} p(\mathbf{z})\, d\mathbf{z}_2}$$

from those with $C = 0$. Once again the first factor can be ignored if $P(C = 0|t, \mathbf{z}_1)$ is independent of t, and the second factor requires either specification or estimation of the marginal distribution of covariate values, $p(\mathbf{z})$. For example, the covariate distribution is known for the genetic factors in Table 8.5. In other circumstances $p(\mathbf{z})$ can be parametrized and parameters in $p(\mathbf{z})$ and $f(t|\mathbf{z})$ can be simultaneously estimated. Some preliminary investigations of this procedure have been made. These calculations suggest that substantial improvements in the estimation of the coefficients of $\mathbf{z}_1$ can be obtained as compared with an estimator based on (8.38). Only relatively simple structures for $p(\mathbf{z})$ and exponential regression models have been considered, however, and even for these the calculations are extensive.

The missing data problems related to Gpd-1 determinations are complicated by the fact that the probability $p(C = 1|t, \mathbf{z})$ clearly does depend on t so that the factor $P(C = 1|t, \mathbf{z})/P(C = 1|\mathbf{z})$ cannot be assumed independent of the failure time parameters that specify the density $f(t|\mathbf{z})$. As indicated previously this happens because Gpd-1 determinations began midway through the study and so are unavailable for any animal that died before an age of 400 days. To the extent that all animals dying beyond 400 days had Gpd-1 determinations the corresponding rank tests (Table 8.5) or regression analysis (Table 8.6) can be justified precisely as above simply by conditioning throughout on $T > 400$. Similar comments apply to the cause-specific analyses simply by replacing t by (t, j) throughout the above discussion, where j is the failure type.

One can go further under the assumption of no missing values for animals that survived more than 400 days, and use the fact that the Gpd-1 phenotype segregates at random to examine the relationship between mortality and Gpd-1 during the first 400 days of life. Define $\theta_1 = P(T < 400|b/b$ phenotype), and $\theta_2 = P(T < 400|b/a$ phenotype). The probability of mortality in the first 400 days is then

$$\begin{aligned} P(T < 400) &= P(T < 400|b/b)P(b/b) + P(T < 400|b/a)P(b/a) \\ &= \frac{\theta_1 + \theta_2}{2}. \end{aligned}$$

The marginal probability for available data on Gpd-1 phenotype and mortality relative to 400 days then gives rise to the trinomial likelihood, which using the sample sizes of Table 8.5 takes value

$$(1-\theta_1)^{30}(1-\theta_2)^{70}(\theta_1+\theta_2)^{104}$$

from which the maximum likelihood estimators are $\hat{\theta}_1 = 1\text{–}2(30)/204 = .71$ and $\hat{\theta}_2 = 1\text{–}2(70)/204 = .31$. A likelihood ratio test gives an approximate χ_2^2 statistic of value

$$60 \log \frac{60}{204} + 140 \log \frac{140}{204} - 200 \log \frac{100}{204} = 16.46$$

under the hypothesis $\theta_1 = \theta_2$. This provides strong evidence for an association between Gpd-1 and mortality in the early time period as well. If mortality rates were assumed proportional during this early time period the instantaneous risk for the b/a phenotype relative to the b/b phenotype would be estimated by $e^{\hat{\beta}} = \log(1-\hat{\theta}_2)/\log(1-\hat{\theta}_1) = .31$, which is reasonably consistent with the data from the later time period.

Unfortunately it was not entirely true that all animals with $T > 400$ received Gpd-1 determinations. The analyses of Tables 8.5 and 8.6 remain valid provided the probability of such determination does not vary with mortality time within the period $T > 400$. This may be approximately the case, though the probability of a missing value appears to decrease somewhat with increasing t. It is appropriate with these data, however, to assume that the $P(C=1|t, \mathbf{z})$ is independent of $\mathbf{z}$. We can therefore denote $g(t) = P(C=1|t, \mathbf{z})$.

The censored data rank tests (Table 8.5) and regression analysis (Table 8.6) can be interpreted even in these circumstances though the meaning of the coefficients in Table 8.6 will be affected by a dependence of $P(C = 1|t, \mathbf{z})$ on t. To see this define

$$\tilde{\lambda}(t|\mathbf{z}, C=1) = \lim_{\Delta t \to 0} \frac{P(t \le T < t+\Delta t | T \ge t, \mathbf{z}, C=1)}{\Delta t}.$$

A conditional likelihood for t, given $\mathbf{z}$ and $C = 1$, can then be written

$$\Pi\left\{\tilde{\lambda}(t_i|\mathbf{z}_i, C_i=1)\exp\left[-\int_0^{t_i} \tilde{\lambda}(u|\mathbf{z}_i, C_i=1)\,du\right]\right\}$$

where the product is over all study subjects with $C = 1$. One can test for association between $\mathbf{z}$ and t from this likelihood using any of the previously described techniques. The censored data rank tests dealing with Gpd-1 are then validated on noting that the hypothesis of no dependence of the ordinary hazard function $\lambda(t|\mathbf{z})$ on $\mathbf{z}$ implies no dependence of $\tilde{\lambda}(t|\mathbf{z}, C=1)$ on $\mathbf{z}$. To see this one can calculate

$$\tilde{\lambda}(t|\mathbf{z}, C=1) = \frac{g(t)f(t|\mathbf{z})}{\int_t^\infty g(u)f(u|\mathbf{z})\,du} \tag{8.39}$$

so that if $\lambda(t|\mathbf{z})$, and therefore $f(t|\mathbf{z})$, is independent of $\mathbf{z}$ so is the whole right side of this expression and hence so is $\tilde{\lambda}(t|\mathbf{z}, C=1)$ independent of $\mathbf{z}$.

The hazard function $\tilde{\lambda}(t|\mathbf{z}, C=1)$ could be modeled in a proportional hazards manner, and it is, in fact, the coefficients from the partial likelihood analysis of such a model that appear in Table 8.6. These coefficients do not then have the usual relative risk interpretation unless $g(u)$ is constant over the time period under consideration ($T > 400$). This happens because the risk sets in the partial likelihood analysis consist only of animals that go on to subsequently receive Gpd-1 determinations. When $g(t)$ varies with t this risk set at any particular time is generally not representative of the overall risk set at that time. For example, if $g(t)$ is monotone increasing the subset of the overall risk set $R(t)$ that gives rise to $C=1$ has an overrepresentation of animals with large failure times. This in turn tends to bias the coefficient for the covariate that has missing values away from zero.

Additional work could be carried out to attempt to assess whether the interpretation of the coefficients in Table 8.6 is seriously affected by the missing data or to obtain alternative estimators of the relative risk parameters. For example, (8.39) could be solved for $f(t|\mathbf{z})$, and hence $\lambda(t|\mathbf{z})$, giving

$$\lambda(t|\mathbf{z}) = \frac{g^{-1}(t)\tilde{\lambda}(t|\mathbf{z})\exp[-\int_0^t \tilde{\lambda}(v|\mathbf{z})\,dv]}{\int_t^\infty g^{-1}(u)\tilde{\lambda}(u|\mathbf{z})\exp[-\int_0^u \tilde{\lambda}(v|\mathbf{z}\,dv]\,du}.$$

A fitted model for $\tilde{\lambda}(t|\mathbf{z})$, for example, based on a proportional hazards model, can then be used in conjunction with known, or estimated values of the function $g(\cdot)$ in order to estimate and compare $\lambda(t|\mathbf{z})$ for various values of $\mathbf{z}$. Such comparisons then provide the desired relative risk estimators. The $g(\cdot)$ function may be estimated by considering the fraction of failures in specific time intervals for which $C=1$, or if $g(t)$ is a fairly smooth function of time by fitting a binary response model to values of C given the corresponding failure times t.

A different approach to relative risk estimation would specify a proportional hazards model for $\lambda(t|\mathbf{z})$ and form a weighted sum of the partial likelihoods corresponding to possible values of the missing $\mathbf{z}_2$ data. The weighting factor would be the binomial coefficient $\binom{n}{x}$ where $n = 204$ and x is the total number of the b/a Gpd-1 phenotype animals corresponding to a particular specification of the missing $\mathbf{z}_2$ values. Such a procedure is unlikely to be computationally feasible, however.

BIBLIOGRAPHIC NOTES

Paired data from the proportional hazards model were considered by Holt and Prentice (1974) in essentially the form given here. Related work is described by Armitage (1959), Downton (1972), and Breslow (1975).

Methods for the analysis of binary response regression data, of the type considered in Section 8.2, have been proposed by many authors. Finney (1971) and Cox (1970) respectively discuss probit and logit models in detail. Most methods of Section 8.2 are given in somewhat greater detail in Prentice (1976a).

The analysis of retrospective studies has also been discussed by many authors. Basic results are given by Cornfield (1951) and Mantel and Haenszel (1959). More philosophical discussion can be found in the references by Miettinen (1974) and Fisher and Patil (1974). Recent methodological work is given by Zelen (1971), Anderson (1972), Prentice (1976b), and Breslow (1976). The approach taken here can be found in Prentice and Breslow (1978).

Nonparametric Bayesian techniques have been considered by several authors. Ferguson (1973) introduced the Dirichlet process and showed that many standard nonparametric techniques could be derived in this way. Doksum (1974) considered a wide class of prior distributions called tail-free or neutral to the right random probability measures, and both the Dirichlet process and the gamma hazard process are special cases of these. Ferguson (1974) gives a survey of work done in this area. The application of these methods to survival data has been considered by Susarla and Van Ryzin (1976), who obtain the Kaplan–Meier estimator from a Dirichlet prior process, by Cornfield and Detre (1977), who consider a gamma type process though there are errors in their work (see Kalbfleisch and MacKay, 1978a), and by Kalbfleisch (1978a). Ferguson and Phadia (1978) have considered the estimation of the survivor function with no covariates for a number of tail-free or neutral to the right prior processes, including the gamma and the Dirichlet cases.

APPENDIX 1

Some Sets of Data

The following sets of data are used for examples and discussion at various places in the text.

Data Set I VETERAN'S ADMINISTRATION LUNG CANCER TRIAL

Data for lung cancer patients: days of survival (t), performance status (x_1), months from diagnosis (x_2), age in years (x_3), and prior* therapy (x_4)

t	x_1	x_2	x_3	x_4	t	x_1	x_2	x_3	x_4	t	x_1	x_2	x_3	x_4
Standard, squamous†					153	60	14	63	10	92	70	10	60	0
72	60	7	69	0	59	30	2	65	0	35	40	6	62	0
411	70	5	64	10	117	80	3	46	0	117	80	2	38	0
228	60	3	38	0	16	30	4	53	10	132	80	5	50	0
126	60	9	63	10	151	50	12	69	0	12	50	4	63	10
118	70	11	65	10	22	60	4	68	0	162	80	5	64	0
10	20	5	49	0	56	80	12	43	10	3	30	3	43	0
82	40	10	69	10	21	40	2	55	10	95	80	4	34	0
110	80	29	68	0	18	20	15	42	0					
314	50	18	43	0	139	80	2	64	0					
100‡	70	6	70	0	20	30	5	65	0	Standard, large				
42	60	4	81	0	31	75	3	65	0	177	50	16	66	10
8	40	58	63	10	52	70	2	55	0	162	80	5	62	0
144	30	4	63	0	287	60	25	66	10	216	50	15	52	0
25‡	80	9	52	10	18	30	4	60	0	553	70	2	47	0
11	70	11	48	10	51	60	1	67	0	278	60	12	63	0
					122	80	28	53	0	12	40	12	68	10
					27	60	8	62	0	260	80	5	45	0
Standard, small					54	70	1	67	0	200	80	12	41	10
30	60	3	61	0	7	50	7	72	0	156	70	2	66	0
384	60	9	42	0	63	50	11	48	0	182‡	90	2	62	0
4	40	2	35	0	392	40	4	68	0	143	90	8	60	0
54	80	4	63	10	10	40	23	67	10	105	80	11	66	0
13	60	4	56	0						103	80	5	38	0
123‡	40	3	55	0	Standard, adeno					250	70	8	53	10
97‡	60	5	67	0	8	20	19	61	10	100	60	13	37	10

Data Set I—*continued*

Data for lung cancer patients: days of survival (t), performance status (x_1), months from diagnosis (x_2), age in years (x_3), and prior* therapy (x_4)

t	x_1	x_2	x_3	x_4	t	x_1	x_2	x_3	x_4	t	x_1	x_2	x_3	x_4
Test, squamous					21	20	4	71	0	73	60	3	70	0
999	90	12	54	10	13	30	2	62	0	8	50	5	66	0
112	80	6	60	0	87	60	2	60	0	36	70	8	61	0
87‡	80	3	48	0	2	40	36	44	10	48	10	4	81	0
231‡	50	8	52	10	20	30	9	54	10	7	40	4	58	0
242	50	1	70	0	7	20	11	66	0	140	70	3	63	0
991	70	7	50	10	24	60	8	49	0	186	90	3	60	0
111	70	3	62	0	99	70	3	72	0	84	80	4	62	10
1	20	21	65	10	8	80	2	68	0	19	50	10	42	0
587	60	3	58	0	99	85	4	62	0	45	40	3	69	0
389	90	2	62	0	61	70	2	71	0	80	40	4	63	0
33	30	6	64	0	25	70	2	70	0					
25	20	36	63	0	95	70	1	61	0	Test, large				
357	70	13	58	0	80	50	17	71	0	52	60	4	45	0
467	90	2	64	0	51	30	87	59	10	164	70	15	68	10
201	80	28	52	10	29	40	8	67	0	19	30	4	39	10
1	50	7	35	0						53	60	12	66	0
30	70	11	63	0	Test, adeno					15	30	5	63	0
44	60	13	70	10	24	40	2	60	0	43	60	11	49	10
283	90	2	51	0	18	40	5	69	10	340	80	10	64	10
15	50	13	40	10	83‡	99	3	57	0	133	75	1	65	0
					31	80	3	39	0	111	60	5	64	0
Test, small					51	60	5	62	0	231	70	18	67	10
25	30	2	69	0	90	60	22	50	10	378	80	4	65	0
103‡	70	22	36	10	52	60	3	43	0	49	30	3	37	0

Source: Prentice (1974). For a discussion of these data see Section 4.5.
*0, no prior therapy; 10, prior therapy.
†Standard therapy, squamous tumor cell type.
‡Censored survival.

Data Set II A CLINICAL TRIAL IN THE TREATMENT OF CARCINOMA OF THE ORO-PHARYNX

Case	Inst.	Sex	Trt Gp.	Grade	Age	Cond.	Site	T	N	Entry Date	Status	Time
1	2	2	1	1	51	1	2	3	1	2468	1	631
2	2	1	2	1	65	1	4	2	3	2968	1	270
3	2	1	1	2	64	2	1	3	3	3368	1	327
4	2	1	1	1	73	1	1	4	0	5768	1	243
5	5	1	2	2	64	1	1	4	3	9568	1	916
6	4	1	2	1	61	1	2	3	0	10668	0	1823
7	4	1	1	2	65	1	2	4	3	10768	1	637
8	4	1	2	3	84	1	4	1	3	12068	1	235
9	6	1	1	2	54	2	1	3	3	13368	1	255
10	3	1	1	2	72	2	4	2	2	15468	1	184
11	3	1	1	2	42	1	4	2	2	15468	1	1064
12	2	1	1	2	61	1	1	4	3	18268	1	414
13	3	1	2	1	71	1	2	3	1	18468	1	216
14	4	1	2	2	83	3	4	3	1	19068	1	324
15	2	1	1	3	43	1	2	4	3	20768	1	480
16	5	1	2	2	52	1	4	4	3	21768	1	245
17	4	2	1	3	68	1	4	2	3	22768	0	1565
18	6	1	2	2	69	1	1	3	0	23368	1	560
19	3	2	2	3	65	3	1	3	0	25968	1	376
20	5	1	1	2	58	1	2	4	3	28068	1	911
21	2	1	2	2	63	1	2	4	3	28068	1	279
22	4	1	1	2	59	3	2	4	3	28268	1	144
23	3	1	1	1	75	1	2	3	1	28268	1	1092
24	6	1	1	1	65	2	1	3	3	28968	1	94
25	4	1	2	3	41	1	2	4	3	29468	1	177
26	3	1	1	2	60	1	4	3	3	29868	0	1472
27	3	1	2	2	72	1	4	1	3	30468	1	526
28	5	1	2	2	51	1	1	4	3	30868	1	173
29	2	2	1	2	72	2	2	3	1	30868	1	575
30	6	1	1	2	49	1	4	3	2	31068	1	222
31	3	1	2	2	82	3	1	3	0	31868	1	167
32	2	2	1	2	64	1	1	2	3	32468	1	1565
33	4	2	2	2	57	2	2	4	3	33568	1	256
34	3	1	2	1	67	2	2	3	3	33368	1	134
35	6	1	2	2	65	2	1	3	0	33868	1	404
36	3	1	2	2	62	1	4	1	2	369	0	1495
37	2	1	1	2	49	1	4	1	3	769	1	162
38	5	1	1	2	60	1	4	3	3	969	1	262
39	3	1	1	2	75	2	2	3	3	1769	1	307
40	2	1	2	2	54	1	2	2	3	2469	1	782
41	3	1	2	2	59	1	4	2	2	2469	1	661
42	5	1	1	2	58	1	1	3	2	3569	1	546
43	3	1	2	2	50	1	1	4	0	4469	0	1766
44	2	1	1	1	60	1	1	3	0	4569	1	374

Data Set II—*continued*

Case	Inst.	Sex	Trt Gp.	Grade	Age	Cond.	Site	T	N	Entry Date	Status	Time
45	3	2	1	1	43	1	2	2	2	4969	0	1489
46	4	1	1	2	48	2	2	3	3	5169	0	1446
47	4	1	2	2	49	3	1	4	3	5669	1	74
48	3	1	1	1	44	1	1	3	1	2769	0	1609
49	2	1	1	1	77	1	1	4	1	8369	1	301
50	2	1	1	1	75	1	2	4	1	9369	1	328
51	3	1	1	1	54	1	1	3	3	11869	1	459
52	3	1	1	1	68	1	1	4	0	12569	1	446
53	6	1	2	3	58	1	4	3	2	12769	0	1644
54	2	1	2	3	66	1	2	4	3	12969	1	494
55	3	1	2	1	47	1	1	3	2	13269	1	279
56	5	1	1	2	60	1	4	3	2	13569	1	915
57	2	1	1	2	66	1	4	4	2	14369	1	228
58	3	1	1	3	51	1	1	3	3	15569	1	127
59	2	2	1	1	49	1	1	3	0	15669	1	1574
60	6	1	1	1	50	1	2	4	0	16669	1	561
61	2	2	1	1	52	1	4	4	3	16769	1	370
62	2	1	2	2	40	1	4	4	3	17869	1	805
63	4	1	2	2	69	1	1	3	3	19969	1	192
64	5	1	2	2	56	1	2	1	3	20469	1	273
65	5	2	2	3	70	2	4	4	3	20469	0	1377
66	3	2	2	3	47	1	4	3	2	23069	1	407
67	3	1	1	3	46	1	2	3	1	24569	1	929
68	3	1	2	1	53	1	4	2	3	26669	1	548
69	3	1	2	1	67	1	4	3	1	27969	0	1317
70	3	2	2	1	68	1	4	3	1	26869	0	1317
71	2	1	1	2	90	1	4	3	3	28069	1	517
72	3	2	1	3	44	1	4	3	2	28969	0	1307
73	5	1	2	2	48	1	1	4	2	29069	1	230
74	4	1	1	2	67	1	2	3	1	30469	1	763
75	5	2	1	2	58	2	4	4	3	30469	1	172
76	4	1	2	2	69	1	1	3	2	32869	0	1455
77	4	1	2	2	75	1	4	3	0	32869	0	1234
78	6	1	2	3	58	1	2	3	3	33069	1	544
79	3	1	1	1	72	1	4	3	3	33269	1	800
80	6	1	1	2	72	1	1	3	0	33569	0	1460
81	6	1	1	3	70	1	4	2	3	33669	1	785
82	6	1	1	2	71	1	2	4	0	34469	1	714
83	1	1	2	2	55	1	1	3	1	35369	1	338
84	3	2	2	1	73	1	1	3	0	36369	1	432
85	1	1	2	2	50	1	4	3	3	870	0	1312
86	6	1	2	2	63	2	1	3	0	4270	1	351
87	2	2	1	1	58	1	2	1	3	4470	1	205
88	1	2	1	2	56	1	4	3	0	4870	0	1219
89	6	2	2	2	62	3	4	4	3	4970	1	11

Data Set II—*continued*

Case	Inst.	Sex	Trt Gp.	Grade	Age	Cond.	Site	T	N	Entry Date	Status	Time
90	3	2	1	1	55	1	4	2	2	5470	1	666
91	2	1	1	1	50	2	2	4	3	5770	1	147
92	3	1	2	1	77	1	4	2	2	7870	0	1060
93	1	2	1	2	67	1	2	3	2	8270	1	477
94	3	1	1	3	53	1	2	3	2	9670	0	1058
95	2	1	2	3	55	1	2	2	2	11070	0	1312
96	6	1	2	1	71	2	2	3	3	11870	1	696
97	2	1	1	1	65	1	1	4	3	12470	1	112
98	1	2	2	2	50	1	2	4	3	13170	1	308
99	5	1	2	2	61	2	4	4	3	14470	1	15
100	5	1	2	1	72	2	1	4	0	14670	1	130
101	4	1	1	1	51	1	2	3	0	15270	1	296
102	4	1	1	2	59	1	4	3	3	15870	1	293
103	2	1	2	2	56	1	1	4	0	16070	1	545
104	3	1	1	2	61	1	1	3	1	16670	0	1086
105	1	1	1	3	61	1	2	2	3	17470	0	1250
106	3	1	2	2	68	2	1	3	3	18770	1	147
107	5	2	1	2	71	2	1	3	3	18970	1	726
108	2	2	2	2	57	1	2	2	2	19070	1	310
109	2	1	1	2	72	1	4	3	1	20570	1	599
110	3	1	1	2	55	1	2	3	0	21170	0	998
111	4	2	2	3	61	1	2	2	2	21970	0	1089
112	5	1	1	1	47	1	4	4	1	23170	1	382
113	4	2	1	3	66	1	2	3	2	24370	0	932
114	1	1	2	2	52	2	4	4	3	25170	1	264
115	1	1	2	1	61	2	4	4	3	25470	1	11
116	5	1	1	1	66	2	2	4	3	25870	0	911
117	2	1	1	3	64	2	4	4	3	28570	1	89
118	5	1	1	2	73	1	1	4	0	28770	1	525
119	2	1	2	2	67	1	1	3	3	31670	0	532
120	3	1	1	2	68	1	1	2	3	32770	1	637
121	6	1	1	3	58	2	2	3	3	33370	1	112
122	6	2	1	1	68	1	4	3	3	33670	0	1095
123	1	1	1	3	85	2	4	2	2	34170	1	170
124	4	1	1	2	74	1	1	3	0	34270	0	943
125	5	1	1	2	53	1	1	4	0	34370	1	191
126	6	1	2	2	60	1	2	1	2	34470	0	928
127	3	1	1	3	58	1	2	3	2	35570	0	918
128	3	1	1	2	66	1	1	1	2	36270	0	825
129	1	1	1	2	58	2	2	2	3	1271	1	99
130	2	1	1	2	39	1	4	3	3	1571	1	99
131	2	1	1	1	54	1	1	4	1	1871	0	933
132	6	1	2	3	49	1	2	2	3	2271	1	461
133	6	1	2	2	52	1	1	3	1	2671	1	347
134	1	1	1	2	35	2	4	3	0	3371	1	372

Data Set II—*continued*

Case	Inst	Sex	Trt Gp.	Grade	Age	Cond.	Site	T	N	Entry Date	Status	Time
135	5	1	2	3	44	1	2	4	3	4371	0	731
136	5	1	1	9	81	1	4	3	3	4971	1	363
137	1	1	2	2	74	2	1	3	0	6771	1	238
138	4	1	2	2	65	1	4	4	3	7571	0	593
139	2	2	1	2	66	2	4	4	3	7771	1	219
140	3	1	1	2	74	2	4	3	2	8871	1	465
141	1	2	2	3	90	0	2	3	0	10571	1	446
142	2	1	2	2	60	1	1	4	1	11371	1	553
143	5	1	2	2	63	1	1	4	1	15371	1	532
144	2	2	2	2	61	1	4	4	1	15471	1	154
145	2	1	2	1	67	1	1	4	3	15971	1	369
146	4	1	1	2	88	1	2	3	0	16171	1	541
147	5	2	2	3	69	2	4	4	0	18371	1	107
148	2	1	2	2	46	1	1	4	1	18871	0	854
149	1	1	1	2	69	1	1	2	2	20171	0	822
150	6	2	1	2	48	1	2	3	0	20271	1	775
151	2	1	1	1	77	1	1	4	0	20271	1	336
152	6	1	2	3	69	1	2	1	3	20271	1	513
153	6	1	2	3	75	1	4	3	3	20971	0	914
154	5	1	1	2	71	1	4	4	3	21671	1	757
155	5	2	1	2	58	1	1	4	3	21871	0	794
156	3	1	2	2	66	2	2	3	2	22171	1	105
157	5	2	1	1	44	1	2	4	1	23771	0	733
158	1	2	2	2	59	1	1	3	1	25371	0	600
159	2	1	1	2	78	9	4	4	3	26371	1	266
160	2	2	2	1	58	2	1	4	3	27371	1	317
161	2	1	2	3	65	1	4	3	3	28071	1	407
162	6	2	2	2	53	2	4	3	2	28471	1	346
163	3	1	2	2	49	1	4	2	2	29471	1	518
164	1	1	2	2	65	1	2	3	2	29971	1	395
165	5	1	1	1	59	1	1	4	1	31471	1	81
166	6	2	2	2	79	1	4	2	2	31971	1	608
167	6	1	1	1	57	1	2	4	3	32171	0	760
168	6	1	1	1	54	1	2	4	0	32371	1	343
169	3	1	2	2	47	1	1	3	0	32671	1	324
170	1	1	1	2	68	1	1	4	1	33071	1	254
171	2	1	1	3	63	1	4	2	2	34071	0	751
172	2	1	1	3	72	1	2	3	3	34271	1	334
173	6	2	2	1	51	2	2	3	1	34771	1	275
174	5	2	2	1	43	1	2	3	3	1272	0	546
175	6	2	2	2	43	2	4	3	3	3572	1	112
176	1	1	2	2	65	4	4	2	3	4672	0	182
177	1	1	2	2	54	1	1	4	3	5472	1	209
178	6	1	2	3	50	1	2	4	3	5572	1	208
179	2	1	2	2	39	1	1	4	3	5672	1	174

Data Set II—*continued*

Case	Inst	Sex	Trt Gp.	Grade	Age	Cond.	Site	T	N	Entry Date	Status	Time
180	2	1	1	1	46	1	1	4	0	5972	0	651
181	2	1	2	3	49	1	2	3	3	8072	0	672
182	1	2	2	2	52	2	1	3	2	8272	1	291
183	1	2	2	2	69	2	4	4	1	13671	0	723
184	4	2	1	2	55	1	2	3	0	14372	1	498
185	4	1	2	2	48	2	2	3	0	14372	0	276
186	1	1	1	2	20	1	4	2	3	15672	0	90
187	4	2	1	2	47	1	4	3	3	15772	1	213
188	5	2	2	2	67	2	4	3	0	20572	1	38
189	4	2	1	2	66	2	2	3	2	20772	1	128
190	2	1	1	1	60	1	2	3	0	20972	0	445
191	2	1	1	1	54	1	2	4	3	22772	1	159
192	5	1	2	2	54	1	1	3	3	24372	1	219
193	4	1	2	2	59	2	1	4	0	24872	1	173
194	5	1	2	3	47	1	1	3	3	27672	0	413
195	3	1	2	2	57	2	1	3	3	12371	1	274

Source: The radiation therapy oncology group. For a discussion of these data see Sections 1.1.2 and 4.5.

Definitions

Sex	1 = male, 2 = female.
Treatment	1 = standard, 2 = test.
Grade	1 = well differentiated, 2 = moderately differentiated, 3 = poorly differentiated.
Age	In years at time of diagnosis.
Condition	1 = no disability, 2 = restricted work, 3 = requires assistance with self care, 4 = bed confined.
Site	1 = faucial arch, 2 = tonsillar fossa, 3 = posterior pillar, 4 = pharyngeal tongue, 5 = posterior wall.
T staging	1 = primary tumor measuring 2 cm or less in largest diameter; 2 = primary tumor measuring 2 cm to 4 cm in largest diameter, minimal infiltration in depth; 3 = primary tumor measuring more than 4 cm; 4 = massive invasive tumor.
N staging	0 = no clinical evidence of node metastases; 1 = single positive node, 3 cm or less in diameter, not fixed; 2 = single positive node, more than 3 cm in diameter, not fixed; 3 = multiple positive nodes or fixed positive nodes.
Date of entry	Day of year and year.
Status	0 = censored, 1 = dead.
Survival	In days from day of diagnosis.

Data Set III STANFORD HEART TRANSPLANT DATA

Patient Ident.	Year of Accept.	Age	Surv. Status	Surv. Time	Surgery	Transplant	Waiting Time	Matching		
								1	2	3
15	68	53	1	1	0	0				
43	70	43	1	2	0	0				
61	71	52	1	2	0	0				
75	72	52	1	2	0	0				
6	68	54	1	3	0	0				
42	70	36	1	3	0	0				
54	71	47	1	3	0	0				
38	70	41	1	5	0	1	5	3	0	.87
85	73	47	1	5	0	0				
2	68	51	1	6	0	0				
103	67	39	1	6	0	0				
12	68	53	1	8	0	0				
48	71	56	1	9	0	0				
102	74	40	0	11	0	0				
35	70	43	1	12	0	0				
95	73	40	1	16	0	1	2	0	0	0
31	69	54	1	16	0	0				
3	68	54	1	16	0	1	1	2	0	1.1
74	72	29	1	17	0	1	5	1	0	.69
5	68	20	1	18	0	0				
77	72	41	1	21	0	0				
99	73	49	1	21	0	0				
20	69	55	1	28	0	1	1	3	1	2.76
70	72	52	1	30	0	1	5	3	1	1.68
101	74	49	0	31	0	0				
66	72	53	1	32	0	0				
29	69	50	1	35	0	0				
17	68	20	1	36	0	0				
19	68	59	1	37	0	0				
4	68	40	1	39	0	1	36	3	0	
100	74	35	0	39	1	1	38	3	0	
8	68	45	1	40	0	0				
44	70	42	1	40	0	0				
16	68	56	1	43	0	1	20	3	0	
45	71	36	1	45	0	1	1	1	0	
1	67	30	1	50	0	0				
22	69	42	1	51	0	1	12	3	0	
39	70	50	1	53	0	1	2	0	0	
10	68	42	1	58	0	1	12	2	1	
35	71	52	1	61	0	1	10	2	1	
37	70	61	1	66	0	1	19	3	0	
68	72	45	1	68	0	1	3	3	0	
60	71	49	1	68	0	1	3	3	0	
62	71	39	1	69	0	0				

Data Set III—*continued*

Patient Ident.	Year of Accept.	Age	Surv. Status	Surv. Time	Surgery	Transplant	Waiting Time	Matching 1	2	3
28	69	53	1	72	0	1	71	2	0	
47	71	47	1	72	0	1	21	3	0	
32	69	64	1	77	0	1	17	4	0	
65	72	51	1	78	0	1	12	2	0	
83	73	53	1	80	0	1	32	4	0	
13	68	54	1	81	0	1	17	3	0	
9	68	47	1	85	0	0				
73	72	56	1	90	0	1	27	3	1	
79	72	53	1	96	0	1	67	2	0	
36	70	48	1	100	0	1	46	2	0	
32	71	41	1	102	0	0				
98	73	28	0	109	0	1	96	4	1	
87	73	46	1	110	0	1	60	2	0	
97	73	23	0	131	0	1	21	3	1	
37	71	41	1	149	0	0				
11	68	47	1	153	0	1	26	1	0	
94	73	43	1	165	1	1	4	3	0	
96	73	26	0	180	0	1	13	2	0	
90	73	52	1	186	1	1	160	3	1	.82
53	71	47	1	188	0	1	41	0	0	0.00
89	73	51	1	207	0	1	139	4	1	1.33
24	69	51	1	219	0	1	83	3	1	1.62
27	69	8	1	263	0	0				
93	73	47	0	265	0	1	28	2	0	.33
51	71	48	1	285	0	1	32	4	1	1.08
67	73	19	1	285	0	1	57	3	0	1.02
16	68	49	1	308	0	1	28	2	0	1.12
84	73	42	1	334	0	1	37	4	0	.60
91	73	47	1	340	0	0				
92	73	44	0	340	0	1	310	1	0	.16
58	71	47	1	342	1	1	21	2	1	1.82
88	73	54	0	370	0	1	31	3	0	.68
86	73	48	0	397	0	1	8	3	1	1.44
82	71	29	0	427	0	0				
81	73	52	0	445	0	1	6	4	1	1.94
80	72	46	0	482	1	1	26	3	0	1.41
78	72	48	0	515	0	1	210	3	0	.81
76	72	52	0	545	1	1	46	3	1	1.70
64	72	48	1	583	1	1	32	1	0	.12
72	72	26	0	596	0	1	4	3	1	1.46
71	72	47	0	620	0	1	31	3	0	.97
69	72	47	0	670	0	1	10	2	0	1.20
7	68	50	1	675	0	1	51	4	0	1.32
23	69	58	1	733	0	1	3	3	0	.96

Data Set III—*continued*

Patient Ident.	Year of Accept.	Age	Surv. Status	Surv. Time	Surgery	Transplant	Waiting Time	Matching		
								1	2	3
63	71	32	0	841	0	1	27	3	1	1.93
30	69	44	1	852	0	1	16	4	0	1.58
59	71	41	0	915	0	1	78	2	0	0.19
56	71	38	0	941	0	1	67	4	0	.98
50	71	45	1	979	1	1	83	0	0	0.00
46	71	48	1	995	1	1	2	2	0	.81
21	69	43	1	1032	0	1	8	2	0	1.13
49	71	36	0	1141	1	1	36	4	0	1.35
41	70	45	0	1321	1	1	58	2	0	.98
14	68	53	1	1386	0	1	37	1	0	.87
26	69	30	0	1400	0	0				
40	70	48	0	1407	1	1	41	4	0	.75
34	69	40	0	1571	0	1	23	2	0	.38
33	69	48	0	1586	0	1	51	3	0	.91
25	69	33	0	1799	0	1	25	2	0	1.06

Source: Crowley and Hu (1977). For a discussion of these data see Sections 1.1.3 and 5.4.

Definitions

Matching	1 = mismatch on alleles, 2 = mismatch on antigen, 3 = mismatch score.
Survival status	1 = dead, 0 = censored.
Previous surgery	1 = yes, 0 = no.
Transplant status	1 = transplanted, 0 = not transplanted.
Time	In days.

Data Set IV DAYS UNTIL OCCURRENCE OF CANCER FOR MALE MICE

Control group

Thymic lymphoma (22%)	159, 189, 191, 198, 200, 207, 220, 235, 245 250, 256, 261, 265, 266, 280, 343, 356, 383 403, 414, 428, 432
Reticulum cell sarcoma (38%)	317, 318, 399, 495, 525, 536, 549, 552, 554 337, 558, 571, 586, 594, 596, 605, 612, 621 628, 631, 636, 643, 647, 648, 649, 661, 663 666, 670, 695, 697, 700, 705, 712, 713, 738 748, 753

Data Set IV—*continued*

Other causes (39%)	40, 42, 51, 62, 163, 179, 206, 222, 228 252, 249, 282, 324, 333, 341, 366, 385, 407 420, 431, 441, 461, 462, 482, 517, 517, 524 564, 567, 586, 619, 620, 621, 622, 647, 651 686, 761, 763
Germ-free group	
Thymic lymphoma (22%)	158, 192, 193, 194, 195, 202, 212, 215, 229 230, 237, 240, 244, 247, 259, 300, 301, 321 337, 415, 434, 444, 485, 496, 529, 537, 624 707, 800
Reticulum cell sarcoma (18%)	430, 590, 606, 638, 655, 679, 691, 693, 696 747, 752, 760, 778, 821, 986
Other causes (46%)	136, 246, 255, 376, 421, 565, 616, 617, 652 655, 658, 660, 662, 675, 681, 734, 736, 737 757, 769, 777, 800, 807, 825, 855, 857, 864 868, 870, 870, 873, 882, 895, 910, 934, 942 1015, 1019

Source: Hoel (1972). For discussion see Sections 1.1.1 and 7.

Data Set V MOUSE LEUKEMIA DATA

T_1	T_2	J	δ	z_1	z_2	z_3	z_4	z_5	z_6
121175	122377	2	2	1	2	2	1	00.0	10000
121175	122377	3	2	1		2	2	07.4	02400
121175	060677	3	1	1		2	2	13.2	00000
121175	121476	2	1	1		2	2	00.0	10000
121175	122377	3	2	2	2	1	1	05.8	00000
121175	010577	2	1	2		1	1	00.0	08800
121175	111677	5	1	1	2	1	1	00.0	08000
121175	101477	1	1	2		1	2	00.0	10000
121175	122377	3	2	2	2	2	1	78.7	00000
121175	122377	3	2	2		2	2	05.3	00080
121175	122377	1	2	2	2	2	2	20.3	00000
121175	040777	1	1	1	1	2	2	04.7	10000
121175	081076	1	1	1		2	2		

Data Set V—*continued*

T_1	T_2	J	δ	z_1	z_2	z_3	z_4	z_5	z_6
121175	050577	1	1	2		1	1	00.0	10000
121175	081077	1	1	1	1	1	1	00.0	10000
121175	091277	2	1	1		1	1	00.0	00080
121175	021777	1	1	2		1	2	00.0	10000
121175	091476	1	1	2		1	2	00.0	
121175	031877	1	1	1	2	1	2	00.0	10000
121275	100676	1	1	2		2	1	00.0	10000
121275	122377	2	2	2	2	2	2	00.0	03600
121275	122377	5	2	2	2	2	2	07.5	02000
121275	062877	2	1	1		2	2	00.4	00000
121275	122377	3	2	2	2	2	2	09.7	00000
121275	081776	1	1	1		2	2	00.0	
121275	080676	3	1	1		1	1		
121275	123377	3	2	1	2	1	1	00.0	10000
121275	041877	3	1	1		1	2	00.0	
121275	122377	3	2	1	2	1	2	00.0	10000
121275	122377	3	2	2	1	2	1	06.1	10000
121275	081977	2	1	1	1	2	1	07.5	10000
121275	122377	3	2	1		2	1	01.7	01000
121275	122377	3	2	1	1	2	1	00.0	10000
121275	102076	1	1	1		1	1	00.0	10000
121275	122377	3	2	2		1	2	18.1	05200
121275	102077	3	1	1	2	1	2	00.0	0300
121275	041177	3	1	1		1	2	00.0	10000
121275	052477	5	1	2	1	1	2	01.6	10000
121275	122377	3	2	1	2	1	2	00.0	05600
121275	070677	1	1	2		1	2	00.0	10000
121575	083076	6	1	1		2	1	00.0	
121575	083076	6	1	2		2	2	00.0	
121575	083076	6	1	2		2	2	00.0	
121575	083076	6	1	1		2	2	00.0	
121575	100776	3	1	2		1	1	00.0	
121575	082677	2	1	2		1	2	00.0	10000
121575	091476	1	1	1		2	1	00.0	
121575	080476	1	1	2		2	1		
121575	082677	1	1	2	2	2	1	68.5	00000
121575	031877	1	1	1	1	2	2	00.0	10000
121575	110477	4	1	2		2	2	00.0	00000
121575	050577	1	1	2	1	2	2	49.7	10000
121575	122377	3	2	2		1	1	02.2	00000
121575	122176	1	1	1		1	1	00.0	10000
121575	122377	3	2	2	2	1	2	00.0	00800

Data Set V—*continued*

T_1	T_2	J	δ	z_1	z_2	z_3	z_4	z_5	z_6
121575	122377	3	1	1	1	1	2	00.0	00360
121575	091277	3	1	1		1	2	00.0	10000
121575	072577	6	1	2		2	1	05.0	00160
121575	072577	6	1	1		2	2	06.2	00480
121575	063077	3	1	1	1	1	1	08.6	00160
121575	101276	1	1	2		1	1	00.0	10000
121575	100477	1	1	2	1	1	1	00.0	10000
121575	100476	1	1	1		1	1	00.0	
121575	052077	1	1	2	1	1	1	06.2	10000
121575	100477	3	1	2	2	1	1	05.1	00000
121575	100477	3	1	1	2	1	1	00.0	02000
121575	072676	3	1	1		1	2	06.7	03100
121575	020977	1	1	1		1	2	00.0	10000
121675	072577	6	1	2		2	1	63.7	00280
121675	072577	6	1	2		2	1	17.7	10000
121675	072577	6	1	1		2	2	06.2	10000
121675	110876	3	1	1		1	1	00.0	
121675	122377	3	2	2		1	1	05.1	02300
121675	122377	3	2	1	2	1	2	00.0	06400
121675	122377	3	2	2	2	1	2	00.0	10000
121775	031877	1	1	2	1	2	1	00.0	10000
121775	081877	1	1	2		2	2	12.8	0040
121775	091477	6	1	1		2	2	47.5	00240
121775	120276	3	1	1		2	2	28.7	
121775	020878	2	2	2	2	2	2	74.7	00040
121775	122377	5	2	2	2	1	1	05.8	00080
121775	122377	2	2	1	2	1	1	00.0	01400
121775	031877	3	1	2	2	1	1	10.7	00000
121775	062077	1	1	2		1	1	06.6	10000
121775	060777	1	1	1	1	1	2	00.0	10000
121775	122377	3	2	2	2	1	2	04.6	00080
121775	102776	1	1	1		1	2	00.0	10000
122275	111577	3	2	1	2	2	1	00.0	10000
122275	111677	1	1	2	1	2	1	00.0	10000
122275	061477	1	1	1		2	2	00.0	10000
122275	111577	3	2	1	2	2	2	00.0	00960
122275	111577	3	2	1	2	2	2	00.0	00000
122275	092476	1	1	1		2	1	00.0	
122275	012677	1	1	1		1	2	00.0	10000
122275	111577	3	1	2	2	1	2	00.0	10000
122275	122377	1	2	1	2	1	2	00.0	04000
122275	042677	2	1	1		1	2	00.0	00100

Data Set V—*continued*

T_1	T_2	J	δ	z_1	z_2	z_3	z_4	z_5	z_6
122275	033077	1	1	1	1	1	2	00.0	10000
122275	010577	1	1	1		2	1	00.0	10000
122275	122377	3	2	2	2	2	1	14.4	00720
122275	070677	1	1	1		2	1	02.2	10000
122275	122377	5	2	1	2	2	1	00.0	02400
122275	122377	3	2	1	2	2	1	00.0	02000
122275	122377	3	2	2	1	2	2	09.2	10000
122275	102776	1	1	1		2	2	00.0	10000
122275	101477	3	1	1	2	1	1	00.0	00880
122275	122377	3	2	1	2	1	1	00.0	10000
122275	122377	3	2	2	1	1	1	12.5	10000
122275	052777	1	1	1		1	1	00.0	10000
122275	091377	2	1	1		1	2	01.3	01500
122275	122377	3	2	2		1	2	87.9	00000
010576	012778	1	1	1		2	1	55.5	00000
010576	072677	1	1	2		2	2	17.4	10000
010576	083077	3	1	2	2	2	2	14.3	00560
010576	020178	3	2	1		1	1	00.0	10000
010576	060777	1	1	1		1	2	06.7	10000
011276	020178	3	2	2	2	2	1	43.4	00000
011276	091376	1	1	2		2	1	00.0	
011276	020178	1	2	1	2	2	1	01.3	00360
011276	020178	2	2	1	2	2	1	02.9	03800
011276	100676	1	1	2		2	1	03.4	10000
011276	122977	2	1	1	2	2	2	00.0	02800
011276	020178	3	2	2	2	2	2	45.4	00000
011276	020178	3	2	2	2	1	1	06.1	01500
011276	020178	3	2	2	2	1	1	02.6	10000
011276	011277	3	1	2		1	2	00.0	
011276	120676	3	1	2		1	2	01.7	
011276	082377	3	1	1		1	2	00.0	10000
011576	020178	3	2	2	2	2	2	27.7	00040
012676	020178	3	2	1	1	2	1	07.1	00040
012676	010577	3	1	2		2	2	02.1	
012676	030877	1	1	2	1	2	2	05.8	10000
012676	102076	1	1	1		2	2	00.0	10000
012676	122977	2	2	1	2	2	2	04.4	00280
012676	020178	3	2	1	2	2	1	00.0	00280
012676	100477	4	1	2		2	2	05.8	10000
012676	012677	6	1	1		2	2	00.0	
012676	020178	1	2	1		2	2	03.3	00080
012676	120776	1	1	2		2	2	27.7	10000
012676	020178	3	2	2	2	1	1	04.0	00040

Data Set V—*continued*

T_1	T_2	J	δ	z_1	z_2	z_3	z_4	z_5	z_6
012676	020178	1	2	2		1	2	00.0	10000
012676	020178	3	2	2	2	1	2	86.6	00040
012676	021777	1	1	2		1	2	05.2	10000
012676	120877	1	2	1	2	2	1	00.0	09200
012676	020178	2	2	1	2	2	1	05.2	10000
012676	020178	3	2	2	2	2	1	06.0	00040
012676	031477	4	1	1		2	2	00.0	
012676	082076	1	1	2		2	2	00.0	
012676	122077	1	2	2		2	2	04.4	10000
012676	021777	1	1	1		1	1	00.0	00000
012676	022176	1	1	1		1	1	00.0	100000
012676	111176	1	1	2		1	2	00.0	10000
012676	020178	3	2	1	2	1	2	00.7	06000
012676	020178	1	2	1		2	1	05.1	00880
012676	020178	3	2	1	2	2	1	04.1	01800
012676	020178	3	2	2		2	1	23.1	00200
012676	011777	2	1	1		2	2	00.0	
012676	091477	3	1	1		2	2	49.5	00400
012676	020178	3	2	2	1	2	2	05.1	10000
012676	111677	2	1	1		1	1	00.0	10000
012676	010177	4	1	2		1	1	05.1	
012676	081177	3	1	2	1	1	1	08.3	10000
012676	020977	1	1	1		1	2	16.6	10000
012676	012778	5	2	1	2	1	2	00.0	01200
012676	112277	1	1	1		1	2	00.0	10000
012776	021777	1	1	2		2	1	06.4	10000
012776	111776	1	1	1		2	1	00.0	10000
012776	111176	1	1	2		2	2	05.5	10000
012776	092077	4	1	2		2	2	18.1	05200
012776	102676	1	1	1		2	2	00.0	
012776	020178	3	2	2	2	1	1	15.2	04000
012776	020178	2	2	2	2	1	1	02.8	10000
012776	062077	1	1	1		1	2	00.0	10000
012776	020878	3	2	2	2	1	2	00.0	10000
012776	033077	1	1	2	2	1	2	00.0	10000
020276	020878	3	2	2	2	2	1	11.5	00040
020276	100477	3	1	2		2	2	05.3	10000
020276	030877	1	1	1	1	2	2	17.3	10000
020276	010178	4	1	2	2	2	2	50.2	00240
020276	090676	1	1	1		2	2	00.0	
020276	020878	3	2	2	1	1	1	03.7	10000
020276	040777	1	1	2	2	1	1	00.0	10000
020276	050677	1	1	1	1	1	1	00.0	10000

Data Set V—*continued*

T_1	T_2	J	δ	z_1	z_2	z_3	z_4	z_5	z_6
020276	101876	2	1	2		1	1	00.0	10000
020276	101876	3	1			1	2		
020276	020878	3	2	1	2	2	1	00.0	10000
020276	050977	3	1	1		2	1	00.0	
020276	020878	3	2	2	1	2	1	05.7	10000
020276	020878	3	2	2	2	2	1	41.1	00040
020276	020878	3	2	1	2	2	2	01.2	03900
020276	050577	1	1	1		2	2	00.0	10000
020276	012177	1	1	1		1	2	00.0	10000
020276	090777	1	1	1	1	1	2	01.2	04200
020276	061677	1	1	1		1	2	00.0	10000
020276	072977	3	1	1	1	1	1	02.9	02900
020976	011078	1	1	1	2	2	1	09.5	09200
020976	020878	3	2	1	2	2	2	02.5	00600
020976	020878	3	2	2	2	2	2	08.6	00800
020976	020878	3	2	1	1	2	2	13.1	10000
020976	020878	3	2	1	2	2	2	00.0	10000
020976	020878	3	2	1	2	2	1	00.0	01000
020976	020878	3	2	2	1	2	1	13.8	10000
020976	020878	3	2	1	2	1	1	01.2	10000
020976	020878	3	2	1	2	1	2	01.6	02200

Source: Mortality data on AKR × (C57BL/6 × AKR)F_1 mice from the laboratories of Dr. Robert Nowinski, Fred Hutchinson Cancer Research Center, Seattle, Washington. For discussion, see Section 8.5.

Definitions

Date of birth (T_1)	Month/day/year.
Date of death (T_2)	Month/day/year.
Disposition of disease at death (J)	1 = thymic leukemia, 2 = nonthymic leukemia, 3 = nonleukemic and no other tumors, 4 = unknown, 5 = other tumors, 6 = accidental death.
"Type" of death (δ)	1 = natural, 2 = terminated.
MHC phenotype (z_1)	1 = k, 2 = b, blank = unknown.
Gpd-1 phenotype (z_2)	1 = b/b, 2 = b/a, blank = unknown.
Sex (z_3)	1 = male, 2 = female.
Coat color (z_4)	1 = c/c, 2 = $c/+$.
Antibody level (z_5)	% gp 70 precipitate; 0 = if % gp 70 < .5, blank = unknown.
Virus level (z_6)	PFL/ml; 0 = if PFL/ml $< 10^{1.6}$, 10000 = if PFU/ml $\geq 10^4$, blank = unknown.

APPENDIX 2

Exercises and Further Results

1. Consider the mouse carcinogenesis data of Appendix 1 from Hoel (1972). Compute the product limit (Kaplan–Meier) estimates (1.10) of the reticulum cell sarcoma survivor functions for the control and the germ free groups by:

(a) Ignoring failures from thymic lymphoma and other causes (i.e., eliminate mice dying by these causes before carrying out calculations).
(b) Regarding failures times from lymphoma or other causes or censored.

Comment on the relative merits of (a) and (b). On the basis of the survivor function plots, does the germ-free environment appear to reduce the risk of reticulum cell sarcoma?

(Chapter 1)

2. Plot on a single graph the logarithms of the life table (1.13), product limit (1.10), and the continuous (1.14) survivor function estimates using the thymic lymphoma data from the germ-free group. Regard failures from reticulum cell sarcoma and other causes as censored. Use grouping intervals of width 50 days for (1.13) and (1.14).

(Chapter 1)

3. An electronic system is at constant risk of failure with hazard λ/hour. In addition, power surges occur each hour (i.e., at times $1, 2, 3, \ldots$), and at each power surge there is a 10% chance that the system will immediately fail. Obtain expressions for the survivor and density functions. (Use a Dirac delta notation for the latter).

(Chapter 1)

4. Let the survival time T be an integer valued random variable with finite mean r_0 and let

$$r_i = E(T - i | T \geq i)$$

be the residual mean life at $T=i$ $(i=1,2,\ldots)$. Show that the survivor function for integer t is

$$F(t)=\frac{1+r_0}{r_t}\prod_0^t\frac{r_i}{1+r_i}.$$

Thus in the discrete case also, the residual mean life time specifies the distribution of T.

(Chapter 1)

5. (a) Show that the exponential distribution is the only continuous distribution for which the mean residual lifetime $r(t)=$ constant for all $t>0$.

(b) Show that if the p.d.f. $f(t)$ is differentiable and $f(t)>0$ $(0<t<\infty)$, then

$$\lim_{t\to\infty} r(t)=\lim_{t\to\infty}\left(-\frac{d}{dt}\log f(t)\right)^{-1}.$$

(c) Examine the form of $r(t)$ for the log-normal and gamma distributions and show that as $t\to\infty$, $r(t)\to\infty$ for the former and $r(t)\to 1/\lambda$ for the latter.

(Chapter 1)

6. As in Section 1.3, let $t_1<t_2<\cdots<t_k$ represent the observed failure times in a sample of size n_0 from a homogeneous population with survivor function $F(t)$. Suppose that d_j items fail at t_j and that n_j items are at risk at t_j-0.

(a) For b a prespecified time $(b>t_1)$ and c a constant $(0\le c\le 1)$, show that, subject to the constraint $F(b)=c$, the nonparametric maximum likelihood estimate of $F(t)$ is

$$\tilde{F}(t)=\prod_{j|t_j<t}(1-\tilde{\lambda}_j)$$

where $t_0=\tilde{\lambda}_0=0$ and $\tilde{\lambda}_j=d_j/(n_j+a)$ if $t_j<b$ and d_j/n_j if $t_j>b$, $j=1,\ldots,k$. The value a is chosen so that $\tilde{F}(b)=c$. Note that if $b\le t_1$, the constrained estimate is not unique for $t<b$. An arbitrary convention would assign a hazard $1-c$ at $t=0$.

(b) Show that the generalized log likelihood ratio statistic for the hypothesis $F(b)=c$ can be written

$$R=\sum_{i|t_i<b}\left[(n_i-d_i)\log\left(1+\frac{a}{n_i-d_i}\right)-n_i\log\left(1+\frac{a}{n_i}\right)\right].$$

Thomas and Grunkemeier (1975) give heuristic arguments that

the usual asymptotic properties apply so that $-2R$ is asymptotically χ^2, under the hypothesis. Use this result to establish a 95% confidence interval for $F(b)$. Compare these results with those obtained in Section 1.3 for the carcinogenesis data of Table 1.1 with $b = 150$.

(Chapter 1)

7. Show that the mean vector and variance matrix for $(d_{1j}, \ldots, d_{rj})$ in the distribution (1.15) are as asserted.

(Chapter 1)

8. Consider again the mouse carcinogenesis data (data set IV, Appendix 1). Use the log-rank test (1.17) to test the hypothesis that germ-free isolation does not affect overall mortality.

(Chapter 1)

9. (a) Let $T_1, \ldots, T_n$ be a random sample from a distribution with survivor function $F(t)$ and suppose that as $t \to 0$

$$F(t) = 1 - \lambda t + o(t)$$

for some $\lambda > 0$. Show that as $n \to \infty$, the survivor function of

$$X_n = n \min(T_1, \ldots, T_n)$$

converges to that of an exponential with failure rate λ. As noted in Chapter 2, this property is often taken as justification for the choice of an exponential model.

(b) Suppose now that for t near 0

$$F(t) = 1 - (\lambda t)^p + o(t^p)$$

where $\lambda > 0$ and $p > 0$. Show that the limiting distribution of

$$Y_n = n^{1/p} \min(T_1, \ldots, T_n)$$

is Weibull with shape p and scale λ.

(Chapter 2; Cox and Hinkley, 1974, p. 472)

10. Show that the moment generating function of the logistic distribution (2.4) is

$$\Gamma(1 + \theta)\Gamma(1 - \theta)$$

and that the $2r$th cumulant is

$$\kappa_{2r} = \frac{(2r)!\zeta(2r)}{r}, \qquad r = 1, 2, \ldots,$$

where $\zeta(n)$ is the Reimann zeta function

$$\zeta(n)=\sum_{i=1}^{\infty} i^{-n}.$$

Hence show that the excess in kurtosis for the extreme value density is 1.2.
[Note that $\zeta(2)=\Pi^2/6$ and $\zeta(4)=\Pi^4/90$.] (Chapter 2)

11. Let W_1 and W_2 be independent with the extreme value density $\exp(w-e^w)$. Show that $V=W_1-W_2$ has the logistic density

$$f(v)=e^v(1+e^v)^{-2}.$$

Derive the variance and kurtosis of the logistic distribution from those of the extreme value distribution.

(Chapter 2)

12. Consider a two sample situation ($z=0,1$) in which the hazard is exponential

$$\lambda(t;z)=\lambda\exp(z\beta)$$

for the continuous failure variable T. The time axis is grouped into disjoint intervals $I_j=[a_{j-1},a_j)$, $j=1,2,\ldots$, where $a_0=0<a_1<\cdots$ and $a_k\to\infty$ as $k\to\infty$. Define the discrete variate $Y=j$ for $T\in I_j$, $j=1,2,\ldots$. Verify that the resulting discrete model is of the form (2.20) and that the same parameter β measures the sample differences. Show also that if the grouping intervals are constant, $a_j-a_{j-1}=c/\lambda$, $j=1,2,\ldots$, where c is a positive constant, say, the discrete model has also the logistic relationship (2.23). For the latter case, note that the log odds ratio is β' where

$$\exp(\beta')=\frac{1-\exp(-ce^{\beta})}{1-\exp(-c)}.$$

This is close to β if c is small.

(Chapter 2)

13. (A) Use expression (3.4) to show that if $t_{(n)}$ is the largest of n independent unit exponential variates, the rth cumulant of $t_{(n)}$ is

$$(r-1)!\sum_{i=1}^{n} i^{-r}.$$

(b) Show that the asymptotic distribution of $X_n=(\log n-t_{(n)})$ is the extreme value distribution with density (2.1). Note also that the m.g.f. of X_n converges uniformly for $\theta \in (-1, 1)$ to the m.g.f. of (2.1). This implies convergence of the moments of X_n to those of (2.1) (c.f. Rao, 1965, p. 101).

(c) By making use of (a) and (b), show that the extreme value density has cumulants

$$\kappa_1 = \psi(1) = \lim_{n\to\infty}\left(\log n - \sum_1^n i^{-1}\right) = -\gamma$$

$$\kappa_r = \psi^{(r-1)}(1) = (-1)^r(r-1)!\zeta(r), \qquad r = 2, 3, 4, \ldots,$$

where $\zeta(r)$ is the Reimann zeta function and

$$\psi^{(r-1)}(x) = \frac{d^r}{dx^r}\log\Gamma(x), \qquad r = 1, 2, \ldots,$$

are the polygamma functions with $\psi(x) = \psi^{(1)}(x)$ the digamma function. Note that this provides an elementary evaluation of the wellknown definite integrals,

$$\int_{-\infty}^{\infty} x \exp(x - e^x)\, dx = -\gamma$$

and

$$\int_{-\infty}^{\infty} x^2 \exp(x - e^x)\, dx = \frac{\Pi^2}{6} + \gamma^2$$

where $\gamma = .5772\ldots$ is Euler's constant. (Chapters 2 and 3)

14. Using the results of Problem 13, determine the expected information matrix for the Weibull distribution with density

$$\lambda p(\lambda t)^{p-1}\exp[-(\lambda t)^p], \qquad 0 < t < \infty.$$

(Chapter 3)

15. Consider the comparison of two type II censored samples where sample i is followed to the observed d_ith failure, $i = 1, 2$.

(a) In the notation of Section 3.3, show that $(\lambda_1 V_1/d_1)/(\lambda_2 V_2/d_2)$ has an F distribution on $2d_1$ and $2d_2$ degrees of freedom.

(b) Suppose $d_1 = 17$ and $d_2 = 19$ and it is observed that $V_1 = 2195$ and $V_2 = 2923$. Compute a 95% confidence interval for $\beta_2 = \log(\lambda_2/\lambda_1)$ and compare with the results given in Section 3.5.2 for the same data but with type I censoring.

(c) For large degrees of freedom, the logarithm of an $F_{(n,m)}$ variate is approximately normal with mean $(n^{-1} - m^{-1})$ and variance $2(n^{-1} +$

m^{-1}) (Atiquallah, 1962). Use this result to obtain an approximate normal distribution for $\hat{\beta}_2$ and compare it with the asymptotic normal distribution of $\hat{\beta}_2$ in Section 3.4.5.

(Chapter 3)

16. Consider again type II censoring in an exponential distribution. As in Section 3.3, suppose n items are placed on test and followed until the dth failure. Let $x_1, \ldots, x_d$ be the normalized spacings defined by $x_i = (n-i+1)(t_{(i)} - t_{(i-1)})$, $i = 1, \ldots, d$.

(a) Obtain the distribution of $y_1, \ldots, y_{d-1}$, given y_d where $y_i = \Sigma_{j=1}^{i} x_j$, $i = 1, \ldots, d$, and show that this distribution is that of the order statistic in a sample size $d-1$ from the uniform distribution on $(0, y_d)$.

(b) Develop a test for large d of the exponential assumption using an approximate distribution for $\Sigma_1^{d-1} y_i$ given y_d. For what kinds of departures from the exponential would you expect this to be a sensitive test? an insensitive test?

(Chapter 3)

17. Consider a type I censored sample from the exponential distribution with censoring time c common to all n individuals on test. Show that the total number of failures, d, has a binomial distribution with parameters n and $p = 1 - \exp(-\lambda c)$. Compare the (asymptotic) efficiency of the m.l. estimator of λ from the marginal distribution of d to that from (V, d). Under what conditions on c is it sensible to base inferences about λ upon d and its distribution alone?

(Chapter 3)

18. A laboratory has n test locations for life testing a particular type of electronic equipment. To conserve time, it is decided to place an item in each test location and test all n items simultaneously. As soon as an item fails, it is immediately replaced by a new item and the system is observed until the dth failure occurs. Suppose that the failure times are exponentially distributed. Let S be the random variable representing the time to cessation of testing. Show that S is sufficient for the failure rate λ. Derive the distribution of S and an exact 95% confidence interval for λ when $n = 25$, $d = 5$ and the fifth failure occurs 407 hours after the start of the experiment.

(Chapter 3)

19. Consider a random censorship model in which failure time T^0 is exponential with failure rate λ and censoring time C is exponential with rate α. Only T, the minimum of T^0 and C, is observed and, as usual, $V = \Sigma_1^n T_i$ and d is the number of failures. Show that (V, d) is sufficient for (λ, α). Show further that V and d are independent, that d is binomial with parameters n and $\lambda(\lambda+\alpha)^{-1}$, and that $2(\lambda+\alpha)V$ has a χ^2 distribution with $2d$ degrees of freedom. Discuss how inference on λ may be carried out.

(Chapter 3)

20. The guaranteed exponential distribution has density function

$$f(t) = \lambda e^{-\lambda(t-G)}, \qquad t > G,$$

where λ, $G > 0$ are unspecified parameters. Let $T_{(1)}, \ldots, T_{(n)}$ be the order statistic on a sample of size n.

(a) Show that $U = \Sigma_2^n T_{(i)}$ and $T_{(1)}$ are jointly sufficient for λ, G and determine the maximum likelihood estimates.
(b) Show that $n(T_{(1)} - G)$, $(n-1)(T_{(2)} - T_{(1)})$, $(n-2)(T_{(3)} - T_{(2)}), \ldots$, $(T_{(n)} - T_{(n-1)})$ are independent exponentials with failure rate λ and hence determine the joint distribution of U and $T_{(1)}$.
(c) Establish methods for the interval estimation of λ and of G.
(d) Apply these results to the group 1 data of Table 1.1. For this purpose, omit the censored data points.

(Chapter 3)

21. Suppose the failure times in Problem 20 are type II censored at $T_{(d)}$ where d is fixed in advance. How would this alter the analysis?

(Chapter 3)

22. Let $T_1, \ldots, T_n$ and $S_1, \ldots, S_m$ be uncensored samples from two guaranteed exponential distributions with parameters (λ_1, G_1) and (λ_2, G_2), respectively.

(a) Outline a test of the hypothesis $\lambda_1 = \lambda_2$.
(b) Supposing that $\lambda_1 = \lambda_2 = \lambda$ is known, develop a test of the hypothesis $G_1 = G_2$. For this purpose, show that $U = T_{(1)} - S_{(1)}$ has a double exponential distribution with density

$$\frac{m}{n+m}\exp(-n\lambda u), \qquad u > 0$$

$$\frac{n}{n+m}\exp(n\lambda u), \qquad u < 0.$$

(c) Generalize this to a test of $G_1 = G_2$ when $\lambda_1 = \lambda_2 = \lambda$ but λ is unknown. Verify that this is the likelihood ratio test of this hypothesis.

(Chapter 3)

23. Freireich et al. (1963) presents the following remission times in weeks from a clinical trial in acute leukemia:

Placebo	1, 1, 2, 2, 3, 4, 4, 5, 5, 8, 8, 8, 8, 11, 11, 12, 12, 17, 22, 23
6-MP	6, 6, 6, 7, 10, 13, 16, 22, 23, 6+, 9+, 10+, 11+, 17+, 19+, 20+, 25+, 32+, 32+, 34+, 35+

(a) Test the hypothesis of equality of remission times in the two groups using Weibull, log-normal, and log-logistic models. Which model appears to fit the data best?
(b) Test for adequacy of an exponential model, relative to the Weibull model.
(c) As a graphical check on the suitability of exponential and Weibull models, compute the Kaplan–Meier estimators $\hat{F}(t)$ of the survivor functions for the two groups. Plot $\log \hat{F}(t)$ versus t and $\log[-\log \hat{F}(t)]$ vs. $\log t$.

(Chapter 3)

24. Suppose uncensored paired failure times (t_{1i}, t_{2i}) have regression variables $z_{1i} = 1$, $z_{2i} = 0$, $i = 1, \ldots, n$. Suppose also that the hazard function for the ith pair can be written

$$\lambda_i \exp(z_{ji}\beta), \qquad j = 1, 2.$$

(a) Show that the m.l.e., $\hat{\beta}$ satisfies

$$n - 2\sum_1^n \frac{w_i e^{\beta}}{1 + w_i e^{\beta}} = 0$$

where $w_i = t_{1i}/t_{2i}$. Does asymptotic likelihood theory apply to $\hat{\beta}$? Show that the usual asymptotic formula, if applicable, would yield an asymptotic variance of $2/n$ for $\hat{\beta}$.
(b) Write down the p.d.f. of w_i, and show that it is independent of λ_i, $i = 1, \ldots, n$. Write down the likelihood function based on w_i, $i = 1, \ldots, n$ and show that the corresponding m.l.e. for β, in this case, satisfies precisely the same equation as that given in (a). Show that asymptotic likelihood results apply to this new (marginal) likelihood function and thereby calculate the asymptotic variance for $\hat{\beta}$. Compare with that given in (a).
(c) Show that $y_i = \log w_i$ arise from a linear model with mean $-\beta$.

Write down the least squares estimator of β and evaluate its efficiency (ratio of asymptotic variance of $\hat{\beta}$ from (b) to least squares variance).

(d) Set $\epsilon_i = \begin{cases} 1, & w_i < 1, \\ 0, & w_i > 1. \end{cases}$

Derive the distribution of ϵ_i and compare the efficiency of the m.l. estimator based on ϵ_i, $i = 1, \ldots, n$ to that given in (b), at $\beta = 0$.
(Chapter 3)

25. Suppose the likelihood function (1.9) arises from a discrete failure time variable with sample space $t_1, \ldots, t_k$. Assume also that censoring can occur only at these discrete times and that a censored failure time t means that the underlying survival time *exceeds* t. Using the asymptotic likelihood methods of Section 3.4 derive the joint asymptotic distribution of the product limit estimators $\hat{F}(t_j)$, $j = 1, \ldots, k$ as defined in (1.10). Compare the asymptotic variance of $\hat{F}(t_j)$ to (1.11).
(Chapter 3)

26. Suppose again that failure time is discrete with sample space $t_1, \ldots, t_k$ and that a censored failure time t means that the underlying uncensored value exceeds t. Suppose that r populations are being compared on the basis of such discrete failure time data. Let λ_{ij} be the conditional probability of failure at t_j in sample i given survival up to t_j. Derive the score test for the hypothesis $\lambda_{ij} = \lambda_j$ all (i, j) and its asymptotic variance. Compare with the log-rank test of Section 1.4.
(Chapter 3)

27. Suppose

$$\lambda(t, z) = \lambda h_0(t) e^{z\beta},$$

where

$$z = \begin{cases} 0, & \text{group 1}, \\ 1, & \text{group 2}. \end{cases}$$

Assuming the ratio of hazard functions between the two samples is 1.5, and assuming no possibility of censoring, calculate the approximate sample size, common to both groups, required to show a difference between groups at the .05 level of significance, with probability (power) .80. How does this sample size depend on $h_0(\cdot)$?
(Chapter 3)

28. A prior density $p(\theta)$ is said to be conjugate to the density $f(x|\theta)$ if, for all θ, $p(\theta)$ is proportional to a possible likelihood function from $f(x|\theta)$. That is,

$$p(\theta) \propto \prod_{i=1}^{n} f(x_i|\theta)$$

for some n and $x_1, \ldots, x_n$.

(a) Show that the class of gamma distributions is conjugate to the exponential density

$$f(t|\lambda) = \lambda e^{-\lambda t}, \qquad t > 0.$$

(b) If the prior distribution for λ is gamma with parameters γ (scale) and ν (shape), show that posterior distribution of λ, given data $t_1, \ldots, t_n$ is gamma with parameters $(\gamma + \Sigma\, t_i)$ and $(n + \nu)$. The gamma family of distributions is said to be closed under sampling from the exponential distribution.

(c) Obtain the predictive distribution of the next observation and also the posterior distribution of the reliability parameter $\rho = e^{-\lambda t} = P(T \geq t/\theta)$ where t is a specified positive number.

(d) Generalize these results to a censored sample $t_1, \ldots, t_n$ with indicators $\delta_1, \ldots, \delta_n$. Note that, in the Bayesian approach, no difficulty is caused by quite complex censoring schemes.

(Chapter 3)

29. Let $(t_1, \mathbf{z}_1), \ldots, (t_n, \mathbf{z}_n)$ be independent observations from the exponential regression model with hazard

$$\lambda(t_i; \mathbf{z}_i) = \lambda \exp(\mathbf{z}_i \boldsymbol{\beta}), \qquad t_i > 0. \tag{1}$$

(a) Show that the estimation problem for $\boldsymbol{\beta}$ is invariant under the group G of scale transformations on the survival time t and thence that

$$a_i = \frac{t_i}{t_1}, \qquad i = 1, \ldots, n,$$

are jointly marginally sufficient for $\boldsymbol{\beta}$.

(b) Obtain the marginal likelihood for $\boldsymbol{\beta}$ and compare the result with that obtained by maximizing the full likelihood over λ.

(Chapter 4)

30. Suppose that n items are placed on test where the survival time t given $\mathbf{z}$ follows the proportional hazards model with hazard

$$\lambda(t; \mathbf{z}) = \lambda_0(t) \exp(\mathbf{z}\boldsymbol{\beta})$$

where $\mathbf{z}$ is time independent. Consider a study design at which m_i items (chosen at random from those at risk) are censored immediately following the ith observed failure time, $i = 1, \ldots, r$, where $r + \sum m_i = n$. Show that for this design, the censored model is invariant under the group of strictly increasing differentiable transformations on t and that the marginal likelihood of $\boldsymbol{\beta}$ is proportional to (4.6).

(Chapter 4)

31. A modified life table estimate for the proportional hazards model. For data from the proportional hazards model with $\mathbf{z}$ time independent, obtain a generalization of the estimate (1.14). Specifically, take the hazard function to be a step function

$$\lambda_0(t) = \lambda_j,\ t \in I_j = [b_0 + \cdots + b_{j-1}, b_0 + \cdots + b_j), \qquad j = 1, \ldots, k,$$

where $b_0 = 0$, $b_k = \infty$ and $b_i > 0$ $(i = 1, \ldots, k-1)$. Take $\boldsymbol{\beta} = \hat{\boldsymbol{\beta}}$ as estimated from the marginal or partial likelihood and show that the maximum likelihood estimate of λ_j is

$$\hat{\lambda}_j = \frac{d_j}{S_j}$$

where d_j is the number of failures in I_j and

$$S_j = b_j \sum_{l \in R_j} e^{\mathbf{z}_l \boldsymbol{\beta}} + \sum_{l \in D_j} (t_l - b_1 - \cdots - b_{j-1}) e^{\mathbf{z}_l \boldsymbol{\beta}},$$

where R_j is the risk set at $b_0 + \cdots + b_j - 0$ and D_j is the set of items failing in I_j. The estimate of the underlying survivor function ($\mathbf{z} = \mathbf{0}$) is now given by $\hat{\hat{F}}(t)$ in (1.14).

(Chapter 4)

32. Compare the estimates in Problem 31 with those obtained from (4.24) for the carcinogenesis data of Pike (Section 1.1.1) (see Figure 4.2).

(Chapter 4)

33. Generalize the results of Problem 31 by allowing $\mathbf{z} = \mathbf{z}(t)$ to be a defined or ancillary time dependent covariate.

(Chapter 5)

34. (a) Assuming that there are no ties in the data, use expression (4.21) to obtain a maximized likelihood for $\boldsymbol{\beta}$ by maximizing over F_0. Write the result in the form

$$L_{\max}(\boldsymbol{\beta}) = L(\boldsymbol{\beta}) R(\boldsymbol{\beta})$$

where $L(\boldsymbol{\beta})$ is the marginal or partial likelihood for $\boldsymbol{\beta}$, (4.6). Examine the form of $R(\boldsymbol{\beta})$ and show that for reasonably large data sets, the effect of $R(\boldsymbol{\beta})$ is small.

(b) Compare the maximized likelihood and the marginal likelihood for the carcinogenesis data of Section 1.1.1.

(Chapter 4)

35. Note that the log-rank statistic (4.19) can be written as

$$\mathbf{S}(\mathbf{0}) = \sum_{i=1}^{k} (\mathbf{s}_i - \bar{\mathbf{s}}_i)'$$

where $\bar{\mathbf{s}}_i = \binom{n_i}{d_i}^{-1} \Sigma_{l \in R_d(t_{(i)})} \mathbf{s}_l$ is the expectation, under the null hypothesis H_0, of $\mathbf{s}_i$ given $R(t_{(i)})$ and the multiplicity d_i. Since under H_0 the $\mathbf{s}_i - \bar{\mathbf{s}}$ can be shown to be uncorrelated, the covariance matrix of $S(0)$ can be estimated by $\Sigma_{i=1}^{k} V_i$. Show that this is identical to the matrix V given in Section 4.2.5 for the $s+1$ sample problem.

(Chapter 4)

36. An estimate of the covariance matrix of the score function statistic $\mathbf{c}$ (4.35) can be obtained by exploiting the relationship $\mathbf{c} = \Sigma_1^k a_i(\mathbf{s}_i - \bar{\mathbf{s}}_i)'$, where $a_i = n_i d_i^{-1} \log(1 - d_i n_i^{-1})$, and the results of the previous question. Show that this gives a variance estimate C^* which is identical to C in (4.36) except for the factor $n_i(n_i - 1)^{-1}$ in the ith term.

(Chapter 4)

37. Suppose there is no censoring and that failure times from (4.1) are grouped into time intervals $I_j = [a_{j-1}, a_j)$, $j = 1, \ldots, k+1$, with $a_0 = 0$ and $a_{k+1} = \infty$ before being observed. Calculate the score statistic from (4.7) for testing the hypothesis $\boldsymbol{\beta} = \mathbf{0}$ and compare it with the grouped data score statistic (4.35).

(Chapter 4)

38. Apply the stratified log-rank test of Section 4.4 to the mouse leukemia data of Appendix 1 (data set V). Define two classes for antibody level ($z_5 < .5$ and $z_5 \geq .5$) and two classes for virus level ($z_6 < 10^4$, $z_6 \geq 10^4$) and consider death by all natural causes. Let the virus level define the strata and test the hypothesis of no dependence of death on antibody level. Compare your results with an unstratified log-rank test of the same hypothesis. See Section 8.5 for some discussion of these data and especially the results in Table 8.6.

(Chapter 4)

39. Consider a two sample problem in which the true hazard for sample z is $\lambda \exp(z\beta)$, $z=0,1$. Suppose that a proportion p of items are in sample 1 and $q=1-p$ are in sample 0. Let each sample be subject to censoring. The time to censoring is exponential with rate θ_i in sample $i=0,1$, where censoring and failure times are independent. Show that the limiting expected information per observation from the rank or partial likelihood analysis based on the proportional model is

$$\lim_{n\to\infty} \frac{I_r(\beta)}{n} = \int_0^\infty \left\{ \frac{\exp[t(\lambda+\theta_0)]}{q} + \frac{\exp[t(\lambda e^\beta + \theta_1)]}{pe^\beta} \right\}^{-1} \lambda\, dt$$

Show also, for the parametric analysis, that the expected information per observation is

$$\frac{I_p(\beta)}{n} = \lambda e^\beta \left(\frac{e^\beta(\lambda+\theta_0)}{q} + \frac{\lambda e^\beta + \theta_1}{p} \right)^{-1}.$$

40. (Continuation)

(a) Suppose that $\beta=0$ and let $\gamma_i = \theta_i/\lambda$, $i=0,1$. Show, in this case, that the asymptotic efficiency of the rank analysis can be expressed as

$$\text{eff}(a) = \left(\frac{1}{q} + \frac{a}{p}\right) \int_0^\infty \left(\frac{e^s}{q} + \frac{e^{as}}{p}\right)^{-1} ds$$

where $a = (1+\gamma_1)/(1+\gamma_0)$. Oakes (1977) gives some evaluations of this for $p=q=\frac{1}{2}$.

(b) The expression eff(a) gives the Pitman efficiency of the log-rank test though the integral, like (4.44), can seldom be evaluated in closed form. Show, however, that if $p=q=\frac{1}{2}$ and $a=2$, the efficiency is $3(1-\log 2) = .92$. For the case $p=q=\frac{1}{2}$, evaluate numerically this efficiency expression for $0<a<4$. Note that for $a=4$, the rate of censorship is more than four times as great in the one sample as compared to the other.

41. Suppose failure times arise in two populations from the same exponential distribution with failure rate λ. Suppose that the first sample ($z=0$) is subject to no censoring whereas the second ($z=1$) is completely censored at t_0. Assuming a proportional hazards model (4.1) calculate the efficiency at $\beta=0$ of the maximum partial likelihood estimator from (4.6) and that of the maximum likelihood estimator from the Weibull submodel as a function of λt_0. Compare these efficiency expressions with each other and with that based on the maximum likelihood estimator of β from a binary response model in which it is only noted whether or not failure time exceeds t_0.

(Chapter 4)

42. Suppose failure times are generated according to the proportional hazards model (4.1) but that observation begins on individual i at a time t_i^0 that may exceed 0. Show that a partial likelihood function can be developed in the manner of Section 4.2.3 and that it has the same form as (4.6) with $R(t_{(i)})$ redefined to include only study subjects at risk and under observation at $t_{(i)}$.

(Chapter 4)

43. Let $(t_i, \delta_i, \mathbf{z}_i)$, $i = 1, \ldots, n$ be a censored sample on a discrete failure time variable with hazard

$$\lambda(t_i; \mathbf{z}_i, \theta) = \sum \lambda(x_j; \mathbf{z}_i, \theta)\delta(t - x_j)$$

$$= \sum \lambda_{ij}(\theta)\delta(t - x_j).$$

Censorings tied with failure times are shifted an infinitesimal amount to the right and so the ties are broken by ordering failure times first. By following the likelihood derivation in Section 5.2, show that the likelihood, or partial likelihood, is $\Pi_1^n L_i(\theta)$ where

$$L_i(\theta) = f(t_i; \mathbf{z}_i, \theta)^{\delta_i} F(t_i + 0; \mathbf{z}_i, \theta)^{1-\delta_i}$$

and f and F are the probability function and the survivor function, respectively.

44. (Continuation) Suppose, for convenience, that θ is a scalar parameter. Note that in general the score contributions $U_i = \partial \log L_i(\theta)/\partial\theta$, $i = 1, \ldots, n$, are not independent. Show, however, by using arguments similar to those in Section 5.4.2 for the partial likelihood that the score contributions are uncorrelated. (Consider the score $U_i(\theta) = \Sigma\, U_{ij}(\theta)$ where $U_{ij}(\theta)$ is the score contribution of the ith item at the jth failure time point and show that the $U_{ij}(\theta)$ are uncorrelated.)

45. Let the failure times of two items to be tested be independent exponentials with failure rate λ. Consider a type II censoring scheme where observation is terminated when the first failure occurs. The contribution to the score of the ith item is

$$U_i = \frac{\delta_i}{\lambda} - T, \qquad i = 1, 2,$$

where δ_i is 1 for a failure and 0 for a censored data point and T is the minimum of the two potential failure times. Show that U_1, U_2 are not independent but that they are, as Problem 44 would suggest, uncorrelated.

(Chapter 5)

46. Suppose that n items are placed on test and as soon as an item fails it is immediately replaced with a new item so that at all times, n items are on test. Suppose further that testing terminates when the nth failure occurs. All items on test at time t are subjected to a temperature $x(t)$ and the failure rate is known to vary with temperature according to the relation

$$\lambda e^{\beta x(t)}.$$

Show that if $t_1 < \cdots < t_r$ are the times at which failures are observed, then the likelihood of λ and β is proportional to

$$\lambda^r \exp\left[\beta \sum x(t_i) - n\lambda \int_0^{t_r} e^{\beta x(t)}\, dt\right].$$

Construct a conditional test of the hypothesis $\beta = 0$.

(Chapter 5)

47. Suppose an individual is at risk of two different types of failure each of which is certain to occur. Let T_1 and T_2 represent the continuous failure times and define

$$\lambda_i(t) = \lim_{\Delta t \to 0+} \frac{P(t \leq T_i < t + \Delta t \mid T_1 \geq t,\, T_2 \geq t)}{\Delta t}, \qquad i = 1, 2,$$

$$\lambda_1(t_1 \mid t_2) = \lim_{\Delta t \to 0+} \frac{P(t_1 \leq T_1 < t_1 + \Delta t \mid T_1 \geq t_1,\, T_2 = t_2)}{\Delta t}, \qquad t_1 > t_2,$$

with a similar definition of $\lambda_2(t_2 \mid t_1)$, $t_2 > t_1$. Show that the joint density function of T_1, T_2 is

$$f(t_1, t_2) = \lambda_2(t_2)\lambda_1(t_1 \mid t_2) \exp\left\{-\int_0^{t_2} [\lambda_1(u) + \lambda_2(u)]\, du - \int_{t_2}^{t_1} \lambda_1(u \mid t_2)\, du\right\}$$

for $t_1 \geq t_2$ and a symmetric part for $t_1 < t_2$. Obtain this result by relating the conditional and marginal densities $f_1(t_1 \mid t_2)$, $t_1 \geq t_2$, and $f_2(t_2)$ to the hazard functions given. Explain why this result would be anticipated from the product integral relationship between the hazard and survivor functions.

(Cox, 1972, Chapters 5 and 7)

48. (Continuation)
Identify $T = T_1$ with failure time and $C = T_2$ with censoring time where censoring and failure times of different individuals are independent. Show that the condition 2 for independent censoring in (5.5) is equivalent to

$$\lambda_1(t) = h_1(t), \qquad 0 < t < \infty,$$

where

$$h_1(t) = \lim_{\Delta t \to 0+} P\{T \in [t, t + \Delta t) | T \geq t\}$$

Thus, the condition requires that the information that an item is uncensored at time t cannot alter the instantaneous failure rate. This is a slightly weaker condition than full independence between T and C.
(Chapter 5)

49. (Continuation)
Williams and Lagakos (1978) obtain a constant-sum condition relating failure and censoring mechanisms and show that under this condition the usual likelihood obtained as a product of density functions and survivor functions is valid. Briefly they define

$$a(u) = P(T < C | T = u), \qquad dB(u) = P[C \in (u, u + du) | T \geq u].$$

and impose the condition

$$a(u) + \int_0^u dB(v) = 1, \qquad u > 0.$$

Show that

$$a(u) = \lambda_1(u) \exp\left[-\int_0^u k(v)\, dv\right] \Big/ h(u)$$

and

$$dB(u) = \lambda_2(u) \exp\left[-\int_0^u k(v)\, dv\right] du$$

where $k(v) = \lambda_1(v) + \lambda_2(v) - h(v)$. Under the assumption that $h(u)$ and $\lambda_1(u)$ are differentiable, verify that a model is constant sum if and only if $h(u) = \lambda_1(u)$ for all $u > 0$.
(Chapter 5)

50. (a) For the bivariate survivor function formulation of Problem 47, let $P(t) = P(T_1 < t,\ T_2 \geq t)$. $P(t)$ is the probability that an item has failed by cause 1 but not by cause 2 at time t and represents the probability of being in remission (e.g.) when T_1 = time to remission and T_2 = time to progression. Show that

$$P(t)=\int_0^t \lambda_1(x)\exp\left[-\int_0^x \lambda_1(u)+\lambda_2(u)\,du\right]\exp\left[-\int_x^t \lambda_2(y|x)dy\right]dx$$

(d) Suppose data of the following four types are available.

I. Censored prior to failure by either cause.
II. Failure by cause 2. Cause I is now not observable.
III. Failure by cause 1, censored prior to failure by cause 2.
IV. Failure by cause 1 and subsequently by cause 2.

Let $h(y)=\lambda_2(y|x)$ and obtain nonparametric maximum likelihood estimates of $R(t)$, $S(t)$, $F(t)$, where

$$-\log R(t)=\int_0^t \lambda_1(u)\,du$$

$$-\log S(t)=\int_0^t \lambda_2(u)\,du$$

$$-\log F(t)=\int_0^t h(u)\,du$$

and consequently of $P(t)$.

(Temkin, 1978, Chapter 5)

51. (Continuation)
Let $\lambda_2(y|x)=h(y-x)$ in Problem 50 and alter the results accordingly.

(Chapters 5 and 7)

52. (Continuation)
Let T_1 represent the waiting time W in the heart transplant data of appendix I and T_2 denote the time to death. Apply the results of Problems 50 and 51 to estimate the probability that an individual has been transplanted and is surviving at a time t after admission to the program.

(Chapters 5 and 7)

53. The log-rank test can be generalized to accommodate defined or ancillary time dependent covariates. The appropriate test statistic is again obtained as the score function based on the partial likelihood for the proportional hazards model. Thus if $\mathbf{z}(t)$ is a vector of time varying covariates, the log-rank statistic is

$$\mathbf{v}=\sum_{i=1}^{k}\left[\mathbf{s}_i(t_{(i)})-n_i^{-1}\sum_{l\in R(t_{(i)})}\mathbf{z}_l(t_{(i)})\right]$$

in the notation of Section 5.4. Variance calculations follow the same lines as before. Apply the log-rank test to the heart transplant data where $z_l(t) = 1$ if the lth patient has been transplanted by time t and 0 otherwise.

(Chapter 5)

54. Consider the model of Section 5.5.2 in which the hazard for the carcinogenesis data is taken to be

$$\lambda_0(t)\exp\{z_1\beta_1 + z_1(\log t - c)\beta_2\}$$

with $z_1 = 0, 1$ a sample indicator. Construct a score function test of $\beta_2 = 0$ and compare the results with the test based on the partial likelihood estimate. Note that such score function tests involve considerably less computation than those tests requiring remaximization of the partial likelihood.

(Chapter 5)

55. For data from the model (5.25) (e.g. the heart transplant data) obtain estimators of the cumulative hazards $\int_0^t \lambda^*(u)\,du$ and $\int_0^t \lambda^{**}(u)\,du$.

(Chapter 5)

56. Show explicitly

(a) The expectation of (6.16) is zero under:
  (1) Censorship that is independent of $\mathbf{z}$.
  (2) Arbitrary independent censoring.

(b) Write the permutation variance for (6.16) assuming censorship independent of $\mathbf{z}$. Compute this variance for both the generalized Wilcoxon (6.9) and the log-rank (6.7) scores for testing $\beta = 0$ with the carcinogenesis data (Table 1.1) as discussed in the illustration of Section 6.6. Compare these with the variance estimates based on the observed Fisher information (6.20).

(c) Show that (6.20) is an estimator of variance for (6.16) under general independent censoring.

(Chapter 6)

57. Calculate explicitly the uncensored scores c_i (Section 6.3.1) based on a double exponential score generating density. Compare these with the sign (median) test scores.

(Chapter 6)

58. Show that the uncensored rank test efficiency (6.5) is unchanged by a location or scale transformation on the data y.

(Chapter 6)

59. The censored data rank tests of Section 6.4 can be based on approximate quantile scores in the manner of (6.12) for the uncensored case. The scores are taken as

$$c_i = \phi(\hat{F}(w_{(i)}))$$
$$C_i = \Phi(\hat{F}(w_{(i)}))$$

where $\hat{F}(w_{(i)}) = \Pi_{j=1}^{i}[n_j/(n_j+1)]$ and the notation is that of Section 6.4. Show that, for the double exponential error density, this leads to a censored version of the sign test with scores

$$c_i = -1 \text{ if } \hat{F}(w_{(i)}) > .5; \quad 0 \text{ if } \hat{F}(w_{(i)}) = .5; \quad 1 \text{ if } \hat{F}(w_{(i)}) < .5$$

$$C_i = \frac{1 - \hat{F}(w_{(i)})}{\hat{F}(w_{(i)})} \quad \text{if } \hat{F}(w_{(i)}) > .5; \quad 1 \text{ if } \hat{F}(w_{(i)}) < .5.$$

(Chapter 6)

60. Derive conditions on the censoring mechanism under which the test statistic (6.16) based on the approximate scores of Problem 55 will be asymptotically equivalent to the same statistic with scores (6.17).

(Chapter 6)

61. Show that the cumulative incidence function (7.10) reduces to one minus the product limit survivor function estimator when there is a single failure type ($m = 1$).

(Chapter 7)

62. Calculate and plot cumulative "incidence" estimators (7.10) for thymic lymphoma mortality for both the control and germ-free groups in the mouse carcinogenesis data (data set IV, Appendix 1). Estimate the relative risk for thymic lymphoma associated with the germ-free environment using (7.12). Test whether this relative risk changes with the age of the animals under study.

(Chapter 7)

63. Denote by Y_1 the time to relapse and by Y_2 the time to death in a clinical trial. Develop a nonparametric estimator of the joint survivor function $Q(y_1, y_2) = P(Y_1 > y_1, Y_2 > y_2)$ under the assumption that the hazard function for death given $Y_1 = y_1$ is independent of y_1.

(Chapter 7)

64. Suppose as in Section 8.3 that m cases ($j = 1$) and n controls ($j = 0$) are selected at failure time t from the multiplicative model (8.18). Suppose the exposure vector $\mathbf{z}(t)$ has finite sample space $\mathbf{z}_1, \ldots, \mathbf{z}_q$.

Show that the induced distribution for $\mathbf{z}(t)$ given (t, j) can be written

$$P[\mathbf{z}(t) = \mathbf{z}_i|(t, j)] = \frac{\exp(\alpha_i + \mathbf{z}_i\boldsymbol{\beta}_j)}{\Sigma_{l=1}^{q} \exp(\alpha_l + \mathbf{z}_l\boldsymbol{\beta}_j)},$$

$i = 1, \ldots, q$, where $\boldsymbol{\beta}_1 = \boldsymbol{\beta}$, $\boldsymbol{\beta}_0 = \mathbf{0}$ and

$$\alpha_i = \log\{P[\mathbf{z}(t) = \mathbf{z}_i|(t, 0)]/P[\mathbf{z}(t) = \mathbf{z}_1|(t, 0)]\}.$$

Compute the asymptotic distribution of the maximum likelihood estimator $\hat{\boldsymbol{\beta}}$ from this induced logistic model. Compare $\hat{\boldsymbol{\beta}}$ and its asymptotic distribution to that obtained from a direct application of the logistic model

$$P[j|t, \mathbf{z}(t)] = \exp\{[\gamma + \mathbf{z}(t)\boldsymbol{\beta}]j\}/\{1 + \exp[\gamma + \mathbf{z}(t)\boldsymbol{\beta}]\}$$

to the case control data, as if the data had been obtained prospectively.

Show that this same result holds even if the sample space for $\mathbf{z}(t)$ is not discrete (Prentice and Pyke, 1979)

(Chapter 8)

65. Suppose that the survivor function $\exp[-\Lambda_0(t)]$ is a realization of a Dirichlet process of Section 8.4.1 with parameters c and $\exp[-\Lambda^*(t)]$. Show that the Kaplan–Meier estimate of Section 1.3 can be obtained as the posterior expectation of $\exp[-\Lambda_0(t)]$ when $c \to 0$.

(Susarla and Van Ryzin, 1976 and Chapter 8)

66. Consider a two-sample problem with $z = 0$ for sample 1 and $z = 1$ for sample 2. In the absence of tied failure times show that the log-rank statistic (6.7), the generalized Wilcoxon statistic (6.9), and the Gehan generalized Wilcoxon statistics (Section 6.5) can all be written in the form

$$\sum_{i=1}^{k} w_i(z_i - n_{1i}n_i^{-1}),$$

where i indexes the ordered failure times $t_1, \ldots, t_k$ in the combined sample, n_i is the size of the risk set just prior to t_i, n_{1i} is the size of the risk set in the second sample ($z_i = 1$) just prior to t_i and w_i are weights that are given respectively by $w_i = n_i$, $w_i \equiv 1$ and $w_i = F_i$ for the three statistics, where

$$F_i = \prod_{j=1}^{i} n_j(n_j + 1)^{-1}$$

is a survivor function estimator. Suggest a generalization for statistics in this form that will accommodate tied failure times. Present a variance estimator for these generalized statistics based on a partial likelihood argument and a hypergeometric distribution at each failure time.

(Chapters 5 and 6)

67. (Continuation)

The Gehan generalized statistic has a weight function; $w_i = n_i$, $i = 1, \ldots, k$, that depends on the censoring mechanism. Suppose n study subjects are followed until the first failure time, t_1, after which a random sample of m randomly selected subjects are followed until failure. Show that with equal initial sample sizes in the two groups, the standardized Gehan test statistic will approach either plus or minus 1 (depending on which sample contains the initial failure) as $n \to \infty$ and regardless of the relationship between the two failure time distributions being compared.

(Prentice and Marek, 1979 and Chapter 6)

APPENDIX 3

Fortran Programs for the Proportional Hazards Model

The following are Fortran subroutines and sample main line programs written to apply the methods of Chapters 4 and 5 based on the proportional hazards model. The principal subroutine FCN carries out the main computations for the case of fixed covariates **z**, and PCN is a more general routine suitable for time varying covariates. The programs are well documented with comment cards so as to make their logic easily understood. Omission of the comment cards reduces greatly the length of the programs.

The routines are completely self-sufficient except that subroutines DECOMP and SOLVE are called in the subprograms FCN and PCN (marked * in listings) to invert a matrix (FPP is the matrix of second partial deriviatives) and to compute an increment DB for the vector $\boldsymbol{\beta}$ of regression coefficients. These subroutines, DECOMP and SOLVE, can be found in the book by Forsythe, Malcolm, and Moler (1977). The only other subroutine not included is the subroutine TIME used in the analysis of the heart transplant data. This subroutine computes the elapsed time between two given dates. The subprograms ORDER and ARRANG are called to arrange all the data on survival time from smallest to largest with failures preceding censorings in the event of ties. On some machines, ARRANG will rearrange only an integer array. In this case, a copy of ARRANG with COVAR specified as real is needed for the real arrays. These could be replaced by any sort routine that accomplishes the same task. In calling either PCN or SCN, this ordering must have been done and the data must be ranked in the vector IRANK.

The initial draft of the FCN subroutine was written by Mr. A. Chang. Mr. P. Wang wrote the initial draft of the PCN routine and Mr. A. McIntosh refined both programs and put them in their final form. We gratefully acknowledge their assistance.

FIXED COVARIATES

This is a sample main line program followed by subroutines to apply the proportional hazards model to post-transplant survival in the heart transplant data of Appendix 1.

```
C
C     COX REGRESSION ANALYSIS FOR HEART TRANSPLANT DATA
C
C
C
C     Z(I,J)        ITH COVARIATE OF JTH PATIENT I=1,2,...M;J=1,2,...N.
C     N             NO OF PATIENTS.
C     M             NO OF COVARIATES.
C     NS            NO. OF STRATA
C     NSTRA         NAME OF STRATA.
C     VN            NAME OF COVARIATES.
C     NAME          PATIENT NAME(#)
C     DB            DATE OF BIRTH.
C     DA            DATE OF ACCEPTANCE.
C     DT            DATE OF TRANSPLANT.
C     DLS           DATE OF LAST SEEN.
C     IND           FAILURE OR CENSORED.
C     T             SURVIVAL TIMES
C     IRANK         RANK OF THE VECTOR T
C     Y             A WORKING VECTOR.
C     IORDER        POINTER VECTOR USED BY SORT ROUTINE.  IT IS
C                   EQUIVALENCED TO Y
C
C

C
      DIMENSION T(103),Z(10,103),NAME(103),IND(103),IRANK(103),Y(103)
      INTEGER DB(103,3),DA(103,3),DT(103,3),DLS(103,3),CAL(103,3),IJ(10)
      DIMENSION EHF(103,1),NP(103)
      DIMENSION NSTRA(10,1),VN(10,7),NEL(103),BETA(10),VAR(10,10)
C
C     THE ARRAY AVG SHOULD BE COMPUTED, NOT INITIALIZED
C
      REAL AVG(10) /1., 50., 500., 48., 48., 0., 2., 0., 1.32, 0./
      INTEGER   CHAR(5)/'A','B','C','D',' '/,ICODE(103,3)
      INTEGER ICAL (3) /10,1,67/
C
C     SWITCH USED BY FCN TO DECIDE ON I/O
C
      LOGICAL OUTPUT / .TRUE. /
C
C     UNIT NUMBERS FOR DATA AND COMMANDS
C
      INTEGER DFILE/8/, CFILE/5/
C
C     DIMENSION CONSTANTS REQUIRED BY FCN
C
```

```
      INTEGER LOFEHF / 1Ø3 / , MB / 1Ø /
C
      INTEGER IORDER (1Ø3)
      EQUIVALENCE  ( IORDER(1) , Y(1) )
C
C
      READ(DFILE,1ØØ)N,M,NS
C
C     READ THE TITLES OF COVARIATES.
C
      DO 4 I=1,NS
4     READ(DFILE,1Ø3)  (NSTRA(J,I),J=1,1Ø)
      DO 5 I=1,M
5     READ(DFILE,1Ø1)  (VN(I,J),J=1,7)
C
C     READ IN DATA
C
      DO 15 I=1,N
         READ(DFILE,1Ø2) NAME(I), (DB(I,J),J=1,3), (DA(I,J),J=1,3) ,
     1   (DT(I,J),J=1,3), (DLS(I,J),J=1,3), (ICODE(I,J),J=1,3), IND(I),
     2   (Z(J,I),J=6,1Ø)
         DO 6 J=1,3
            CAL(I,J)=ICAL(J)
6        CONTINUE
15    CONTINUE
C
C     COMPUTE POST-TRANSPLANT SURVIVAL TIMES ( DLS - DT )
C
      CALL TIME(DT,DLS,T,N,Ø)
C
C     COMPUTE WAITING TIMES.
C
      CALL TIME(DA,DT,Y,N,2)
      DO 2Ø I=1,N
         Z(2,I)=Y(I)
C
C        IF NAME(I) NEVER GET A HEART AT ALL,NEL(I)=Ø,Z(1,I)=Ø.Ø
C        IF NAME(I) HAS CODE C,NEL(I)=Ø
C
         NEL(I)=1
         Z(1,I)=1.Ø
         IF(ICODE(I,1).EQ.CHAR(3)) NEL(I) = Ø
C
         IF(Y(I).GE.Ø.Ø)GO TO 2Ø
         NEL(I)=Ø
         Z(1,I)=Ø.Ø
2Ø    CONTINUE
C
C     CALENDAR TIME AT TRANSPLANT
C
      CALL TIME(CAL,DT,Y,N,Ø)
      DO 21 I=1,N
21    Z(3,I)=Y(I)
C
C     COMPUTE AGE AT ACCEPTANCE INTO THE PROGRAM.
```

```
C
      CALL TIME(DB,DA,Y,N,1)
      DO 22 I=1,N
22    Z(4,I)=Y(I)
C
C     COMPUTE AGE AT TRANSPLANT.
C
      CALL TIME(DB,DT,Y,N,1)
      DO 25 I=1,N
      Z(5,I)=Y(I)
25    CONTINUE
C
C-----------------------------------------------------------------------
C
C     COVARIATE MATRIX IS OBTAINED,WITH
C     ROW ONE        TRANSPLANT STATUS.
C     ROW TWO        WAITING TIME TO TRANSPLANT.
C     ROW THREE      CALENDAR TIME AT TRANSPLANT
C     ROW FOUR       AGE AT ACCEPTANCE INTO THE PROGRAM.
C     ROW FIVE       AGE AT TRANSPLANT.
C     ROW SIX        PREVIOUS SURGERY.
C     ROW SEVEN      MISMATCH ON ALLELES.
C     ROW EIGHT      MISMATCH ON ANTIGEN.
C     ROW NINE       CONTINUIOUS MISMATCH SCORE
C     ROW TEN        DEATH BY REJECTION.
C
C-----------------------------------------------------------------------
C
C     COMPUTE RANK VECTOR.
C
C
C     FIRST, CALL ORDER TO GET AN ORDERED SET OF POINTERS TO THE
C     SURVIVAL TIMES.  NEXT, CALL ARRANG TO RE-ARRANGE THE TIMES, AND
C     THE OTHER COVARIATES, INTO SORTED ORDER.
C
      MN=M*N
      CALL ORDER( T, IND, N, IORDER )
C
      CALL ARRANG( IORDER, N, 1, N, T )
      CALL ARRANG( IORDER, N, 1, N, IND )
      CALL ARRANG( IORDER, N, 1, N, NEL )
      CALL ARRANG( IORDER, N, 1, N, NAME )
      CALL ARRANG( IORDER, N, M, MN, Z )
C
C     COMPUTATION OF RANK VECTOR
C
      JRANK = 1
      IRANK (1) = JRANK
      DO 40 I = 2, N
         IF( IND(I) .EQ. 0 ) GO TO 41
         IF( T(I) .NE. T(I-1) )  JRANK = JRANK + 1
41       IRANK (I) = JRANK
40    CONTINUE
C
C     T IS NOW REORDERED SMALLEST TO LARGEST.
```

```
C     IRANK VECTOR CONTAINS RANKS OF ORDERED DATA.
C
C     READ COVARIATE INDICATOR IJ - IJ SPECIFIES COVARIATES INCLUDED AND
C     EXCLUDED FOR THIS RUN
C
26       READ(CFILE,300,END=99) (IJ(I),I=1,M)
         DO 27 I=1,M
27       BETA(I)=0.0
         IO=1
         CALL FCN(IO,M,N,NS,Z,T,IJ,NEL,NSTRA,VN,IND,IRANK,Y,BETA,F,VAR,
     1   EHF,NP,LOFEHF,AVG,MB,OUTPUT)
C
C          MAXIMUM LIKELIHOOD ESTIMATES HAVE BEEN OUTPUT.
C        BETA CONTAINS THE MAXIMUM LIKELIHOOD ESTIMATES.
C        F CONTAINS THE MAX LOG LIKELIHOOD.
C        VAR CONTAINS THE COVARIATE MATRIX OF THE ESTIMATES.
C
C
C        IF IO=1,NO CONVERGENCE REACHED AFTER 8 ITERATIONS.
C
         IF(IO.EQ.1)GO TO 29
         GO TO 26
29       WRITE(6,110)
         WRITE(6,301) (BETA(I),I=1,M)
      GO TO 26
C
99    STOP
C
C*********************************************************************
C
C     FORMAT STATEMENTS
C
100   FORMAT(3I4)
101   FORMAT(7A2)
102   FORMAT(I3,1X,3I3,3I3,3I3,3I3,3A1,1X,I2,3F2.0,F5.2,F2.0)
103   FORMAT(10A2)
110   FORMAT('0','NO CONVERGENCE')
300   FORMAT(10I3)
301   FORMAT(' ',10F10.6/)
C
C
      END


      SUBROUTINE ORDER( T, IND, N, PTRS )
C
C PURPOSE
C     THIS ROUTINE SORTS THE SURVIVAL TIMES IN ARRAY T FROM SMALLEST TO
C     LARGEST.  DEATHS PROCEED LOSSES IN THE EVENT OF A TIE.
C
C APPROACH
C     A HEAPSORT ALGORITHM IS USED TO DO THE SORT.  SORTING IS DONE
C     INDIRECTLY.  ON RETURN, THE VECTOR P CONTAINS POINTERS TO THE
C     ORDERED DATA, THAT IS, P(1) IS INDEX OF SMALLEST TIME, P(2) SECOND
C     SMALLEST, ETC.
```

```
C
C     HERE WE ARE DOING A SORT ON MULTIPLE KEY FIELDS, WHERE THE FIRST
C     FIELD IS TO BE IN ASCENDING ORDER, AND THE SECOND IN DESCENDING
C     ORDER WITHIN THE FIRST.  TO SIMPLIFY MY DOCUMENTATION, I HAVE
C     WRITTEN IT AS IF I WERE DOING AN ASCENDING SORT ON KEYS
C
C
C DESCRIPTION OF PARAMETERS
C     N       NUMBER OF TIMES TO BE SORTED
C     T       VECTOR OF SURVIVAL TIMES, LENGTH N
C     IND     INDICATOR VARIABLE - 1 MEANS DEATH, 0 MEANS LOSS (IE
C             CENSORED OBSERVATION)
C     PTRS    POINTER VECTOR
C
      INTEGER PTRS(N), IND(N)
      REAL T(N)
      INTEGER R, P, PJ, PJ1, HTOP, HEND
C
C
C     INITIALIZE POINTERS
C
      DO 10 I = 1, N
10      PTRS(I) = I
C
C
      HTOP = N/2 + 1
      HEND = N
C
C
2000     IF( HTOP .LE. 1 ) GO TO 2500
C
C        WE ARE CREATING A HEAP
C
         HTOP = HTOP - 1
         P = PTRS(HTOP)
         GO TO 3000
C
C        WE ARE SORTING THE HEAP.  TAKE THE TOP ELEMENT, AND PUT IT IN
C        THE SORTED SECTION.  THIS DISPLACES AN ELEMENT, WHICH WE WILL
C        PUT BACK IN THE HEAP
C
2500     P = PTRS(HEND)
         PTRS(HEND) = PTRS(1)
         HEND = HEND - 1
C
C        DONE IF HEND IS NOW ONE
C
         IF ( HEND .EQ. 1 ) GO TO 9000
C
C
3000     J = HTOP
C
C
4000        I = J
            J = J + J
```

```
C
C           PICK THE LARGEST SON OF WHAT USED TO BE KEY(PTRS(J))
C           PROVIDED THAT A SON EXISTS
C
            IF ( J .GT. HEND ) GO TO 8000
            PJ = PTRS(J)
            IF ( J .EQ. HEND ) GO TO 6000
C
C
5000        PJ1 = PTRS(J+1)
            IF ( T(PJ) - T(PJ1) ) 5050, 5010, 6000
5010        IF ( IND(PJ) .LE. IND(PJ1) ) GO TO 6000
C
C           KEY(PTRS(J)) < KEY(PTRS(J+1))
C
5050        J = J + 1
            PJ = PJ1
C
C
C           IS LARGEST SON LARGER THAN THE THING WE ARE TRYING TO
C           ADD TO THE HEAP ??
C           IF IT IS, PROMOTE IT TO THE ITH POSITION, AND REPEAT THE
C           PROCESS WITH ITS SONS.  IF NOT, THE KEY WE AE TRYING TO
C           ADD GOES HERE.
C
6000        IF ( T(P) - T(PJ) ) 7000, 6010, 8000
6010        IF ( IND(P) .LE. IND(PJ) ) GO TO 8000
C
C           KEY(PTRS(P)) < KEY(PTRS(PJ))
C
7000        PTRS(I) = PTRS(J)
         GO TO 4000
C
C
8000     PTRS(I) = P
      GO TO 2000
C
C
9000  PTRS(1) = P
      RETURN
      END


      SUBROUTINE ARRANG ( P, N, M, NM, COVAR )
C
C PURPOSE
C     THIS SUBROUTINE ARRANGES AN ARRAY OF COVARIATES, 'COVAR', IN THE
C     ORDER SPECIFIED BY THE POINTER ARRAY P.
C
C APPROACH
C     THE RE-ARRANGEMENT IS DONE IN PLACE, AND THE POINTER VECTOR IS
C     RETURNED INTACT.
C
C DESCRIPTION OF PARAMETERS
C     P      POINTER ARRAY, LENGTH N
```

```
C     COVAR COVARIATE ARRAY, DIMENSIONED (M,N) IN THE MAIN PROGRAM.
C           THE ARRAY IS DECLARED AS TYPE INTEGER HERE, BUT THE DECLARED
C           TYPE IN THE CALLING PROGRAM IS IMMATERIAL.
C     N     LENGTH OF P AND COVAR
C     M     WIDTH OF COVAR
C     NM    M*N
C
      INTEGER P(N), COVAR(NM)
C
C
C LOCAL VARIABLES
C     KEEP  AN ARRAY OF WORKING STORAGE.  ITS LENGTH SHOULD BE >=M
C
      INTEGER KEEP (20)
C
C
      DO 60 I = 1, N
C
C        IF ITH ELEMENT IS IN ITS PROPER PLACE, GO TO END OF LOOP
C        OTHERWISE, SAVE THE ELEMENT, AND CHASE AROUND THE CYCLE,
C        PUTTING ELEMENTS IN THEIR CORRECT PLACES.  THINK OF THE POINTER
C        ARRAY AS EXPRESSING A PERMUTATION OF THE COVARIATE ELEMENTS.
C        THIS PERMUTATION CAN BE EXPRESSED AS THE COMPOSITION OF
C        DISJOINT CYCLES.  THIS LOOP PERFORMS ONE SUCH CYCLE.
C
C        WHEN AN ELEMENT IS IN ITS PROPER PLACE, IT IS MARKED BY MAKING
C        THE POINTER NEGATIVE.
C
         IF( ( P(I) .LT. 0 )  .OR.  ( P(I) .EQ. I ) )  GO TO 50
C
C           SAVE THE ITH ELEMENT
C
            L1 = M*(I-1)
            DO 10 L = 1, M
               KEEP(L) = COVAR (L1+L)
10          CONTINUE
C
            J = I
C
C              MOVE ELEMENT TO ITS PROPER PLACE, AMD MARK IT
C
20             K = P(J)
               P(J) = -K
               LJ = M*(J-1)
               LK = M*(K-1)
               DO 30 L = 1, M
                  COVAR (LJ+L)  =  COVAR (LK+L)
30             CONTINUE
C
               J = K
            IF( P(J) .NE. I ) GO TO 20
C
C           MOVE ITH ELEMENT TO ITS PROPER PLACE, AND MARK IT
C
            P(J) = -I
```

```
          L1 = M*(J-1)
          DO 40 L = 1, M
             COVAR (L1+L)  = KEEP (L)
40        CONTINUE
C
50       P(I) = IABS (P(I))
60    CONTINUE
C
      RETURN
      END


      SUBROUTINE FCN(IO,M,N,NS,Z,T,IJ,NEL,NSTRA,VN,IND,IRANK,X,BETA,F,
     1VAR,EHF,NP,LOFEHF,AVG,MB,OUTPUT)
C
C PURPOSE -----------------------------------------------------------C
C     THIS SUBROUTINE COMPUTES:                                      C
C             1. MAXIMUM LIKELIHOOD ESTIMATES                        C
C             2. MAXIMUM LIKELIHOOD                                  C
C             3. ESTIMATED COVARIANCE MATRIX                         C
C             4. ESTIMATED HAZARDS                                   C
C     FROM THE MARGINAL LIKELIHOOD OF BETA ARISING OUT OF THE JOINT  C
C     DISTRIBUTION OF THE RANK VECTORS FROM THE STRATA               C
C--------------------------------------------------------------------C
C
C DESCRIPTION OF PARAMETERS: -----------------------------------------
C     IO      OPTION INDICATOR AND CONVERGENCE RETURN CODE.
C             IO=1 FIND MLE AND EHF,AND NO CONVERGENCE
C             IO=2 FIND EHF AND MAX LIKELIHOOD,MLE ARE INPUT VALUES
C             IO=3 FIND MLE ONLY
C             IO=4 FIND MAXIMUM LIKELIHOOD ONLY
C     M       NUMBER OF COVARIATES(DIM. OF IJ,VN,Z ETC)
C     N       NUMBER OF POINTS
C     NS      NUMBER OF STRATA
C     Z       MATRIX OF BASIC COVARIABLES(MBXN)
C     T       VECTOR OF SURVIVAL TIMES OF LENGTH N
C     IJ      INPUT VECTOR OF CODES FOR EACH COVARIATES.
C             IJ(K)=1   COV. K IS INCLUDED
C             IJ(K)=0   COV. K IS USED IN COMPUTATION,BUT NOT IN MAXIMIZ
C             IJ(K)=-1  COV. K IS OUT
C     NEL     STRATUM INDICATORS
C             NEL(I)=0    ITH RECORD IS OMITTED
C             NEL(I)=J    ITH RECORD IS IN THE JTH STRATA;J=1,2,...NS
C     NSTRA   STRATA NAMES
C     VN      VARIABLE NAMES
C     IND     INDICATOR VECTOR FOR DEATH OR CENSORED
C     IRANK   RANK VECTOR FROM THE ORDER ROUTINE
C     X       A WORKING VECTOR OF LENGTH N.
C     BETA    INPUT AND OUTPUT VECTOR OF LENGTH M OF BETA VALUES.
C     F       OUTPUT VALUE OF THE LOG. LIKELIHOOD.
C     VAR     OUTPUT VALUE OF THE ESTIMATED COVARIANCE MATRIX (MXM).
C     EHF     ESTIMATED HAZARD FUNCTION
C     NP      WORKING VECTOR WHOSE SIZE MUST BE >= THE MAX. RANK
C     LOFEHF  SIZE OF FIRST DECLARED DIMENSION OF EHF
C     AVG     APPROX. AVERAGES OF THE COVARIATES
```

```
C     MB      DIMENSION OF Z
C     OUTPUT  SWITCH, SET TO .TRUE. IF FCN IS TO PRINT ITS RESULTS
C-----------------------------------------------------------------------
C
C
      DIMENSION Z(MB,N),IRANK(N),IND(N),IJ(M),T(N),X(N),BETA(M),VAR(M,M)
      DIMENSION EHF(LOFEHF,NS),NP(LOFEHF)
      DIMENSION NEL(N),NSTRA(10,NS),VN(M,7),AVG(M)
      DIMENSION IK(20),ID(20),B(20),A(20)
      DIMENSION FP(20),FPP(20,20),TEMP(20),IPVT(20),DB(20)
      DIMENSION SI1(20),SI2(20,20),SS(20)
      INTEGER SSN(15)
      DOUBLE PRECISION FP, SI1, SI2, T1, T2, SI, SS, A, S, DXMI, DXI
C
      INTEGER RETNUM
      LOGICAL OUTPUT, CONVER
C
C
C OTHER REQ'D ROUTINES------------------------------------------C
C     DECOMP, SOLVE                                             C
C---------------------------------------------------------------C
C
C
C
C********************************C
C     START THE COMPUTATIONS     C
C********************************C
C
C
C     FIND MAXIMUM RANK
C
      MAX = IRANK (N)
C
C
C---------------------C
C     CHECK RANGE OF IJ C
C---------------------C
C
      DO 10 I=1,M
        IF( IABS ( IJ (I) ) .LE. 1 ) GO TO 10
        WRITE(6,1002) I,IJ(I)
        GO TO 9999
10    CONTINUE
C
C-----------------------------------------------------------------------
C     COMPUTE POSITIONS OF FAILURE TIMES
C     NP      POSITION OF FAILURE POINTS(AND FIRST POSITION IS RECORDED
C             IN THE EVENT OF TIES)
C-----------------------------------------------------------------------
C
      J=1
      DO 20 I=1,N
        IF(IRANK(I).NE.J)GO TO 20
        I1 = IRANK(I)
        NP(I1)=I
```

```
       J=J+1
20    CONTINUE
C
C---------------------------------------------------C
C     DETERMINE MC AND POSITION OF EACH COVARIATE  C
C     MC      NO. OF COVARIATES WITH CODE 1        C
C---------------------------------------------------C
C
      J1=1
      DO 30 I=1,M
        IF(IJ(I).NE.1)GO TO 30
        ID(J1)=I
        B(J1)=BETA(I)
        J1=J1+1
30    CONTINUE
      MC=J1-1
C
C-------------------------------------------------C
C     DETERMINE MD                                C
C     MD       NO. OF COVARIATES WITH CODE Ø OR 1 C
C-------------------------------------------------C
C
      DO 4Ø I=1,M
        IF(IJ(I).NE.Ø)GO TO 4Ø
        ID(J1)=I
        B(J1)=BETA(I)
        J1=J1+1
4Ø    CONTINUE
      MD=J1-1
C
C-----------------------------------------------------------C
C      PRINT INPUT COVARIATES AND INTIAL VALUES.            C
C-----------------------------------------------------------C
C
      IF ( .NOT. OUTPUT ) GO TO 7Ø
C
      WRITE(6,1ØØØ)
      WRITE(6,2ØØØ) N
      WRITE(6,1Ø7Ø)
      DO 5Ø I=1,M
        IF(IJ(I).EQ.1)WRITE(6,1ØØ1) I,(VN(I,L),L=1,7),BETA(I)
5Ø    CONTINUE
      DO 6Ø I=1,M
        IF(IJ(I).EQ.Ø)WRITE(6,1ØØ3) I,(VN(I,L),L=1,7),I,BETA(I)
6Ø    CONTINUE
      WRITE(6,1Ø71)
C
7Ø    CONTINUE
C
C---------------------------C
C     CHECK STRATUM INFRO.  C
C---------------------------C
C
      DO 8Ø J=1,NS
8Ø      SSN(J)=Ø
```

```
C
      DO 100 I=1,N
        J=NEL(I)
        IF(J.EQ.0)GO TO 100
        IF(J.LT.0.OR.J.GT.NS)GO TO 90
        SSN(J)=SSN(J)+1
        GO TO 100
90      WRITE(6,2005) I
100   CONTINUE
C
C-----------------------------------C
C     PRINT OUT STRATUM INFORMATION  C
C-----------------------------------C
C
      IF ( .NOT. OUTPUT ) GO TO 120
C
      WRITE(6,1004) NS
      WRITE(6,1006)
      DO 110 J=1,NS
110     WRITE(6,1005) J,(NSTRA(I,J),I=1,10),SSN(J)
C
120   CONTINUE
C
C     SPECIAL CASE FOR IO = 2
C
      IF ( IO .EQ. 2 ) GO TO 5000
C
C-+++++++++++++++++++++++++++++++++++++++++C
C     START ITERATIONS TO FIND MLE'S AND ML   C
C-+++++++++++++++++++++++++++++++++++++++++C
C
      ITER=1
4000    CONTINUE
        F = 0.0
C
C-----------------------------------------C
C     DEFINE X(I) (EXPONENTIAL CONTRIBUTES) C
C-----------------------------------------C
C
      DO 4002 I=1,N
         X(I)=0.0
         DO 4001 J=1,MD
            K=ID(J)
4001        X(I)=X(I)+B(J)*(Z(K,I)-AVG(K))
4002     X(I)=EXP(X(I))
C
C
C
C------------------------------------------------------------------C
C     INITIALIZE FIRST AND SECOND PARTIAL VECTOR AND MATRIX RESPT.   C
C------------------------------------------------------------------C
C
       IF( IO .EQ. 4 ) GO TO 4020
       DO 4010 I = 1, MC
         FP (I) = 0.0D0
```

```
C------------------------------C
C       INITIALIZE SI,SI1,SI2  C
C------------------------------C
C
        SI=0.0D0
        IF ( IO .EQ. 4 ) GO TO 4032
        DO 4031 I=1,MC
          SI1(I)=0.0D0
          DO 4031 J=1,MC
            SI2(I,J)=0.0D0
4031    CONTINUE
C
C
4032    CONTINUE
C
C-------------------------------------------------------------C
C     ENTER RISK SET LOOP AND                                 C
C     COMPUTE LOG LIKELIHOOD,FIRST AND SECOND PARTIALS.       C
C-------------------------------------------------------------C
C
4040       CONTINUE
C
C          CALL COMMON ROUTINE TO FIND EXPONENTIAL CONTRIBUTES
C
           ASSIGN 4070 TO RETNUM
           GO TO 3000
4070       CONTINUE
C
C
           IF(MI.EQ.0)GO TO 4100
           S=0.0D0
           DO 4080 J=1,MD
             S = S + SS(J)*DBLE(B(J))
4080       CONTINUE
C
C          ADD TO
C          F   LOG LIKELIHOOD
C          FP  FIRST PARTIAL
C          FPP NEGATIVE OF SECOND PARTIAL MATRIX
C
           XMI=MI
           DXMI = MI
           F=F+S-XMI*DLOG(SI)
           IF( IO .EQ. 4 ) GO TO 4100
           DO 4090 K=1,MC
             FP(K)=FP(K)+SS(K)-DXMI*SI1(K)/SI
             DO 4090 L=K,MC
               T1=SI2(K,L)/SI
               T2=SI1(K)*SI1(L)/SI**2
               FPP(K,L)=FPP(K,L)+XMI*SNGL(T1-T2)
               FPP(L,K) = FPP(K,L)
4090       CONTINUE
4100       CONTINUE
C
C
```

```
          DO 4010 L = 1, MC
            FPP(I,L) = 0.0
4010    CONTINUE
4020    CONTINUE
        LL = 1
C
C----------------------C
C     ENTER STRATA LOOP C
C----------------------C
C
4030      II = MAX
C
            II = II - 1
            IF ( II .GE. 1 ) GO TO 4040
C
C---------------------------C
C     END OF RISK SET LOOP   C
C---------------------------C
C
          LL = LL + 1
          IF ( LL .LE. NS ) GO TO 4030
C
C---------------------------C
C     END OF STRATA LOOP     C
C---------------------------C
C
        IF ( IO .NE. 4 ) GO TO 6000
C
        IF ( .NOT. OUTPUT ) GO TO 9999
C
        WRITE(6,1063)F
        GO TO 9999
C
C
6000    CONTINUE
C
C       SOLVE THE EQUATION FPP*DB = FP
C       FOR DB
C
        DO 6010 I = 1, MC
           DB(I) = FP(I)
6010    CONTINUE
        CALL DECOMP ( 20, MC, FPP, COND, IPVT, TEMP )
        IF ( COND+1.0   .EQ.  COND ) GO TO 6050
        CALL SOLVE ( 20, MC, 20, 1, FPP, DB, IPVT )
C
C       DO ONE NEWTON-RAPHSON ITERATION
C       CHECK FOR CONVERGENCE AT THE SAME TIME
C
        CONVER = .TRUE.
        DO 6020 I=1,MC
          B(I)=B(I)+DB(I)
          I1 = ID(I)
          BETA(I1)=B(I)
          CONVER = CONVER  .AND.  (ABS(DB(I)) .LT. ABS(B(I))*1.E-04)
```

```
6020    CONTINUE
C
C       IF WE HAVE CONVERGENCE, EXIT FROM ITER LOOP
C
        IF ( CONVER ) GO TO 7000
C
C       CHALK UP ANOTHER ITERATION, AND TRY AGAIN
C       BUT DONT DO TOO MANY
C
        ITER=ITER+1
        IF(ITER.LE.10)GO TO 4000
      ITER = 10
      GO TO 6060
C
C       NEARLY SINGULAR SECOND PARTIAL MATRIX - SAY SO
C
6050    WRITE ( 6, 1081 )
C
C--------------------------------------------C
C     NO CONVERGENCE AFTER 10 ITERATIONS     C
C--------------------------------------------C
C
6060  IO = 1
      WRITE ( 6, 1080 ) ITER, ( FP(I), I=1, MC )
      WRITE ( 6, 1082 ) ( DB(I), I=1, MC )
      DO 6070 I = 1, MC
        WRITE ( 6, 1066 ) (FPP(I,K), K=1,MC)
6070  CONTINUE
      WRITE ( 6, 1090 ) COND
      GO TO 9999
C
C     CONVERGENCE REACHED
C     FIND INVERSE OF NEGATIVE OF SECOND PARTIAL MATRIX
C
7000  CONTINUE
C
      DO 7061 I = 1, MC
        DO 7060 J = 1, MC
          VAR(I,J) = 0.0
7060    CONTINUE
        VAR(I,I) = 1.0
7061  CONTINUE
      CALL SOLVE ( 20, MC, M, MC, FPP, VAR, IPVT )
C
C
C
      IF ( .NOT. OUTPUT ) GO TO 7055
C
      WRITE(6,1070)
      WRITE(6,1069)
      DO 7010 I=1,M
        IF(IJ(I).EQ.0)WRITE(6,1003) I,(VN(I,L),L=1,7),I,BETA(I)
7010  CONTINUE
      DO 7020 I=1,M
        IF(IJ(I).EQ.1)WRITE(6,1061) I,(VN(I,L),L=1,7),BETA(I)
```

```
7020  CONTINUE
      WRITE(6,1071)
      WRITE(6,901) ITER
      WRITE(6,1063) F
      WRITE(6,1065)
C
C
C
      DO 7050 I=1,MC
        WRITE(6,1066) (VAR(I,K),K=1,MC)
7050  CONTINUE
      WRITE ( 6, 1090 ) COND
C
7055  CONTINUE
C
C----------------------------------------C
C     COMPUTE ASYMPTOTIC NORMAL DEVIATES   C
C----------------------------------------C
C
      IF ( .NOT. OUTPUT ) GO TO 7075
C
      WRITE(6,2218)
      WRITE(6,2219)
      DO 7070 K=1,MC
        AYND=B(K)/SQRT(VAR(K,K))
        AX=ABS(AYND)
        Y=1.0/(1.0+0.2316419*AX)
        D=0.3989423*EXP(-AYND*AYND/2.0)
        PR=1.0-D*Y*((((1.330274*Y-1.821256)*Y+1.781478)*Y-0.3565638)*Y+
     1     0.3193815)
        PR=1.0-PR
        WRITE(6,2220) ID(K),AYND,PR
7070  CONTINUE
      WRITE(6,1070)
C
7075  IF ( IO .EQ. 3 ) GO TO 9999
C
C     IF WE COME HERE
C     1) WE HAVE CONVERGENCE
C     2) IO = 1  => WANT EHF.  SET IO TO 2, AND FALL INTO CODE FOR
C        IO = 2
C
      IO=2
C
      GO TO 5003
C
C
C     SECTION OF PROGRAM TO DEAL WITH THE CASE IO = 2
C
C
5000  CONTINUE
C
C----------------------------------------C
C     DEFINE X(I) (EXPONENTIAL CONTRIBUTES) C
C----------------------------------------C
```

```
C
      DO 5002 I=1,N
         X(I)=0.0
         DO 5001 J=1,MD
            K=ID(J)
5001        X(I)=X(I)+B(J)*(Z(K,I)-AVG(K))
5002     X(I)=EXP(X(I))
C
C-----------------------------------------------------------------C
C     INITIALIZE EHF                                              C
C     EHF     ESTIMATED HAZARD VALUES AT EACH FAILURE POINT,AND FOR C
C             EACH STRATA(MAX*NS)                                 C
C-----------------------------------------------------------------C
C
5003  CONTINUE
C
C
      DO 5010 J=1,NS
        DO 5010 I=1,MAX
          EHF(I,J)=0.0
5010  CONTINUE
C
      LL = 1
C
C     ENTER STRATA LOOP
C
5020    II = MAX
        SI = 0.0D0
C
C       ENTER RISK SET LOOP
C
5021      CONTINUE
C
C         CALL INTERNAL SUBROUTINE
C
          ASSIGN 5030 TO RETNUM
          GO TO 3000
5030      CONTINUE
C
C
          IF ( MI .EQ. 0 ) GO TO 5080
C
C
C-----------------------------------------------------------------C
C     COMPUTE ESTIMATE OF HAZARD AT T(II)                         C
C                                                                 C
C     IF A SINGLE FAILURE OCCURS AT T(II),THEN THE HAZARD IS SOLVED C
C     ANALYTICALLY                                                C
C-----------------------------------------------------------------C
C
          IF(MI.GT.1)GO TO 5040
          J2=IK(MI)
          XI = X (J2)
          EHF ( II, LL ) = 1.0 - ( 1.0 - XI/SI ) ** (1./XI)
          GO TO 5080
```

```
C
C-----------------------------------------------C
C     OTHERWISE AN ITERATIVE SOL'N IS REQUIRED.  C
C-----------------------------------------------C
C
5040      CONTINUE
C
C-------------------------------------------------C
C     A SUITABLE STARTING VALUE FOR THE ITERATION IS: C
C-------------------------------------------------C
C
          ALPH=MI/SI
          ALPH=EXP(-ALPH)
C
C---------------------------------------------------------C
C     FIND ESTIMATE OF HAZARD,BY NEWTON-RAPHSON METHOD     C
C---------------------------------------------------------C
C
          DO 5050 ITER = 1, 10
            GI=0.0
            GI1=0.0
            DO 5060 K=1,MI
              J1=IK(K)
              XI=X(J1)
              XII=ALPH**XI
              GI=GI+XI/(1.0-XII)
              GI1=GI1+XII*XI**2/(ALPH*(1.0-XII)**2)
5060        CONTINUE
            STEP=(GI-SI)/GI1
            ALPH=ALPH-STEP
C
C           CONVERGENCE TEST
C
            IF(ABS(STEP).LE.1.E-03)GO TO 5070
5050      CONTINUE
C
C         NO CONVERGENCE - SAY SO
C
          I1 = NP(II)
          TT=T(I1)
          WRITE(6,1072) II,LL,TT
          GO TO 5080
C
C         CONVERGENCE - SUPPLY EHF VALUE
C
5070      EHF(II,LL)=1.0-ALPH
5080      CONTINUE
C
          II = II - 1
          IF ( II .GE. 1 ) GO TO 5021
C
C       END OF RISK SET LOOP
C
        LL = LL + 1
        IF ( LL .LE. NS ) GO TO 5020
```

```
C
C     END OF STRATA LOOP
C
C
C--------------------------------C
C     OUTPUT ESTIMATED HAZARDS   C
C--------------------------------C
C
      IF ( .NOT. OUTPUT ) GO TO 5100
C
      WRITE(6,400)
      DO 5090 I=1,MAX
        I1 = NP(I)
        TT=T(I1)
        WRITE(6,401) I,TT,(EHF(I,LL),LL=1,NS)
5090  CONTINUE
C
5100  CONTINUE
C
C
9999  RETURN
C
C*********************************************************************C
C
C     THIS ROUTINE IS CALLED TO PERFORM CALCULATIONS COMMON TO THE CODE
C     FOR IO = 2 AND IO ~= 2.  THE RETURN ADDRESS MUST BE PLACED IN
C     VARIABLE RETNUM BY MEANS OF AN ASSIGN STATEMENT
C
C
3000  CONTINUE
C
C------------------------------------------C
C     INITIALIZE MI: NO. OF DEATH AT T(II). C
C------------------------------------------C
C
      MI=0
C
C-------------------------------------------------------------------C
C     INITIALIZE SS,SUM OF COVARIATES OF THOSE DYING AT T(II)       C
C-------------------------------------------------------------------C
C
C
      DO 3010 I=1,MD
3010    SS(I)=0.0D0
      IP = NP (II)
      DO 3060 I=IP,N
         IF(IRANK(I).NE.II)GO TO 3070
         IF(NEL(I).NE.LL)GO TO 3060
C
C        CENTRE COVARIATE VALUES
C
         DO 3015 J = 1, MC
            L1 = ID(J)
            A(J) = Z(L1,I) - AVG (L1)
3015     CONTINUE
```

```
C
C-----------------------------------C
C         CALCULATE SI SI1 SI2      C
C-----------------------------------C
C
          DXI = X(I)
          SI = SI + DXI
          GO TO (3020,3040,3020,3040),IO
 3020     DO 3030 K=1,MC
             SI1(K)=SI1(K)+DXI*A(K)
             DO 3030 J=K,MC
                SI2(K,J)=SI2(K,J)+DXI*A(K)*A(J)
 3030     CONTINUE
C
 3040     CONTINUE
C
          IF(IND(I).EQ.0)GO TO 3060
          MI=MI+1
          IK(MI)=I
          DO 3050 J=1,MD
             SS(J) = SS(J) + A(J)
 3050     CONTINUE
C
 3060  CONTINUE
C
C
 3070  CONTINUE
C
C      RETURN TO CALLER
C
       GO TO RETNUM,(4070,5030)
C
C*********************************************************************C
C
C-----------------------------C
C      FORMAT STATEMENTS      C
C-----------------------------C
C
400    FORMAT('1','RANK',5X,'TIME',5X,'HAZARD ONE')
401    FORMAT(' ',I3,3X,F8.2,4X,F9.7)
1002   FORMAT('0',5X,'***IJ IS OUT OF RANGE:IJ(',I5,')=',I5,'***')
1000   FORMAT('1',5X,'INPUT COVARIATES AND INITIAL VALUES')
2000   FORMAT('0','TOTAL NUMBER OF POINTS IS',I5)
1070   FORMAT('0',55(2H *))
1001   FORMAT(' ','VARIABLE NUMBER',I3,2X,7A2,2X,'IS ITERATED',5X,'INITIA
      1L VALUE=',F7.4)
1071   FORMAT(' ',55(2H *))
2005   FORMAT('0',5X,I4,'TH STRATUM INDICATOR IS OUT OF RANGE')
1004   FORMAT('0','TOTAL NUMBER OF STRATA IS',I4)
1006   FORMAT(' ','STRATA NUMBER',10X,'STRATA NAME ',10X,'NO.OF POINTS')
1005   FORMAT(' ',4X,I4,10X,10A2,8X,I5)
1072   FORMAT('0','NO CONVERGENCE AFTER 10 ITERATIONS FOR EHF(',I3,',',I3
      1,') AT TIME=',F10.3)
1069   FORMAT(' ',28HMAXIMUM LIKELIHOOD ESTIMATES)
1061   FORMAT(1H ,23HCOEFFICIENT OF VARIABLE,I3,2H  ,7A2,2HIS,F9.6)
```

```
901   FORMAT('Ø','***CONVERGENCE REACHED IN',I3,' ITERATIONS***')
1Ø63  FORMAT(1HØ,25HTHE MAX LOG LIKELIHOOD IS,F16.8)
1Ø65  FORMAT(1HØ,27HESTIMATED COVARIANCE MATRIX/)
1Ø66  FORMAT(1H ,1ØF13.6)
1Ø8Ø  FORMAT('- MAXIMUM LIKELIHOOD ESTIMATES DID NOT CONVERGE' /
    1       '  COMPUTATION HALTED AFTER ', I2, ' ITERATIONS' /
    2       'Ø SCORE FUNCTION' /   7(1X, 1PD15.6) )
1Ø81  FORMAT('- SINGULAR SECOND PARTIAL MATRIX; COMPUTATION ABANDONED')
1Ø82  FORMAT('Ø LAST BETA INCREMENT WAS' / 7(1X, 1PE15.6) )
1Ø9Ø  FORMAT('Ø CONDITION NUMBER OF SECOND PARTIAL MATRIX: ', 1PE15.6 )
2218  FORMAT('Ø','SUMMARY:'/1X,'-------')
2219  FORMAT(' ','COEFFICIENT NO.',4X,'N.D.',6X,'P-VALUES(1-SIDED)')
222Ø  FORMAT(' ',6X,I2,8X,F9.6,8X,F6.4)
1ØØ3  FORMAT(' ','VARIABLE NUMBER ',I3,3X,7A2,' IS INCLUDED',5X,'BETA(',
     1I2,')=',F1Ø.6)
      END
```

```
     INPUT COVARIATES AND INITIAL VALUES

TOTAL NUMBER OF POINTS IS  1Ø3

 * * * * * * * * * * * * * * * * * * * * * * * * * * * * * * * * * * *
VARIABLE NUMBER  2  WAIT TIME       IS ITERATED     INITIAL VALUE= Ø.Ø
VARIABLE NUMBER  3  YEAR ACCEPT     IS ITERATED     INITIAL VALUE= Ø.Ø
 * * * * * * * * * * * * * * * * * * * * * * * * * * * * * * * * * * *

TOTAL NUMBER OF STRATA IS    1
STRATA NUMBER             STRATA NAME             NO.OF POINTS
      1               POST-TRANSPLANT                   65

 * * * * * * * * * * * * * * * * * * * * * * * * * * * * * * * * * * *
MAXIMUM LIKELIHOOD ESTIMATES
COEFFICIENT OF VARIABLE  2  WAIT TIME      IS-Ø.ØØ3494
COEFFICIENT OF VARIABLE  3  YEAR ACCEPT    IS-Ø.ØØØ429
 * * * * * * * * * * * * * * * * * * * * * * * * * * * * * * * * * * *

***CONVERGENCE REACHED IN  4 ITERATIONS**

THE MAX LOG LIKELIHOOD IS    -143.72969Ø55

ESTIMATED COVARIANCE MATRIX

     Ø.ØØØØ31    -Ø.ØØØØØØ
    -Ø.ØØØØØØ     Ø.ØØØØØØ

 CONDITION NUMBER OF SECOND PARTIAL MATRIX:     3.925251E Ø2

SUMMARY:
-------
COEFFICIENT NO.    N.D.       P-VALUES(1-SIDED)
       2        -Ø.625348        Ø.2659
       3        -1.57918Ø        Ø.Ø571

 * * * * * * * * * * * * * * * * * * * * * * * * * * * * * * * * * * *
```

RANK	TIME	HAZARD ONE
1	-10.00	0.0
2	1.00	0.0158992
3	2.00	0.0160963
4	4.00	0.0163664
5	11.00	0.0169216
6	13.00	0.0174290
7	15.00	0.0
8	16.00	0.0177837
9	24.00	0.0184249
10	26.00	0.0191655
11	27.00	0.0195527
12	30.00	0.0198385
13	40.00	0.0203452
14	45.00	0.0206033
15	47.00	0.0208085
16	48.00	0.0213448
17	49.00	0.0225350
18	51.00	0.0237341
19	52.00	0.0492408
20	55.00	0.0262929
21	61.00	0.0275478
22	64.00	0.0293194
23	65.00	0.0313782
24	66.00	0.0659770
25	67.00	0.0355136
26	69.00	0.0369074
27	128.00	0.0386057
28	137.00	0.0401403
29	148.00	0.0
30	162.00	0.0415239
31	229.00	0.0424898
32	254.00	0.0449110
33	281.00	0.0470756
34	298.00	0.0489459
35	323.00	0.0517525
36	552.00	0.0745389
37	625.00	0.0878078
38	731.00	0.1125866
39	837.00	0.1455337
40	898.00	0.0
41	995.00	0.1832393
42	1025.00	0.2071036
43	1351.00	0.3092646

INPUT COVARIATES AND INITIAL VALUES

TOTAL NUMBER OF POINTS IS 103

```
* * * * * * * * * * * * * * * * * * * * * * * * * * * * * * * * * * *
VARIABLE NUMBER  2  WAIT TIME      IS ITERATED     INITIAL VALUE= 0.0
VARIABLE NUMBER  5  AGE TRANS      IS ITERATED     INITIAL VALUE= 0.0
VARIABLE NUMBER  6  PRE SURGERY    IS ITERATED     INITIAL VALUE= 0.0
VARIABLE NUMBER  9  MIS SCORE      IS ITERATED     INITIAL VALUE= 0.0
* * * * * * * * * * * * * * * * * * * * * * * * * * * * * * * * * * *
```

```
TOTAL NUMBER OF STRATA IS    1
STRATA NUMBER            STRATA NAME              NO.OF POINTS
      1               POST-TRANSPLANT                  65

 * * * * * * * * * * * * * * * * * * * * * * * * * * * * * * * * * *
MAXIMUM LIKELIHOOD ESTIMATES
COEFFICIENT OF VARIABLE  2  WAIT TIME      IS-0.003092
COEFFICIENT OF VARIABLE  5  AGE TRANS      IS 0.051244
COEFFICIENT OF VARIABLE  6  PRE SURGERY    IS-0.817645
COEFFICIENT OF VARIABLE  9  MIS SCORE      IS 0.465959
 * * * * * * * * * * * * * * * * * * * * * * * * * * * * * * * * * *

***CONVERGENCE REACHED IN  4 ITERATIONS**

THE MAX LOG LIKELIHOOD IS    -138.32466125

ESTIMATED COVARIANCE MATRIX

      0.000025      0.000002     -0.000098      0.000141
      0.000002      0.000480      0.000795     -0.000110
     -0.000098      0.000795      0.235130      0.008243
      0.000141     -0.000110      0.008243      0.085972

 CONDITION NUMBER OF SECOND PARTIAL MATRIX:     5.088250E 03

SUMMARY:
-------
COEFFICIENT NO.     N.D.        P-VALUES(1-SIDED)
       2         -0.622264          0.2669
       5          2.339540          0.0097
       6         -1.686206          0.0459
       9          1.589165          0.0560

 * * * * * * * * * * * * * * * * * * * * * * * * * * * * * * * * * *
```

TIME DEPENDENT COVARIATES

This is a sample main line program followed by subroutines to apply the proportional hazards model with time varying covariates to the heart transplant data of Appendix 1.

```
C
C PURPOSE ---------------------------------------------------------------C
C     THE PROGRAM GIVES ANALYSES OF THE HEART TRANSPLANT DATA,
C     USING THE METHOD OF PARTIAL LIKELIHOOD,UNDER THE ASSUMPTION    C
C     OF PROPORTIONAL HAZARD.                                         C
C-----------------------------------------------------------------------C
C
C
C DESCRIPTION OF PARAMETERS: -------------------------------------------
C     CONSULT THE DESCRIPTIONS IN SUBROUTINES PCN,COVPCN
C-----------------------------------------------------------------------
```

```
C
C
      DIMENSION T(103),Z(9,103),NAME(103),IND(103),IRANK(103),Y(103)
      DIMENSION EHF(62,1),NP(62)
      DIMENSION NSTRA(10,1),VN(20,7),NEL(103),BETA(20),VAR(20,20)
      DIMENSION IJ(20),A(20),ZO(9),B(20)
C
C
C OTHER REQ'D ROUTINES--------------------------------------C
C     PCN,COVPCN                                            C
C---------------------------------------------------------C
C
C     UNIT NUMBERS FOR DATA AND COMMANDS
C
      INTEGER DFILE/8/, CFILE/5/
C
C     SWITCH USED BY PCN TO DECIDE ON I/O
C
      LOGICAL OUTPUT / .TRUE. /
C
C
      EXTERNAL COVPCN
C
C
      READ ( DFILE, 100 ) N, MB
      READ ( CFILE, 100 ) M, NS
100   FORMAT(4I4)
C
C----------------------------------------C
C     READ THE TITLES OF COVARIATES ETC.  C
C     NOTE STRATA NAMES AND COVARIATE NAMES C
C     READ FROM COMMAND FILE              C
C----------------------------------------C
C
      DO 4 I=1,NS
4     READ(CFILE,103) (NSTRA(J,I),J=1,10)
103   FORMAT(10A2)
      DO 5 I=1,M
5     READ(CFILE,101) (VN(I,J),J=1,7)
101   FORMAT(7A2)
C
C-----------------------------C
C     READ IN THE ORDERED DATA    C
C-----------------------------C
      DO 15 I=1,N
      READ(DFILE,102) NAME(I),T(I),IND(I),IRANK(I),(Z(J,I),J=1,MB)
      NEL(I)=1
15    CONTINUE
102   FORMAT(I3,1X,F6.1,1X,I1,1X,I2,4F9.2,5F5.2)
C
C-----------------------------------------------------------------C
C     Z CONTAINS BASIC COVARIATES                                 C
C     SEE DESCRIPTION IN COVPCN                                   C
C     FOR DEFINITIONS OF WORKING COVARIATES SEE DESCRIPTION IN COVPCN C
C-----------------------------------------------------------------C
C
```

```
      LOFEHF = 62
      MAX = IRANK (N)
C
C
C-----------------------------------------------------------------------C
C     READ COVARIATE INDICATOR IJ.  DATA CARDS ARE:                      C
C     -1 -1 -1 -1 -1 -1 -1 -1 -1 -1 -1 -1 -1 -1 -1  1 -1 -1 -1           C
C      1 -1 -1 -1 -1 -1 -1 -1  1 -1 -1 -1 -1 -1 -1 -1  1 -1 -1 -1        C
C-----------------------------------------------------------------------C
C
26    READ(CFILE,300,END=99) (IJ(I),I=1,M)
300   FORMAT(20I3)
C
C-------------------------------C
C     SET INITIAL VALUE OF BETA  C
C-------------------------------C
C
      DO 27 I=1,M
27    BETA(I)=0.0
      IO=1
      CALL PCN(IO,M,N,NS,T,IND,Z,MB,IJ,NEL,NSTRA,VN,IRANK,Y,BETA,F,VAR,
     1EHF,LOFEHF,NP,COVPCN,OUTPUT)
C
C-------------------------------------------------------------C
C       MAXIMUM LIKELIHOOD ESTIMATES HAVE BEEN OUTPUT.         C
C     BETA CONTAINS THE MAXIMUM LIKELIHOOD ESTIMATES.          C
C     F CONTAINS THE MAX LOG LIKELIHOOD.                       C
C     VAR CONTAINS THE COVARIATE MATRIX OF THE ESTIMATES.      C
C     EHF CONTAINS THE ESTIMATED HAZARDS                       C
C     NP CONTAINS THE POSITION OF FAILURE POINTS               C
C-------------------------------------------------------------C
C
C
C-------------------------------------------------------------C
C     IF IO=1,NO CONVERGENCE REACHED AFTER 10 ITERATIONS       C
C-------------------------------------------------------------C
C
      IF(IO.EQ.1)GO TO 29
      GO TO 26
29    WRITE(6,110)
110   FORMAT('0','NO CONVERGENCE')
      WRITE(6,311) (BETA(I),I=1,M)
311   FORMAT(' ',10F10.6/)
      GO TO 26
C
C
99    STOP
      END

      SUBROUTINE PCN(IO,M,N,NS,T,IND,Z,MB,IJ,NEL,NSTRA,VN,IRANK,X,BETA,
     1F,VAR,EHF,LOFEHF,NP,COVFUN,OUTPUT)
C
C PURPOSE ---------------------------------------------------------------C
C     THIS SUBROUTINE COMPUTES:                                          C
C              1. MAXIMUM LIKELIHOOD ESTIMATES                           C
```

```
C             2. MAXIMUM LIKELIHOOD                                        C
C             3. ESTIMATED COVARIANCE MATRIX                               C
C             4. ESTIMATED HAZARDS                                         C
C     FROM THE PARTIAL LIKELIHOOD OF BETA,UNDER THE ASSUMPTION OF          C
C     PROPORTIONAL HAZARD MODEL.                                           C
C--------------------------------------------------------------------------C
C
C DESCRIPTION OF PARAMETERS: ----------------------------------------------
C     IO      OPTION INDICATOR AND CONVERGENCE RETURN CODE.
C             IO=1 FIND MLE,EHF AND F,AND NO CONVERGENCE
C             IO=2 FIND EHF AND MAX LIKELIHOOD,MLE ARE INPUT VALUES
C             IO=3 FIND MLE ONLY
C             IO=4 FIND MAXIMUM LIKELIHOOD ONLY
C     M       NUMBER OF COVARIATES(DIM. OF IJ,VN,Z ETC)
C     N       NUMBER OF POINTS
C     NS      NUMBER OF STRATA
C     T       VECTOR OF SURVIVAL TIMES OF LENGTH N
C     IND     INDICATOR VECTOR FOR DEATH OR CENSORED
C     Z       MATRIX OF BASIC COVARIABLES(MBXN)
C     MB      DIMENSION OF Z
C     IJ      INPUT VECTOR OF CODES FOR EACH COVARIATES.
C             IJ(K)=1   COV. K IS INCLUDED
C             IJ(K)=Ø   COV. K IS USED IN COMPUTATION,BUT NOT IN MAXIMIZ
C             IJ(K)=-1  COV. K IS OUT
C     NEL     STRATUM INDICATORS
C             NEL(I)=Ø   ITH RECORD IS OMITTED
C             NEL(I)=J   ITH RECORD IS IN THE JTH STRATA;J=1,2,...NS
C     NSTRA   STRATA NAMES
C     VN      VARIABLE NAMES
C     IRANK   RANK VECTOR FROM THE ORDER ROUTINE
C     X       A WORKING VECTOR OF LENGTH N.
C     BETA    INPUT AND OUTPUT VECTOR OF LENGTH M OF BETA VALUES.
C     F       OUTPUT VALUE OF THE LOG. LIKELIHOOD.
C     VAR     OUTPUT VALUE OF THE ESTIMATED COVARIANCE MATRIX (MXM).
C     EHF     ESTIMATED HAZARD FUNCTION
C     LOFEHF  SIZE OF THE FIRST DECLARED DIMENSION OF EHF
C     NP      WORKING VECTOR WHOSE SIZE MUST BE >= THE MAX. RANK
C     COVFUN  FUNCTION CALLED TO FIND COVARIATE VALUES AT PARTICULAR
C             TIMES
C     OUTPUT  A LOGICAL VARIABLE, SET TO .TRUE. IF THE SUBROUTINE IS
C             TO OUTPUT ITS RESULTS
C--------------------------------------------------------------------------
C
C
      DIMENSION Z(MB,N),IRANK(N),IND(N),IJ(M),T(N),X(N),BETA(M),VAR(M,M)
      DIMENSION EHF(LOFEHF,NS),NP(LOFEHF)
      DIMENSION NEL(N),NSTRA(1Ø,NS),VN(M,7)
      DIMENSION IK(2Ø),ID(2Ø),B(2Ø),A(2Ø)
      DIMENSION FP(2Ø),FPP(2Ø,2Ø),TEMP(2Ø),IPVT(2Ø),DB(2Ø)
      DIMENSION SI1(2Ø),SI2(2Ø,2Ø),SS(2Ø)
      INTEGER SSN(15)
      INTEGER RETNUM
      LOGICAL OUTPUT, CONVER
      DOUBLE PRECISION FP, SI1, SI2, T1, T2, SI, SS, A, S, DXMI, DXI
C
```

```
C
C OTHER REQ'D ROUTINES-----------------------------------C
C     DECOMP, SOLVE, COV                                  C
C-------------------------------------------------------C
C
C********************************C
C     START THE COMPUTATIONS     C
C********************************C
C
C
C     FIND MAXIMUM RANK
C
      MAX = IRANK (N)
C
C
C-------------------C
C     CHECK RANGE OF IJ C
C-------------------C
C
      DO 1Ø I=1,M
        IF( IABS ( IJ (I) ) .LE. 1 ) GO TO 1Ø
        WRITE(6,1ØØ2) I,IJ(I)
        GO TO 9999
1Ø    CONTINUE
C
C-------------------------------------------------------------------------
C     COMPUTE POSITIONS OF FAILURE TIMES
C     NP      POSITION OF FAILURE POINTS(AND FIRST POSITION IS RECORDED
C             IN THE EVENT OF TIES)
C-------------------------------------------------------------------------
C
      J=1
      DO 2Ø I=1,N
        IF(IRANK(I).NE.J)GO TO 2Ø
        I1 = IRANK(I)
        NP(I1)=I
        J=J+1
2Ø    CONTINUE
C
C----------------------------------------------C
C     DETERMINE MC AND POSITION OF EACH COVARIATE  C
C     MC      NO. OF COVARIATES WITH CODE 1        C
C----------------------------------------------C
C
      J1=1
      DO 3Ø I=1,M
        IF(IJ(I).NE.1)GO TO 3Ø
        ID(J1)=I
        B(J1)=BETA(I)
        J1=J1+1
3Ø    CONTINUE
       C=J1-1
C
C---------------------------------------------C
C     DETERMINE MD                               C
```

```
C     MD       NO. OF COVARIATES WITH CODE Ø OR 1 C
C---------------------------------------------C
C
      DO 4Ø I=1,M
        IF(IJ(I).NE.Ø)GO TO 4Ø
        ID(J1)=I
        B(J1)=BETA(I)
        J1=J1+1
4Ø    CONTINUE
      MD=J1-1
C
C------------------------------------------------------C
C      PRINT INPUT COVARIATES AND INTIAL VALUES.        C
C------------------------------------------------------C
C
      IF ( .NOT. OUTPUT ) GO TO 7Ø
C
      WRITE(6,1ØØØ)
      WRITE(6,2ØØØ) N
      WRITE(6,1Ø7Ø)
      DO 5Ø I=1,M
        IF(IJ(I).EQ.1)WRITE(6,1ØØ1) I,(VN(I,L),L=1,7),BETA(I)
5Ø    CONTINUE
      DO 6Ø I=1,M
        IF(IJ(I).EQ.Ø)WRITE(6,1ØØ3) I,(VN(I,L),L=1,7),I,BETA(I)
6Ø    CONTINUE
      WRITE(6,1Ø71)
C
7Ø    CONTINUE
C
C--------------------------C
C     CHECK STRATUM INFO.  C
C--------------------------C
C
      DO 8Ø J=1,NS
8Ø      SSN(J)=Ø
C
      DO 1ØØ I=1,N
        J=NEL(I)
        IF(J.EQ.Ø)GO TO 1ØØ
        IF(J.LT.Ø.OR.J.GT.NS)GO TO 9Ø
        SSN(J)=SSN(J)+1
        GO TO 1ØØ
9Ø      WRITE(6,2ØØ5) I
1ØØ   CONTINUE
C
C----------------------------------C
C     PRINT OUT STRATUM INFORMATION C
C----------------------------------C
C
      IF ( .NOT. OUTPUT ) GO TO 12Ø
C
      WRITE(6,1ØØ4) NS
      WRITE(6,1ØØ6)
      DO 11Ø J=1,NS
```

```
110      WRITE(6,1005) J,(NSTRA(I,J),I=1,10),SSN(J)
C
120   CONTINUE
C
C     SPECIAL CASE FOR IO = 2
C
      IF ( IO .EQ. 2 ) GO TO 5000
C
C-++++++++++++++++++++++++++++++++++++++++++++C
C     START ITERATIONS TO FIND MLE'S AND ML   C
C-++++++++++++++++++++++++++++++++++++++++++++C
C
      ITER=1
4000    CONTINUE
        F = 0.0
C
C---------------------------------------------------------------------C
C     INITIALIZE FIRST AND SECOND PARTIAL VECTOR AND MATRIX RESPT.    C
C---------------------------------------------------------------------C
C
        IF( IO .EQ. 4 ) GO TO 4020
        DO 4010 I = 1, MC
          FP (I) = 0.0D0
          DO 4010 L = 1, MC
            FPP(I,L) = 0.0
4010    CONTINUE
4020    CONTINUE
        LL = 1
C
C---------------------C
C     ENTER STRATA LOOP C
C---------------------C
C
4030      II = MAX
C
C------------------------------------------------------------C
C     ENTER RISK SET LOOP AND                                C
C     COMPUTE LOG LIKELIHOOD,FIRST AND SECOND PARTIALS.      C
C------------------------------------------------------------C
C
4040        CONTINUE
C-------------------------C
C     INITIALIZE SI1,SI2    C
C-------------------------C
C
            IF ( IO .EQ. 4  ) GO TO 4060
            DO 4050 I=1,MC
              SI1(I)=0.0D0
              DO 4050 J=1,MC
                SI2(I,J)=0.0D0
4050        CONTINUE
4060        CONTINUE
C
C           CALL COMMON ROUTINE TO FIND EXPONENTIAL CONTRIBUTES
C
```

```
            ASSIGN 4070 TO RETNUM
            GO TO 3000
4070        CONTINUE
            IF(MI.EQ.0)GO TO 4100
            S=0.0D0
            DO 4080 J=1,MD
              S=S+SS(J)*DBLE(B(J))
4080        CONTINUE
C
C
C           ADD TO
C           F   LOG LIKELIHOOD
C           FP  FIRST PARTIAL
C           FPP NEGATIVE OF SECOND PARTIAL MATRIX
C
C
            XMI=MI
            DXMI = MI
            F=F+S-XMI*DLOG(SI)
            IF( IO .EQ. 4 ) GO TO 4100
            DO 4090 K=1,MC
              FP(K)=FP(K)+SS(K)-DXMI*SI1(K)/SI
              DO 4090 L=K,MC
                T1=SI2(K,L)/SI
                T2=SI1(K)*SI1(L)/SI**2
                FPP(K,L)=FPP(K,L)+XMI*SNGL(T1-T2)
                FPP(L,K) = FPP(K,L)
4090        CONTINUE
4100        CONTINUE
C
C
            II = II - 1
            IF ( II .GE. 1 ) GO TO 4040
C
C--------------------------C
C     END OF RISK SET LOOP   C
C--------------------------C
C
          LL = LL + 1
          IF ( LL .LE. NS ) GO TO 4030
C
C--------------------------C
C     END OF STRATA LOOP     C
C--------------------------C
C
        IF ( IO .NE. 4 ) GO TO 6000
C
        IF ( .NOT. OUTPUT ) GO TO 9999
C
        WRITE(6,1063)F
        GO TO 9999
C
C
6000    CONTINUE
C
```

```
C
C     IF WE COME HERE
C     1) WE HAVE CONVERGENCE
C     2) IO = 1  => WANT EHF.  SET IO TO 2, AND FALL INTO CODE FOR
C        IO = 2
C
      IO=2
C
C
C     SECTION OF PROGRAM TO DEAL WITH THE CASE IO = 2
C
C
5000  CONTINUE
C
C-----------------------------------------------------------------C
C     INITIALIZE EHF                                              C
C     EHF     ESTIMATED HAZARD VALUES AT EACH FAILURE POINT,AND FOR C
C             EACH STRATA(MAX*NS)                                 C
C-----------------------------------------------------------------C
C
      DO 5010 J=1,NS
        DO 5010 I=1,MAX
          EHF(I,J)=0.0
5010  CONTINUE
C
      LL = 1
C
C     ENTER STRATA LOOP
C
5020    II = MAX
C
C       ENTER RISK SET LOOP
C
5021      CONTINUE
C
C         CALL INTERNAL SUBROUTINE
C
          ASSIGN 5030 TO RETNUM
          GO TO 3000
5030      CONTINUE
C
C
          IF ( MI .EQ. 0 ) GO TO 5080
C
C
C-----------------------------------------------------------------C
C     COMPUTE ESTIMATE OF HAZARD AT T(II)                         C
C                                                                 C
C     IF A SINGLE FAILURE OCCURS AT T(II),THEN THE HAZARD IS SOLVED C
C     ANALYTICALLY                                                C
C-----------------------------------------------------------------C
C
          IF(MI.GT.1)GO TO 5040
          J2 = IK (MI)
          XI = X (J2)
```

```
      EHF ( II, LL ) =  1.0 -  ( 1.0 - XI/SI ) ** (1./XI)
      GO TO 5080
C
C------------------------------------------C
C     OTHERWISE AN ITERATIVE SOL'N IS REQUIRED.  C
C------------------------------------------C
C
5040  CONTINUE
C
C----------------------------------------------C
C     A SUITABLE STARTING VALUE FOR THE ITERATION IS: C
C----------------------------------------------C
C
      ALPH=MI/SI
      ALPH=EXP(-ALPH)
C
C--------------------------------------------------C
C     FIND ESTIMATE OF HAZARD,BY NEWTON-RALPHSON METHOD    C
C--------------------------------------------------C
C
      DO 5050 ITER = 1, 10
        GI=0.0
        GI1=0.0
        DO 5060 K=1,MI
          J1=IK(K)
          XI=X(J1)
          XII=ALPH**XI
          GI=GI+XI/(1.0-XII)
          GI1=GI1+XII*XI**2/(ALPH*(1.0-XII)**2)
5060    CONTINUE
        STEP=(GI-SI)/GI1
        ALPH=ALPH-STEP
C
C       CONVERGENCE TEST
C
        IF(ABS(STEP).LE.1.E-03)GO TO 5070
5050  CONTINUE
C
C     NO CONVERGENCE - SAY SO
C
      I1 = NP(II)
      TT=T(I1)
      WRITE(6,1072) II,LL,TT
      GO TO 5080
C
C     CONVERGENCE - SUPPLY EHF VALUE
C
5070  EHF(II,LL)=1.0-ALPH
5080  CONTINUE
C
      II = II - 1
      IF ( II .GE. 1 ) GO TO 5021
C
C    END OF RISK SET LOOP
C
```

```
        LL = LL + 1
        IF ( LL .LE. NS ) GO TO 5020
C
C      END OF STRATA LOOP
C
C
C-------------------------------C
C      OUTPUT ESTIMATED HAZARDS     C
C-------------------------------C
C
      IF ( .NOT. OUTPUT ) GO TO 5100
C
      WRITE(6,400)
      DO 5090 I=1,MAX
        I1 = NP(I)
        WRITE(6,401) I,T(I1),(EHF(I,LL),LL=1,NS)
5090  CONTINUE
C
5100  CONTINUE
C
C
9999  RETURN
C
C*******************************************************************C
C
C      THIS ROUTINE IS CALLED TO PERFORM CALCULATIONS COMMON TO THE CODE
C      FOR IO = 2 AND IO ~= 2.  THE RETURN ADDRESS MUST BE PLACED IN
C      VARIABLE RETNUM BY MEANS OF AN ASSIGN STATEMENT
C
C
3000  CONTINUE
C
C-----------------------------------------C
C      INITIALIZE MI: NO. OF DEATH AT T(II). C
C-----------------------------------------C
C
      MI=0
C
C-----------------------------------------------------------------C
C      INITIALIZE SS,SUM OF COVARIATES OF THOSE DYING AT T(II)       C
C-----------------------------------------------------------------C
C
C
      SI = 0.0D0
C
      DO 3010 I=1,MD
3010    SS(I)=0.0D0
C
C
C-----------------------------------------------------------------C
C      FIND T(II) AND DEFINE X(I) (EXPONENTIAL CONTRIBUTES)FOR RISK  C
C      SET R(T(II)).                                                 C
C-----------------------------------------------------------------C
C
      IP=NP(II)
```

```
      FAIL=T(IP)
      DO 3070 I=IP,N
        IF(NEL(I).NE.LL)GO TO 3070
        DXI = 0.0D0
        DO 3020 J=1,MD
          A(J)=COVFUN(ID(J),FAIL,Z(1,I),MB)
          DXI=DXI+A(J)*DBLE(B(J))
 3020   CONTINUE
        DXI = DEXP(DXI)
        X(I) = DXI
C
C
C---------------------------C
C     CALCULATE SI SI1 SI2  C
C---------------------------C
C
        SI=SI+DXI
        GO TO(3030,3050,3030,3050),IO
 3030   DO 3040 K=1,MC
          SI1(K)=SI1(K)+A(K)*DXI
           DO 3040 J=K,MC
             SI2(K,J)=SI2(K,J)+DXI*A(K)*A(J)
 3040   CONTINUE
 3050   IF(IRANK(I).NE.II.AND.MI.EQ.0)GO TO 3080
        IF(IRANK(I).NE.II.OR.IND(I).EQ.0)GO TO 3070
        MI=MI+1
        IK(MI)=I
C
        IF( IO .EQ. 2) GO TO 3070
        DO 3060 J=1,MD
          SS(J)=SS(J)+A(J)
 3060   CONTINUE
 3070 CONTINUE
 3080 CONTINUE
C
C     NOW RETURN TO CALLER
C
      GO TO RETNUM,(4070,5030)
C
C*******************************************************************C
C
C--------------------------C
C     FORMAT STATEMENTS    C
C--------------------------C
C
 400  FORMAT('1','RANK',5X,'TIME',5X,'HAZARD')
 401  FORMAT(' ',I3,F8.2,3X,10(1X,F9.7) / (18X,10(1X,F9.7)) )
 901  FORMAT('0','***CONVERGENCE REACHED IN',I3,' ITERATIONS***')
 1000 FORMAT('1',5X,'INPUT COVARIATES AND INITIAL VALUES')
 1001 FORMAT(' ','VARIABLE NUMBER',I3,2X,7A2,2X,'IS ITERATED',5X,'INITIA
     1L VALUE=',F7.4)
 1002 FORMAT('0',5X,'***IJ IS OUT OF RANGE:IJ(',I5,')=',I5,'***')
 1003 FORMAT(' ','VARIABLE NUMBER',I3,3X,7A2,' IS INCLUDED',5X,'BETA('
     1,I2,')=',F10.6)
 1004 FORMAT('0','TOTAL NUMBER OF STRATA IS',I4)
```

```
1005  FORMAT(' ',4X,I4,10X,10A2,8X,I5)
1006  FORMAT(' ','STRATA NUMBER',10X,'STRATA NAME ',10X,'NO.OF POINTS')
1061  FORMAT(1H ,23HCOEFFICIENT OF VARIABLE,I3,2H  ,7A2,2HIS,F9.6)
1063  FORMAT(1H0,25HTHE MAX LOG LIKELIHOOD IS,F16.8)
1065  FORMAT(1H0,27HESTIMATED COVARIANCE MATRIX/)
1066  FORMAT(1H ,10F13.6)
1069  FORMAT(' ',28HMAXIMUM LIKELIHOOD ESTIMATES)
1070  FORMAT('0',55(2H *))
1071  FORMAT(' ',55(2H *))
1072  FORMAT('0','NO CONVERGENCE AFTER 10 ITERATIONS FOR EHF(',I3,',',I3
     1,') AT TIME=',F10.3)
1080  FORMAT('- MAXIMUM LIKELIHOOD ESTIMATES DID NOT CONVERGE' /
     1       '  COMPUTATION HALTED AFTER ', I2, ' ITERATIONS' /
     2       '0 SCORE FUNCTION' /   7(1X, 1PD15.6) )
1081  FORMAT('- SINGULAR SECOND PARTIAL MATRIX; COMPUTATION ABANDONED')
1082  FORMAT('0 LAST BETA INCREMENT WAS' / 7(1X, 1PE15.6) )
1090  FORMAT('0 CONDITION NUMBER OF SECOND PARTIAL MATRIX: ', 1PE15.6 )
2000  FORMAT('0','TOTAL NUMBER OF POINTS IS',I5)
2005  FORMAT('0',5X,I4,'TH STRATUM INDICATOR IS OUT OF RANGE')
2218  FORMAT('0','SUMMARY:'/1X,'-------')
2219  FORMAT(' ','COEFFICIENT NO.',4X,'N.D.',6X,'P-VALUES(1-SIDED)')
2220  FORMAT(' ',6X,I2,8X,F9.6,8X,F6.4)
      END


      REAL FUNCTION COVPCN ( NC, T, B, MB )
C
C*************************************************************************
C     FUNCTION TO COMPUTE THE VALUE OF THE JTH COVARIATE OF THE ITH
C     POINT AT TIME T
C*************************************************************************
C
C     PARAMETERS:
C
C     T       TIME AT WHICH COVARIATES ARE TO BE EVALUATED
C     MB      DIMENSION OF B
C
C     B CONTAINS THE BASIC COVARIABLES OF THE ITH POINT AS FOLLOWS:
C
C             1. TRANSPLANT STATUS
C             2. WAITING TIME FOR TRANSPLANT
C             3. YEAR OF ACCEPTANCE TO PROGRAM - OCT. 1/67
C             4. AGE AT ACCEPTANCE
C             5. PREVIOUS SURGERY
C             6. NO. OF MISMATCHES
C             7. MISMATCH ON HL-A2
C             8. MISMATCH SCORE
C             9. REJECTION INDICATOR VARIABLE
C
C
C     NC      COVARIATE NUMBER AS FOLLOWS:
C
C     CONSTANT COVARIATES
C
C      1      AGE AT ACCEPTANCE - 48.0
```

```
C       2      YEAR AT ACCEPTANCE - OCT. 1, 1967.
C       3      PREVIOUS SURGERY (1=YES,Ø=NO)
C       4      (1)*(2)
C       5      UNUSED
C       6      UNUSED
C       7      UNUSED
C       8      UNUSED
C
C     TIME DEPENDENT COVARIATES (Ø.Ø UNTIL TIME OF TRANSPLANT)
C
C       9-16  AS 1-8
C      17      TRANSPLANT STATUS (Ø=NONE)
C      18      WAITING TIME TO TRANSPLANT
C      19      MISMATCH SCORE
C      2Ø      LN(T-W+1)  WHERE W IS WAITING TIME TO TRANSPLANT
C
      REAL B(MB)
C
      COVPCN = Ø.Ø
      IF ( NC .GT. 2Ø ) GO TO 99
      IF ( NC .GT. 8 ) GO TO 5Ø
C
C     CONSTANT COVARIABLES
C     STATEMENT NUMBER CORRESPONDS TO VALUE OF NC
C
      GO TO ( 1, 2, 3, 4, 99, 99, 99, 99 ), NC
C
C
C
C     TIME DEPENDENT COVARIABLES
C
5Ø    J = NC - 8
      IF ( B(1).EQ.Ø  .OR.  B(2).GE.T ) GO TO 99
      GO TO ( 9, 1Ø, 11, 12, 99, 99, 99, 99, 17, 18, 19, 2Ø ) ,J
C
C
C     AGE AT ACCEPTANCE
9     CONTINUE
1     COVPCN = B(4) - 48.Ø
      GO TO 99
C
C     YEAR OF ACCEPTANCE
C
1Ø    CONTINUE
2     COVPCN = B(3)
      GO TO 99
C
C     PREVIOUS SURGERY
11    CONTINUE
3     COVPCN = B(5)
      GO TO 99
C
C     AGE * YEAR ACCEPT.
12    CONTINUE
4     COVPCN = B(3) * (B(4)-48.Ø)
```

```
      GO TO 99
C
C     TRANSPLANT STATUS
17    COVPCN = B(1)
      GO TO 99
C
C     TRANSPLANT WAITING TIME
18    COVPCN = B(2)
      GO TO 99
C
C     MISMATCH SCORE
19    COVPCN = B(8)
      GO TO 99
C
C     LN(T-W+1)
20    COVPCN = ALOG( T-B(2)+1. )
C
C
99    RETURN
      END
```

```
     INPUT COVARIATES AND INITIAL VALUES

TOTAL NUMBER OF POINTS IS  103

 * * * * * * * * * * * * * * * * * * * * * * * * * * * * * * * * * * *
VARIABLE NUMBER 17  TRNS STATUS(T)  IS ITERATED     INITIAL VALUE= 0.0
 * * * * * * * * * * * * * * * * * * * * * * * * * * * * * * * * * * *

TOTAL NUMBER OF STRATA IS   1
STRATA NUMBER            STRATA NAME           NO.OF POINTS
      1              COMPLETE SAMPLE              103

 * * * * * * * * * * * * * * * * * * * * * * * * * * * * * * * * * * *
MAXIMUM LIKELIHOOD ESTIMATES
COEFFICIENT OF VARIABLE 17  TRNS STATUS(T)IS 0.125667
 * * * * * * * * * * * * * * * * * * * * * * * * * * * * * * * * * * *

***CONVERGENCE REACHED IN  3 ITERATIONS**

THE MAX LOG LIKELIHOOD IS   -298.23657227

ESTIMATED COVARIANCE MATRIX

     0.090647

 CONDITION NUMBER OF SECOND PARTIAL MATRIX:    1.000000E 00

SUMMARY:
-------
COEFFICIENT NO.    N.D.       P-VALUES(1-SIDED)
      17         0.417391        0.3382

 * * * * * * * * * * * * * * * * * * * * * * * * * * * * * * * * * * *
```

RANK	TIME	HAZARD
1	1.00	0.0097087
2	2.00	0.0293286
3	3.00	0.0300928
4	5.00	0.0205303
5	6.00	0.0209168
6	8.00	0.0106677
7	9.00	0.0107517
8	12.00	0.0109557
9	16.00	0.0330742
10	17.00	0.0114141
11	18.00	0.0115191
12	21.00	0.0231951
13	28.00	0.0117162
14	30.00	0.0118363
15	32.00	0.0120938
16	35.00	0.0121820
17	36.00	0.0123322
18	37.00	0.0124446
19	39.00	0.0125485
20	40.00	0.0258055
21	43.00	0.0132117
22	45.00	0.0134126
23	50.00	0.0135579
24	51.00	0.0137569
25	53.00	0.0139228
26	58.00	0.0141194
27	61.00	0.0142942
28	66.00	0.0145297
29	68.00	0.0294513
30	69.00	0.0152073
31	72.00	0.0308757
32	77.00	0.0159855
33	78.00	0.0162806
34	80.00	0.0165501
35	81.00	0.0168666
36	85.00	0.0170969
37	90.00	0.0174147
38	96.00	0.0177655
39	100.00	0.0180868
40	102.00	0.0184426
41	110.00	0.0192230
42	149.00	0.0200182
43	153.00	0.0204553
44	165.00	0.0208825
45	186.00	0.0219206
46	188.00	0.0224794
47	207.00	0.0230673
48	219.00	0.0236120
49	263.00	0.0242219
50	285.00	0.0512629
51	308.00	0.0271624
52	334.00	0.0279208
53	340.00	0.0287775
54	343.00	0.0307241

```
55  584.00    0.0423547
56  675.00    0.0524267
57  733.00    0.0557406
58  852.00    0.0638071
59  980.00    0.0814991
60  996.00    0.0897992
61 1032.00    0.0999822
62 1387.00    0.1515459


     INPUT COVARIATES AND INITIAL VALUES

TOTAL NUMBER OF POINTS IS  103

 * * * * * * * * * * * * * * * * * * * * * * * * * * * * * * * * * * *
VARIABLE NUMBER  1  AGE ACCEPT        IS ITERATED     INITIAL VALUE= 0.0
VARIABLE NUMBER  9  AGE ACCEPT(T)     IS ITERATED     INITIAL VALUE= 0.0
VARIABLE NUMBER 17  TRNS STATUS(T)    IS ITERATED     INITIAL VALUE= 0.0
 * * * * * * * * * * * * * * * * * * * * * * * * * * * * * * * * * * *

TOTAL NUMBER OF STRATA IS   1
STRATA NUMBER          STRATA NAME              NO.OF POINTS
      1              COMPLETE SAMPLE                103

 * * * * * * * * * * * * * * * * * * * * * * * * * * * * * * * * * * *
MAXIMUM LIKELIHOOD ESTIMATES
COEFFICIENT OF VARIABLE  1  AGE ACCEPT     IS 0.011859
COEFFICIENT OF VARIABLE  9  AGE ACCEPT(T)  IS 0.041301
COEFFICIENT OF VARIABLE 17  TRNS STATUS(T) IS 0.074535
 * * * * * * * * * * * * * * * * * * * * * * * * * * * * * * * * * * *

***CONVERGENCE REACHED IN  4 ITERATIONS**

THE MAX LOG LIKELIHOOD IS    -294.68847656

ESTIMATED COVARIANCE MATRIX

     0.000336     -0.000330     -0.001948
    -0.000330      0.000804      0.001401
    -0.001948      0.001401      0.103245

 CONDITION NUMBER OF SECOND PARTIAL MATRIX:     8.226162E 02

SUMMARY:
-------
COEFFICIENT NO.      N.D.        P-VALUES(1-SIDED)
      1            0.647110          0.2588
      9            1.456894          0.0726
     17            0.231965          0.4083

 * * * * * * * * * * * * * * * * * * * * * * * * * * * * * * * * * * *
```

Bibliography

Aalen, O. (1976). Nonparametric inference in connection with multiple decrement models. *Scand. J. Stat.*, **3**, 15–27.

Abramowitz, M. and I. A. Stegun (Eds.) (1965). *Handbook of Mathematical Functions*, New York: Dover.

Adichie, J. N. (1967). Estimates of regression parameters based on rank tests. *Ann. Math. Stat.*, **38**, 894–904.

Altshuler, B. (1970). Theory for the measurement of competing risks in animal experiments. *Math. Biosc.*, **6**, 1–11.

Anderson, J. A. (1972). Separate sample logistic discrimination. *Biometrika*, **59**, 19–35.

Anderson, J. R. (1978). Robust linear regression based on residual ranks. Unpublished Ph.D. thesis, Department of Biostatistics, University of Washington.

Andrews, D. F., P. J. Bickel, F. R. Hampel, P. J. Huber, W. H. Rogers, and J. W. Tukey (1972). *Robust Estimates of Location: Survey and Advances*, Princeton, New Jersey: Princeton University.

Andrews, D. F. (1974). A robust method for multiple linear regression. *Technometrics*, **16**, 523–531.

Anscombe, F. J. (1964). Normal likelihood functions. *Annu. Inst. Stat. Math.* (*Tokyo*), **16**, 1–19.

Armitage, P. and R. Doll (1954). The age distribution of cancer and a multistage theory of carcinogenesis. *Br. J. Cancer*, **8**, 1–12.

Armitage, P. (1959). The comparison of survival curves. *J. R. Stat. Soc. A*, **122**, 279–292.

Armitage, P. and R. Doll (1961). Stochastic models for carcinogenesis. *Proc. Fourth Berkeley Symposium in Mathematical Statistics*, **IV**, Berkeley: University of California Press, 19–38.

Atiquallah, M. (1962). The estimation of residual variance in quadratically balanced least squares problems and the robustness of the F-test. *Biometrika*, **49**, 83–91.

Atkinson, A. C. (1970). A method for discriminating between models (with discussion). *J. R. Stat. Soc. B*, **32**, 323–353.

Ayer, M., H. D. Brunk, G. M. Ewing, W. T. Reid, and E. Silverman (1955). An empirical distribution function for sampling with incomplete information. *Ann. Math. Stat.*, **26**, 641–647.

Barlow, R. E. and F. Proschan (1976). Asymptotic theory of total time on test processes, with applications to life testing. *Multivariate Analysis IV* (P. R. Krishnaiah, Ed.). Amsterdam: N. Holland, 227–237.

Barnard, G. A. (1963). Some aspects of the fiducial argument. *J. R. Stat. Soc. B*, **25**, 111–114.

Barndorff-Nielsen, O. (1973). On M ancillarity. *Biometrika*, **60**, 447–455.

Basu, A. P. (1971). Bivariate failure rate. *J. Am. Statist. Assoc.*, **66**, 103–104.

Bartholomew, D. J. (1957). A problem in life testing. *J. Am. Stat. Assoc.*, **52**, 350–355.

Batchelor, J. R. and M. Hackett (1970). HLA matching in treatment of burned patients with skin allografts. *Lancet*, **2**, 581–583.

Berkson, J. and R. P. Gage (1952). Survival curve for cancer patients following treatment. *J. Am. Stat. Assoc.*, **47**:, 501–515.

Berkson, J. and L. Elveback (1960). Competing exponential risks, with particular reference to the study of smoking and lung cancer. *J. Am. Stat. Assoc.*, **55**, 415–428.

Billingsley, P. (1968). *Convergence of Probability Measures*, New York: Wiley.

Birnbaum, A. and E. Laska (1967). Efficiency robust two-sample rank tests. *J. Am. Stat. Assoc.*, **62**, 1241–1251.

Bliss, C. I. (1935). The calculation of the dosage-mortality curve. *Ann. Appl. Biol.*, **22**, 134–167.

Breslow, N. E. (1970). A generalized Kruskal–Wallis test for comparing K samples subject to unequal patterns of censorship. *Biometrika*, **57**, 579–594.

Breslow, N. E. (1974). Covariance analysis of censored survival data. *Biometrics*, **30**, 89–99.

Breslow, N. E. and J. Crowley (1974). A large sample study of the life table and product limit estimates under random censorship. *Ann. Stat.*, **2**, 437–453.

Breslow, N. E. (1975). Analysis of survival data under the proportional hazards model. *Int. Stat. Rev.*, **43,** 45–58.

Breslow, N. E. (1976). Regression analysis of the log odds ratio: a method for retrospective studies. *Biometrics*, **32**, 409–416.

Brown, B. W., M. Hollander, and R. M. Korwan, (1974). Nonparametric tests of independence for censored data with applications to heart transplant studies. *Reliability and Biometry*: *Statistical Analysis of Life Length* (F. Prochan and R. J. Serfling, Eds.). Philadelphia: SIAM.

Byar, D. P., R. Huse, and J. C. Bailar, III (1974). An exponential model relating censored survival data and concomitant information for prostrate cancer patients. *J. Natl. Cancer Inst.*, **52**, 321–326.

Chiang, C. L. (1960). A stochastic study of life table and its applications: I. Probability distribution of the biometric functions. *Biometrics*, **16**, 618–635.

Chiang, C. L. (1961). On the probability of death from specific causes in the presence of competing risks. *Proc. Fourth Berkeley Symposium in Mathematical Statistics*, **IV** (L. M. Le Cam et al., Eds.), Berkeley: University of California Press, 169–180.

Chiang, C. L. (1968). *Introduction to Stochastic Processes in Biostatistics*, New York: Wiley.

Chiang, C. L. (1970). Competing risks and conditional probabilities. *Biometrics*, **26**, 767–776.

Clark, D. A., E. B. Stinson, R. B. Griepp, J. S. Schroeder, N. E. Shumway, and D. C. Harrison (1971). Cardiac transplantation in men, VI. Prognosis of patients selected for cardiac transplantation. *Ann. Intern. Med.*, **75**, 15–21.

Conover, W. J. (1971). *Practical Nonparametric Statistics*, New York: Wiley.

Cornfield, J. (1951). A method of estimating comparative rates from clinical data. Applications to cancer of the lungs, breast, and cervix. *J. Nat. Canc. Inst.*, **11**, 1269–1275.

Cornfield, J. (1957). The estimation of the probability of developing a disease in the presence of competing risks. *Am. J. Public Health*, **47**, 601–607.

Cornfield, J. (1971). The university group diabetes program. A further statistical analysis of the mortality findings. *J. Am. Med. Assoc.*, **217**, 1676–1687.

Cornfield, J. and K. Detre (1977). Bayesian analysis of life tables. *J. R. Stat. Soc. B*, **39**, 86–94.

Cox, D. R. (1953). Some simple tests for Poisson variates. *Biometrika*, **40**, 354–360.

Cox, D. R. (1959). The analysis of exponentially distributed life-times with two types of failure. *J. R. Stat. Soc. B*, **21**, 411–421.

Cox, D. R. (1961). Tests of separate families of hypotheses. *Proc. Fourth Berkeley Symposium in Mathematical Statistics*, **I**, Berkeley: University of California Press, 105–123.

Cox, D. R. (1962a). Further results on tests of separate families of hypotheses. *J. R. Stat. Soc. B*, **24**, 406–424.

Cox, D. R. (1962b). *Renewal Theory*, London: Methuen.

Cox, D. R. (1964). Some applications of exponential ordered scores. *J. R. Stat. Soc. B*, **26**, 103–110.

Cox, D. R. and P. A. Lewis (1966). *The Statistical Analysis of a Series of Events*, London: Methuen.

Cox, D. R. and E. J. Snell (1968). A general definition of residuals (with discussion). *J. R. Stat. Soc. B*, **30**, 248–275.

Cox, D. R. (1970). *The Analysis of Binary Data*, London: Methuen.

Cox, D. R. (1972). Regression models and life tables (with discussion). *J. R. Stat. Soc. B*, **34**, 187–220.

Cox, D. R. (1973). The statistical analysis of dependencies in point processes. *Symposium on Point Processes* (P. A. W. Lewis, Ed.), New York: Wiley.

Cox, D. R. and D. V. Hinkley (1974). *Theoretical Statistics*, London: Chapman and Hall.

Cox, D. R. (1975). Partial likelihood. *Biometrika*, **62**, 269–276.

Crowley, J. (1974a). Asymptotic normality of a new nonparametric statistic for use in organ transplant studies. *J. Am. Stat. Assoc.*, **69**, 1006–1011.

Crowley, J. (1974b). A note on some recent likelihoods leading to the log rank test. *Biometrika*, **61**, 533–538.

Crowley, J. and N. Breslow (1975). Remarks on the conservatism of $\Sigma(0-E)^2/E$ in survival data. *Biometrics*, **31**, 957–961.

Crowley, J. and D. R. Thomas (1975). Large sample theory for the log rank test. *Technical Report No. 415*, Department of Statistics, University of Wisconsin.

Crowley, J. and M. Hu (1977). Covariance analysis of heart transplant data. *J. Am. Stat. Assoc.*, **72**, 27–36.

Crump, K. F., D. G. Hoel, C. H. Langley, R. Peto (1976). Fundamental carcinogenic processes and their implications for low dose risk assessment. *Canc. Res.*, **36**, 2973–2979.

Crump, K. S. (1977). Response to open query: theoretical problems in the modified Mantel–Bryan procedure. *Biometrics*, **33**, 752–755.

Cutler, S. J. and F. Ederer (1958). Maximum utilization of the life table in analyzing survival. *J. Chronic Dis.*, 699–712.

David, H. A. (1970). *Order Statistics*, New York: Wiley.

David, H. A. and M. Moeschberger (1978). *Theory of Competing Risks*, London: Griffin.

Doksum, K. A. (1974). Tailfree and neutral random probability measures and their posterior distributions. *Ann. Probab.*, **2**, 183–201.

Downton, F. (1972). Contribution to the discussion of paper by D. R. Cox. *J. R. Stat. Soc. B*, **34**, 202–205.

Dumonceaux, R. and C. E. Antle (1973). Discrimination between the lognormal and Weibull distributions. *Technometrics*, **15**, 923–926.

Efron, B. (1967). The two sample problem with censored data. *Proc. Fifth Berkeley Symposium in Mathematical Statistics*, **IV**, New York: Prentice-Hall, 831–853.

Efron, B. (1977). Efficiency of Cox's likelihood function for censored data. *J. Am. Stat. Assoc.*, **72**, 557–565.

Elveback, L. (1958). Estimation of survivorship in chronic disease; the actuarial method. *J. Am. Stat. Assoc.*, **53**, 420–440.

Epstein, B. and M. Sobel (1953). Life testing. *J. Am. Stat. Assoc.*, **48**, 486–502.

Farewell, V. T. and R. L. Prentice (1977). A study of distributional shape in life testing. *Technometrics*, **19**, 69–76.

Feigl, P. and M. Zelen (1965). Estimation of exponential survival probabilities with concomitant information. *Biometrics*, **21**, 826–838.

Feller, W. (1966). *An Introduction to Probability Theory*, Vol. II, 1st ed., New York: Wiley (2nd ed. 1971).

Ferguson, T. S. (1973). A Bayesian analysis of some nonparametric problems. *Ann. Stat.*, **1**, 209–230.

Ferguson, T. S. (1974). Prior distributions on spaces of probability measures. *Ann. Stat.*, **2**, 615–629.

Ferguson, T. S. and E. G. Phadia (1979). Bayesian nonparametric estimation based on censored data. *Ann. Stat.*, **7**, 163–186.

Finney, D. J. (1971). *Probit Analysis*, London: Cambridge University Press.

Fisher, L. and P. Kanarek (1974). Presenting censored survival data when censoring and survival times may not be independent. *Reliability and Biometry*: *Statistical Analysis of Lifelength* (F. Prochan and R. J. Serfling, Eds.), Philadelphia: SIAM, 303–326.

Fisher, L. and K. Patil (1974). Matching and unrelatedness. *Am. J. Epidemiol.*, **100**, 347–349.

Fisher, R. A. (1922). On the mathematical foundations of theoretical statistics. *Phil. Trans. R. Soc.* (*London*) *A*, **222**, 309–368.

Fisher, R. A. (1925). Theory of statistical estimation. *Proc. Camb. Phil. Soc.*, **22**, 700–725.

Fisher, R. A. and F. Yates (1938). *Statistical Tables for Biological, Agricultural and Medical Research*, London: Oliver and Boyd (1st ed. 1938, 6th ed. 1963).

Fisher, R. A. (1956). *Statistical Methods and Scientific Inference*, New York: Hafner.

Forsythe, G. E., M. A. Malcolm, and C. B. Moler (1977). *Computer Methods for Mathematical Computations*, Englewood Cliffs, N.J.: Prentice-Hall.

Fraser, D. A. S. (1968). *The Structure of Inference*, New York: Wiley.

Freireich, E. O. et al. (1963). The effect of 6-mercaptopmine on the duration of steroid induced remission in acute leukemia. *Blood*, **21**, 699–716.

Gail, M. H. (1972). Does cardiac transplantation prolong life? A reassessment. *Ann. Inter. Med.*, **76**, 815–817.

Gail, M. H. (1975). A review and critique of some models used in competing risk analysis. *Biometrics*, **31**, 209–222.

Gastwirth, J. L. (1970). On robust rank tests. *Nonparametric Techniques in Statistical Inference* (M. L. Puri, Ed.), London: Cambridge University Press.

Gehan, E. A. (1965a). A generalized Wilcoxon test for comparing arbitrarily singly-censored samples. *Biometrika*, **52**, 203–223.

Gehan, E. A. (1965b). A generalized two-sample Wilcoxon test for doubly censored data. *Biometrika*, **52**, 650–652.

Gehan, E. (1969). Estimating survivor functions from the life table. *J. Chronic Dis.*, **21**, 629–644.

Gilbert, J. P. (1962). Random censorship. Ph.D. thesis, University of Chicago.

Gilbert, J. P. et al. (1975). Report of the committee for the assessment of biometric aspects of controlled trials of hypoglycemic agents. *J. Am. Med. Assoc.*, **231**, 583–608.

Glasser, M. (1967). Exponential survival with covariance. *J. Am. Stat. Assoc.*, **62**, 561–568.

Gompertz, B. (1825). On the nature of the function expressive of the law of human mortality. *Phil. Trans. R. Soc. (London)*, **115**, 513–583.

Greenwood, M. (1926). The natural duration of cancer. *Reports on Public Health and Medical Subjects*, **33**, London: Her Majesty's Stationery Office, 1–26.

Gross, A. J. and V. A. Clark (1975). *Survival Distributions: Reliability Applications in the Biomedical Sciences*, New York: Wiley.

Hagar, H. W. and L. J. Bain (1970). Inferential procedures for the generalized gamma distribution. *J. Am. Stat. Assoc.*, **65**, 1601–1609.

Hájek, J. and Z. Sĭdák (1967). *Theory of Rank Tests*, New York: Academic Press.

Hájek, J. (1969). *A Course in Nonparametric Statistics*, San Francisco: Holden-Day.

Hampel, F. (1974). The influence curve and its role in robust estimation. *J. Am. Stat. Assoc.*, **69**, 383–393.

Harter, H. L. (1967). Maximum likelihood estimation of the parameters of four parameter generalized gamma population from complete and censored samples. *Technometrics*, **9**, 159–165.

Hartley, H. O. and R. L. Sielken, Jr. (1977). Estimation of "safe doses" in carcinogenesis experiments. *Biometrics*, **33**, 1–30.

Hettmansperger, T. P. and J. W. McKean (1977). A robust alternative based on ranks to least squares in analyzing linear models. *Technometrics*, **19**, 275–284.

Hodges, J. L. and E. L. Lehmann (1963). Estimates of location based upon rank tests. *Ann. Math. Stat.*, **34**, 598–611.

Hoel, D. G. (1972). A representation of mortality data by competing risks. *Biometrics*, **28**, 475–488.

Hogg, R. V. (1974). Adaptive robust procedures: A partial review and some suggestions for future applications and theory (with discussion). *J. Am. Stat. Assoc.*, **69**, 909–927.

Holford, T. R. (1976). Life tables with concomitant information. *Biometrics*, **32**, 587–598.

Hollander, M. and D. A. Wolfe (1973). *Nonparametric Statistical Methods*, New York: Wiley.

Holt, J. D. and R. L. Prentice (1974). Survival analysis in twin studies and matched pair experiments. *Biometrika*, **61**, 17–30.

Holt, J. D. (1978). Competing risk analysis with special reference to matched pair experiments. *Biometrika*, **65**, 159–166.

Huber, P. J. (1972). Robust statistics: a review. *Ann. Math. Stat.*, **43**, 1041–1067.

Huber, P. J. (1973). Robust regression: asymptotics, conjectures and Monte Carlo. *Ann. Stat.*, **1**, 799–821.

Irwin, J. O. (1942). The distribution of the logarithm of survival times when true law is exponential, *J. Hyg.*, **42**, 328–333.

Johnson, N. L. and S. Kotz (1970a). *Distributions in Statistics. Continuous Univariate Distributions*, Vol. 1, Boston: Houghton Mifflin.

Johnson, N. L. and S. Kotz (1970b). *Distributions in Statistics. Continuous Univariate Distributions*, Vol. 2, Boston: Houghton Mifflin.

Johnson, N. L. and S. Kotz (1975). A vector multivariate hazard rate. *J. Mult. Anal.*, **5**, 53–66.

Johnson, R. A. and K. G. Mehrotra (1972). Locally most powerful rank tests for the two-sample problem with censored data. *Ann. Math. Stat.*, **43**, 823–831.

Jurečková, J. (1969). Asymptotic linearity of a rank statistic in regression parameter. *Ann. Math. Stat.*, **40**, 1889–1900.

Jurečková, J. (1971). Nonparametric estimate of regression coefficients. *Ann. Math. Stat.*, **42**, 1328–1338.

Kalbfleisch, J. D. and D. A. Sprott (1970). Application of likelihood methods to models involving a large number of parameters (with discussion). *J. R. Stat. Soc. B*, **32**, 175–208.

Kalbfleisch, J. D. and R. L. Prentice (1973). Marginal likelihoods based on Cox's regression and life model. *Biometrika*, **60**, 267–278.

Kalbfleisch, J. D. (1974). Some efficiency calculations for survival distributions. *Biometrika*, **61**, 31–38.

Kalbfleisch, J. D. and A. A. McIntosh (1977). Efficiency in survival distributions with time dependent covariables. *Biometrika*, **64**, 47–50.

Kalbfleisch, J. D. (1978a). Nonparametric Bayesian analysis of survival time data. *J. R. Stat. Soc. B*, **40**, 214–221.

Kalbfleisch, J. D. (1978b). Likelihood methods and nonparametric tests. *J. Am. Stat. Assoc.*, **73**, 167–170.

Kalbfleisch, J. D. and R. J. MacKay (1978a). Some remarks on a paper by Cornfield and Detre. *J. R. Stat. Soc. B*, **40**, 175–177.

Kalbfleisch, J. D. and R. J. MacKay (1978b). Censoring and the immutable likelihood. Technical Report 78–09, Department of Statistics, University of Waterloo.

Kalbfleisch, J. D. and R. J. MacKay (1979). On constant-sum models for censored survival data. *Biometrika*, **66**, 87–90.

Kaplan, E. L. and P. Meier (1958). Nonparametric estimation from incomplete observations. *J. Am. Stat. Assoc.*, **53**, 457–481.

Kay, R. (1977). Proportional hazard regression models and the analysis of censored survival data. *J. R. Stat. Soc. C*, **26**, 227–237.

Kay, R. (1979). Some further asymptotic efficiency calculations for survival data regression models. *Biometrika*, **66**, 91–96.

Kimball, A. W. (1958). Disease incidence estimation in populations subject to multiple causes of death. *Bull. Int. Inst. Stat.*, **36**, 193–204.

Kimball, A. W. (1969). Models for the estimation of competing risks from grouped data. *Biometrics*, **25**, 329–337.

Kingman, J. F. C. (1975). Random discrete distributions (with discussion). *J. R. Stat. Soc. B*, **37**, 1–22.

Koziol, J. and N. Reid (1977). On multiple comparisons among K samples subject to unequal patterns of censorship. *Commun. Stat. Theor. Math. A*, **6**(12), 1149–1164.

Lagakos, S. (1977). A covariate model for partially censored data subject to competing causes of failure. *J. R. Stat. Soc. C*, **27**, 235–241.

Lagakos, S., C. J. Sommer, and M. Zelen (1978). Semi-Markov models for partially censored data. *Biometrika*, **65**, 311–317.

Lagakos, S. (1979). General right censoring and its impact on the analysis of survival data. *Biometrics*, **35**, 139–156.

Lawless, J. F. (1973). Conditional versus unconditional confidence intervals for the parameters of the Weibull distribution. *J. Am. Stat. Ass.*, **68**, 665–669.

Lawless, J. F. (1978). Confidence interval estimation for the Weibull and extreme value distributions. *Technometrics*, **20**, 355–364.

Le Cam, L. (1970). On the assumptions used to prove asymptotic normality of maximum likelihood estimates. *Ann. Math. Stat.*, **41**, 802–828.

Lehmann, E. L. (1953). The power of rank tests. *Ann. Math. Stat.*, **24**, 23–43.

Lehmann, E. L. (1959). *Testing Statistical Hypotheses*, New York: Wiley.

Lehmann, E. L. (1975). *Nonparametrics: Statistical Methods Based on Ranks*, San Francisco: Holden Day.

Liu, P. Y. and J. Crowley (1978). Large sample theory for the MLE based on Cox's regression model for survival data. *Technical Report 1, Biostatistics,* Wisconsin Clinical Career Center, University of Wisconsin, Madison.

McKean, J. W. and T. P. Hettmansperger (1978). A robust analysis of the general linear model based on one step *r*-estimates. *Biometrika*, **65**, 571–579.

Makeham, W. M. (1860). On the law of mortality and the construction of annuity tables. *J. Inst. Actuaries (London)*, **8**.

Makeham, W. M. (1874). On an application of the theory of the composition of decremental forces. *J. Inst. Actuaries (London)*, **18**, 317–322.

Mann, N. R., R. E. Schafer, and N. D. Singpurwalla (1974). *Methods for Statistical Analysis of Reliability and Life Data*, New York: Wiley.

Mantel, N. and W. Haenszel (1959). Statistical aspects of the analysis of data from retrospective studies of disease. *J. Natl. Cancer Inst.*, **22**, 719–748.

Mantel, N. and W. R. Bryan (1961). Safety testing for carcinogens, *J. Natl. Cancer Inst.*, **27**, 455–470.

Mantel, N. (1963). Chi-square tests with one degree of freedom: extensions of the Mantel–Haenszel procedure. *J. Am. Stat. Assoc.*, **58**, 690–700.

Mantel, N. (1966). Evaluation of survival data and two new rank order statistics arising in its consideration. *Cancer Chemother. Rep.*, **50**, 163–170.

Mantel, N. and M. Myers (1971). Problems of convergence of maximum likelihood iterative procedures in multiparameter situations. *J. Am. Stat. Assoc.*, **66**, 484–491.

Mantel, N. and D. P. Byar (1974). Evaluation of response time data involving transient states: an illustration using heart transplant data. *J. Am. Stat. Ass.*, **69**, 81–86.

Mantel, N., N. R. Bohidar, C. C. Brown, J. L. Ciminera, and J. W. Tukey (1975). An improved Mantel Bryan procedure for "safety" testing of carcinogens. *Cancer Res.*, **35**, 865–872.

Marshall, A. W. and I. Olkin (1967). A multivariate exponential distribution. *J. Am. Stat. Assoc.*, **62**, 30–44.

Meier, P. (1977). Estimation of a distribution function from incomplete observations. *Perspectives in Probability and Statistics* (J. Gani, Ed.), New York: Academic Press, 67–87.

Miettinen, O. S. (1974). Confounding and effect modification. *Am. J. Epidemiol.*, **100**, 350–353.

Moeschberger, M. L. and H. A. David (1971). Life tests under competing causes of failure and the theory of competing risks. *Biometrics*, **27**, 909–933.

Moeschberger, M. L. (1974). Life tests under competing causes of failure. *Technometrics*, **16**, 39–47.

Moran, P. A. P. (1959). *The Theory of Storage*, London: Methuen.

Moran, P. A. P. (1971). Maximum likelihood estimation in non-standard conditions. *Proc. Camb. Phil. Soc.*, **70**, 441–450.

Myers, M., B. F. Hankey, and N. Mantel (1973). A logistic-exponential model for use with response-time data involving regressor variables. *Biometrics*, **29**, 257–269.

Nádas, A. (1971). The distribution of the identified minimum of a normal pair determines the distribution of the pair. *Technometrics*, **13**, 201–202.

Nelson, W. B. (1970). Statistical methods for accelerated life test data—the inverse power law model. *General Electric Corporate Research and Development T15 Report 71-C-011.*

Nelson, W. B. and G. J. Hahn (1972). Linear estimation of a regression relationships from censored data, part 1—simple methods and their applications (with discussion). *Technometrics*, **14**, 247–276.

Oakes, D. (1977). The asymptotic information in censored survival data. *Biometrika*, **64**, 441–448.

Osborn, D. and R. Madley (1968). The incomplete beta function and its ratio to the complete beta function. *Math. of Comp.*, **22**, 159–162.

Peterson, A. V. (1975). Nonparametric estimation in the competing risks problem. Ph.D. thesis, Department of Statistics, Stanford University.

Peterson, A. V. (1976). Bounds for a joint distribution function with fixed sub-distribution functions: application to competing risks. *Proc. Natl. Acad. Sci. U.S.*, **73**, 11–13.

Peto, R. (1972a). Rank tests of maximal power against Lehmann-type alternatives. *Biometrika*, **59**, 472–475.

Peto, R. (1972b). Contribution to the discussion of paper by D. R. Cox. *J. R. Stat. Soc. B*, **34**, 205–207.

Peto, R. and J. Peto (1972). Asymptotically efficient rank invariant test procedures (with discussion). *J. R. Stat. Soc. A*, **135**, 185–206.

Peto, R. and P. Lee (1973). Weibull distributions for continuous carcinogenesis experiments. *Biometrics*, **29**, 457–470.

Peto, R. and M. C. Pike (1973). Conservatism of the approximation $(0-E)^2/E$ in the log rank test for survival data or tumor incidence data. *Biometrics*, **29**, 579–584.

Peto, R., M. C. Pike, P. Armitage, N. E. Breslow, D. R. Cox, S. V. Howard, N. Mantel, K. McPherson, J. Peto, and P. G. Smith (1977). Design and analysis of randomized clinical trials requiring prolonged observation of each patient. Part 2. Analysis and examples. *Br. J. Cancer*, **35**, 1–39.

Pike, M. C. (1966). A method of analysis of certain class of experiments in carcinogenesis, *Biometrics*, **22**, 142–161.

Pike, M. C. (1970). A note on Kimball's paper "models for the estimation of competing risks from grouped data." *Biometrics*, **26**, 579–581.

Prentice, R. L. (1973). Exponential survivals with censoring and explanatory variables. *Biometrika*, **60**, 279–288.

Prentice, R. L. (1974). A log-gamma model and its maximum likelihood estimation. *Biometrika*, **61**, 539–544.

Prentice, R. L. (1975). Discrimination among some parametric models. *Biometrika*, **62**, 607–614.

Prentice, R. L. and E. R. Shillington (1975). Regression analysis of Weibull data and the analysis of clinical trials. *Util. Math.*, **8**, 257–276.

Prentice, R. L. (1976a). Use of the logistic model in retrospective studies. *Biometrics*, **32**, 599–606.

Prentice, R. L. (1976b). A generalization of the probit and logit models for dose response curves. *Biometrics*, **32**, 761–768.

Prentice, R. L. (1978). Linear rank tests with right censored data. *Biometrika*, **65**, 167–179.

Prentice, R. L. and N. E. Breslow (1978). Retrospective studies and failure time models. *Biometrika*, **65**, 153–158.

Prentice, R. L. and L. A. Gloeckler (1978). Regression analysis of grouped survival data with application to breast cancer data. *Biometrics*, **34**, 57–67.

Prentice, R. L., J. D. Kalbfleisch, A. V. Peterson, Jr., N. Flournoy, V. T. Farewell, and N. E. Breslow (1978). The analysis of failure times in the presence of competing risks. *Biometrics*, **34**, 541–554.

Prentice, R. L. and J. D. Kalbfleisch (1979). Hazard rate models with covariates. *Biometrics*, **35**, 25–39.

Prentice, R. L. and P. Marek (1979). A qualitative discrepancy between censored data rank tests. *Biometrics*, **35**, in press.

Prentice, R. L. and R. Pyke (1979). Logistic disease incidence models and case-control studies. *Biometrika*, **66**, in press.

Puri, M. L. (ed.) (1970). *Nonparametric Techniques in Statistical Inference*, London: Cambridge University Press.

Puri, M. L. and P. K. Sen (1971). *Nonparametric Methods in Multivariate Analysis*, New York: Wiley.

Rao, C. R. (1965). *Linear Statistical Inference and Its Applications*, New York: Wiley.

Sampford, M. R. and J. Taylor (1959). Censored observations in randomized block experiments. *J. R. Stat. Soc. B*, **21**, 214–237.

Samuels, S. (1978). Robustness of survival estimators. Unpublished Ph.D. thesis. Department of Biostatistics, University of Washington.

Sarhan, A. E. and B. G. Greenberg, (Eds.) (1962). *Contributions to Order Statistics*, New York: Wiley.

Savage, I. R. (1956). Contributions to the theory of rank order statistics—the two sample case. *Ann. Math. Stat.*, **27**, 590–615.

Savage, I. R. (1957). Contributions to the theory of rank order statistics. The "trend" case. *Ann. Math. Stat.*, **28**, 968–977.

Seal, H. L. (1954). The estimation of mortality and other decremental probabilities. *Skand. Akt.*, **37**, 137–162.

Seal, H. L. (1977). Studies in the history of probability and statistics XXXV. Multiple decrements or competing risks. *Biometrika*, **64**, 429–439.

Snyder, D. L. 1975). *Random Point Processes*, New York: Wiley.

Sprott, D. A. and J. D. Kalbfleisch (1969). Examples of likelihoods and comparison with point estimates and large sample approximations. *J. Am. Stat. Assoc.*, **64**, 468–484.

Sprott, D. A. (1975a). Marginal and conditional sufficiency. *Biometrika*, **62**, 599–605.

Sprott, D. A. (1975b). Application of maximum likelihood methods for finite samples. *Sankhya B*, **37**, 259–270.

Stacy, E. W. (1962). A generalization of the gamma distribution. *Ann. Math. Stat.*, **33**, 1187–1192.

Stacy, E. W. and G. A. Mihram (1965). Parameter estimation for a generalized gamma distribution. *Technometrics*, **7**, 349–358.

Susarla, V. and J. Van Ryzin (1976). Nonparametric Bayesian estimation of survival curves from incomplete observations. *J. Am. Stat. Assoc.*, **71**, 897–902.

Tarone, R. and J. Ware (1977). On distribution-free tests for equality of survival distributions. *Biometrika*, **64**, 156–160.

Temkin, N. R. (1978). An analysis for transient states with application to tumor shrinkage. *Biometrics*, **34**, 571–580.

Thomas, D. R. (1969). Conditional locally most powerful rank tests for the two-sample problem with arbitrarily censored data. *Technical Report No. 7,* Department of Statistics, Oregon State University.

Thomas, E. D., R. Storb, R. A. Clift, A. Fefer, F. L. Johnson, P. E. Neiman, K. G. Lerner, H. Glucksberg, and C. D. Buckner (1975a). Bone-marrow transplantation (first of two parts). *N. Engl. J. Med.*, **292**, 832–843.

Thomas, E. D., R. Storb, R. A. Clift, A. Fefer, F. L. Johnson, P. E. Neiman, K. G. Lerner, H. Glucksberg, and C. D. Buckner (1975b). Bone-marrow transplantation (second of two parts). *N. Engl. J. Med.*, **292**, 895–902.

Thomas, D. R. and G. L. Grunkemeier (1975). Confidence interval estimation of survival probabilities for censored data. *J. Am. Stat. Ass.*, **70**, 865–871.

Thompson, M. E. and V. P. Godambe (1974). Likelihood ratio vs. most powerful rank test: a two sample problem in hazard analysis. *Sankhya*, **36**, 13–40.

Thompson, W. A. (1977). On the treatment of grouped observations in life studies. *Biometrics*, **33**, 463–470.

Tsiatis, A. (1975). A nonidentifiability aspect of the problem of competing risks. *Proc. Natl. Acad. Sci.*, **72**, 20–22.

Tsiatis, A. (1978). A large sample study of the estimate for the integrated hazard function in Cox's regression model for survival data. *Technical Report No. 562,* Department of Statistics, University of Wisconsin, Madison.

Turnbull, B. W., B. W. Brown, and M. Hu (1974). Survivorship analysis of heart transplant data (A). *J. Am. Stat. Assoc.*, **69**, 74–80.

Turnbull, B. W. (1974). Nonparametric estimation of a survivorship function with doubly censored data. *J. Am. Stat. Ass.*, **69**, 169–173.

Turnbull, B. W. (1976). The empirical distribution function with arbitrarily grouped censored and truncated data. *J. R. Stat. Soc. B*, **38**, 290–295.

van der Waerden, B. L. (1953). Ein neuer Test für das Problem der zwei Stichproben. *Math. Ann.*, **126**, 93–107.

Ware, J. H. and D. P. Byar (1979). Methods for the analysis of censored survival data. To appear in *Perspectives in Biometrics*, Vol. 2 (R. M. Elashoff, Ed.).

Whittemore, A. and J. B. Keller (1978). Asthma and air pollution a quantitative theory. To appear in *Proc. SIMS Conference on Environmental Health*, SIAM.

Williams, J. S. and S. W. Lagakos (1977). Models for censored survival analysis: Constant sum and variable sum models. *Biometrika*, **64**, 215–224.

Williams, J. S. (1978). Efficient analysis of Weibull survival data from experiments on heterogeneous patient populations. *Biometrics*, **34**, 209–222.

Zelen, M. (1971). The analysis of several 2×2 contingency tables. *Biometrika*, **38**, 129–137.

Zippin, C. and P. Armitage (1966). Use of concomitant variables and incomplete survival information in the estimation of an exponential survival parameter. *Biometrics*, **22**, 665–672.

Author Index

Subject Index

Applied Probability and Statistics (Continued)

DODGE and ROMIG • Sampling Inspection Tables, *Second Edition*

DRAPER and SMITH • Applied Regression Analysis

DUNN • Basic Statistics: A Primer for the Biomedical Sciences, *Second Edition*

DUNN and CLARK • Applied Statistics: Analysis of Variance and Regression

ELANDT-JOHNSON • Probability Models and Statistical Methods in Genetics

ELANDT-JOHNSON and JOHNSON • Survival Models and Data Analysis

FLEISS • Statistical Methods for Rates and Proportions

GALAMBOS • The Asymptotic Theory of Extreme Order Statistics

GIBBONS, OLKIN, and SOBEL • Selecting and Ordering Populations: A New Statistical Methodology

GNANADESIKAN • Methods for Statistical Data Analysis of Multivariate Observations

GOLDBERGER • Econometric Theory

GOLDSTEIN and DILLON • Discrete Discriminant Analysis

GROSS and CLARK • Survival Distributions: Reliability Applications in the Biomedical Sciences

GROSS and HARRIS • Fundamentals of Queueing Theory

GUPTA and PANCHAPAKESAN • Multiple Decision Procedures: Theory and Methodology of Selecting and Ranking Populations

GUTTMAN, WILKS, and HUNTER • Introductory Engineering Statistics, *Second Edition*

HAHN and SHAPIRO • Statistical Models in Engineering

HALD • Statistical Tables and Formulas

HALD • Statistical Theory with Engineering Applications

HARTIGAN • Clustering Algorithms

HILDEBRAND, LAING, and ROSENTHAL • Prediction Analysis of Cross Classifications

HOEL • Elementary Statistics, *Fourth Edition*

HOLLANDER and WOLFE • Nonparametric Statistical Methods

HUANG • Regression and Econometric Methods

JAGERS • Branching Processes with Biological Applications

JESSEN • Statistical Survey Techniques

JOHNSON and KOTZ • Distributions in Statistics
Discrete Distributions
Continuous Univariate Distributions—1
Continuous Univariate Distributions—2
Continuous Multivariate Distributions

JOHNSON and KOTZ • Urn Models and Their Application: An Approach to Modern Discrete Probability Theory

JOHNSON and LEONE • Statistics and Experimental Design in Engineering and the Physical Sciences, Volumes I and II, *Second Edition*

JUDGE, GRIFFTHS, HILL • The Theory and Practice of Econometrics

KALBFLEISCH and PRENTICE • The Statistical Analysis of Failure Time Data

KEENEY and RAIFFA • Decisions with Multiple Objectives

LANCASTER • An Introduction to Medical Statistics

LEAMER • Specification Searches: Ad Hoc Inference with Nonexperimental Data

McNEIL • Interactive Data Analysis